Chemical Constituents and Biological Activities of Natural Sources

Chemical Constituents and Biological Activities of Natural Sources

Guest Editors

Míchel Frédérich
Anne Bialecki

MDPI

Basel • Beijing • Wuhan • Barcelona • Belgrade • Novi Sad • Cluj • Manchester

Guest Editors
Míchel Frédérich
Department of Pharmacy
Université de Liège
Liège
Belgium

Anne Bialecki
Faculté des Sciences et
Technologies
Université de la Réunion
Saint-Denis
France

Editorial Office
MDPI AG
Grosspeteranlage 5
4052 Basel, Switzerland

This is a reprint of the Special Issue, published open access by the journal *Molecules* (ISSN 1420-3049), freely accessible at: https://www.mdpi.com/journal/molecules/special_issues/43655TNB67.

For citation purposes, cite each article independently as indicated on the article page online and as indicated below:

Lastname, A.A.; Lastname, B.B. Article Title. *Journal Name* **Year**, *Volume Number*, Page Range.

ISBN 978-3-7258-8073-7 (Hbk)
ISBN 978-3-7258-8074-4 (PDF)
https://doi.org/10.3390/books978-3-7258-8074-4

Contents

About the Editors

Míchel Frédérich

Michel Frédérich is a Belgian pharmacist born in Verviers, Belgium. He holds a degree in Pharmaceutical Sciences from the University of Liège and completed postdoctoral training at Leiden University. He is a Full Professor of Pharmacognosy at ULiège. His expertise is recognized through several roles in European and Belgian pharmacopoeial and regulatory bodies. He served as Director of the Interdisciplinary Center for Drug Research at ULiège (CIRM) from 2020 to 2026. His research focuses on natural products with antimalarial, anticancer and anti-inflammatory properties, as well as on the quality control of herbal medicinal products. He has supervised numerous doctoral researchers and led several nationally and internationally funded research projects. His scientific output includes many peer-reviewed international publications and reflects a strong academic impact in pharmacognosy and natural product research.

Anne Bialecki

Anne Bialecki is a Full Professor of Chemistry at the University of La Réunion, where she also serves as director of the Master's programme in Chemical and Biotechnological Valorisation of Biodiversity. Her research focuses on the exploration of bioactive molecules derived from both marine biodiversity—including marine invertebrates and associated microorganisms—and terrestrial biodiversity, particularly plants. Her work encompasses the extraction, purification, and structural analysis of natural compounds with applications in perfumery, cosmetics, pharmacy, and chemical ecology. She served as Director of the Chembiopro laboratory for a decade, from 2014 to 2024, leading a research team dedicated to the valorisation of natural resources from the Indian Ocean region.

Preface

Natural products have long served as an invaluable source of inspiration for drug discovery and the development of bioactive compounds. The extraordinary chemical diversity found in terrestrial plants, marine organisms, microorganisms, and other biological matrices continues to reveal molecules of significant therapeutic interest, driving forward research in pharmacology, medicinal chemistry, and related disciplines.

This Reprint, published in the journal *Molecules* within the Natural Products Chemistry section, brings together a collection of original research articles and comprehensive reviews united under the theme "Chemical Constituents and Biological Activities of Natural Sources." It is addressed to the scientific community working at the interface of natural product chemistry, pharmacognosy, biochemistry, and biomedical sciences.

The contributions gathered here reflect the breadth and vitality of current research in the field, spanning the metabolomic profiling of endemic plant species, the structural elucidation of novel bioactive molecules, mechanistic reviews on alkaloids and bioactive peptides, and pharmacological investigations targeting inflammation, cancer, and viral infections. Together, they illustrate the remarkable potential of natural sources as reservoirs of therapeutic leads.

We hope this Reprint will serve as a valuable reference for researchers engaged in the exploration and valorisation of natural chemical diversity.

Míchel Frédérich and Anne Bialecki
Guest Editors

molecules MDPI

ticle

ynthesis and Biological Evaluation of Curvularin-Type)erivatives with Potential Anti-Inflammatory, Anticancer, nd Antimicrobial Activities

yung Hee Kim [1,†], Tai Kyoung Kim [2,†], Ju-Mi Hong [1,3], Jin A Kim [1,3], Min Ju Kim [1], Jin-Hyoung Kim [1,4], ung Han Yim [1,2,*], Il-Chan Kim [1,*] and Se Jong Han [1,4,*]

[1] Division of Life Sciences, Korea Polar Research Institute, Incheon 21990, Republic of Korea; happy_vu@naver.com (K.H.K.); wnal5555@kopri.re.kr (J.-M.H.); kimja611@kopri.re.kr (J.A.K.); mjkim1113@kopri.re.kr (M.J.K.); kimjh@kopri.re.kr (J.-H.K.)

[2] CRYOTECH Inc., 2F-211-3, 71 Mieumsandan 5-ro 41beon-gil, Gangseo-gu, Busan 46744, Republic of Korea; tkkim@cryotech.co.kr

[3] Department of Marine Sciences, Incheon National University, Incheon 22012, Republic of Korea

[4] Department of Polar Science, University of Science and Technology, Daejeon 34113, Republic of Korea

[*] Correspondence: jhyim@kopri.re.kr (J.H.Y.); ickim@kopri.re.kr (I.-C.K.); hansj@kopri.re.kr (S.J.H.); Tel.: +82-32-760-5540 (J.H.Y.); +82-760-5541 (I.-C.K.); +82-760-5521 (S.J.H.); Fax: +82-32-760-5509 (J.H.Y. & I.-C.K. & S.J.H.)

[†] These authors contributed equally to this work.

Abstract

Curvularins, a class of macrocyclic lactones, have cytotoxic, antimicrobial, and anti-inflammatory properties. Curvularin, a 12-membered macrolactone, was used as a scaffold to design and synthesize structurally modified analogues to investigate structure–activity relationships and improve biological efficacy. Three series of curvularin-based analogues, Cur-5H-OMe, Cur-4P-OMe, and Cur-OMe, were synthesized with the same core structure but different substituent sizes and positions. Nine representative derivatives were evaluated for anti-inflammatory, anticancer, antibacterial, and antifungal activities. In LPS-stimulated RAW 264.7 macrophages, most compounds inhibited nitric oxide (NO) production in a concentration-dependent manner but exhibited cytotoxicity at high concentrations. Cytotoxicity assays against HaCaT cells and human cancer cell lines (HCT116, HeLa, and A375) revealed limited selectivity toward cancer cells. Antimicrobial evaluation indicated selective activity against the Gram-positive bacteria, *Staphylococcus aureus*. Compound **23** exhibited superior antibacterial potency compared with kanamycin and notable antifungal activity against *Candida albicans*. This study provides a versatile synthetic platform and identifies key structural features of curvularin derivatives, demonstrating their potential as anti-inflammatory and antimicrobial lead compounds.

Keywords: curvularin analogues; anti-inflammatory; anticancer; antimicrobial; *Staphylococcus aureus*; *Candida albicans*

1. Introduction

The exploration of extreme environments, particularly in polar marine ecosystems, has revealed several microorganisms that were previously unknown [1–3]. Among these, marine-derived fungi have emerged as a prolific and promising source of structurally unique and biologically active secondary metabolites [4,5]. Unlike terrestrial and mesophilic fungal species, marine psychrophilic fungi have adapted to survive and thrive under extreme environmental stressors, such as sub-zero temperatures, high salinity, and limited nutrient availability. These harsh conditions have driven the evolution of distinct metabolic pathways that enable the production of novel bioactive compounds [6–8]. Consequently, psychrophilic fungi have attracted growing interest in the fields of natural product chemistry, biotechnology, and pharmaceutical research.

Marine fungi isolated from polar environments are particularly noteworthy for their ability to biosynthesize metabolites with both unprecedented chemical scaffolds and a wide range of physiological functions. Many of these compounds have demonstrated antimicrobial, antifungal, anticancer, and anti-inflammatory properties [9–12]. Such characteristics render polar marine fungi an invaluable and relatively untapped reservoir for drug discovery and development. The fungi have biosynthetic gene clusters that differ significantly from those of more commonly studied fungi, opening new avenues for identifying structurally novel compounds with desirable pharmacological properties. Moreover, the increasing global demand for alternative therapies to treat chronic diseases, such as inflammation-related disorders, has further underscored the need to explore new and sustainable biological resources [13–15].

Among the marine-derived psychrophilic fungi investigated by our research group, the strain *Penicillium* sp. SF-5859 has shown particularly promising potential [16]. This strain was isolated from a marine sponge collected in the frigid waters of the Antarctic Ocean, an environment characterized by extreme cold and ecological isolation. During preliminary studies, this fungal isolate was found to produce a range of curvularin-type secondary metabolites. Curvularins are a class of macrocyclic lactones known for their cytotoxic, antimicrobial, and anti-inflammatory properties. In this case, the secondary metabolites isolated from *Penicillium* sp. SF-5859 demonstrated strong anti-inflammatory activity, particularly through the suppression of inflammatory mediators such as nitric oxide (NO) and prostaglandin E2 (PGE2).

Given the demonstrated efficacy of the natural curvularin-type metabolites in reducing key inflammatory markers, expanding upon this discovery through chemical modification is crucial. Derivatives of natural products often display improved pharmacokinetic profiles, enhanced potency, and increased selectivity compared with their parent compounds. Therefore, in this study, we aimed to synthesize a series of curvularin-based analogues to investigate the relationship between molecular structure and anti-inflammatory activity. Moreover, we explored how specific functional group modifications affect the biological activity of these compounds.

2. Results and Discussion

2.1. *Synthesis of Curvularin Analogues*

To systematically investigate the structure–activity relationships (SARs) of curvularin, three series of analogues, namely Cur-5H-OMe (**8**), Cur-4P-OMe (**16**), and Cur-OMe (**23**), were synthesized. The synthesis of all three series followed a unified, convergent strategy comprising esterification, Friedel–Crafts acylation, dehydrohalogenation, ring-closing metathesis (RCM), and selective demethylation (Schemes 1–3).

Initially, the esterification of 3,5-dimethoxyphenylacetic acid (**1**) with the respective alcohols (**2**, **10**, **17**) was attempted using DCC and DMAP. However, the complete removal of the dicyclohexylurea (DCU) byproduct posed difficulty, which reduced the overall purification efficiency [17]. By switching to the Mukaiyama reagent (2-chloro-1-methylpyridinium iodide) in the presence of triethylamine (TEA), the esterification proceeded smoothly [18]. This optimized condition successfully afforded the ester intermediates **3** (76%), **11** (65%), and **18** (65%).

https://doi.org/10.3390/molecules3106106

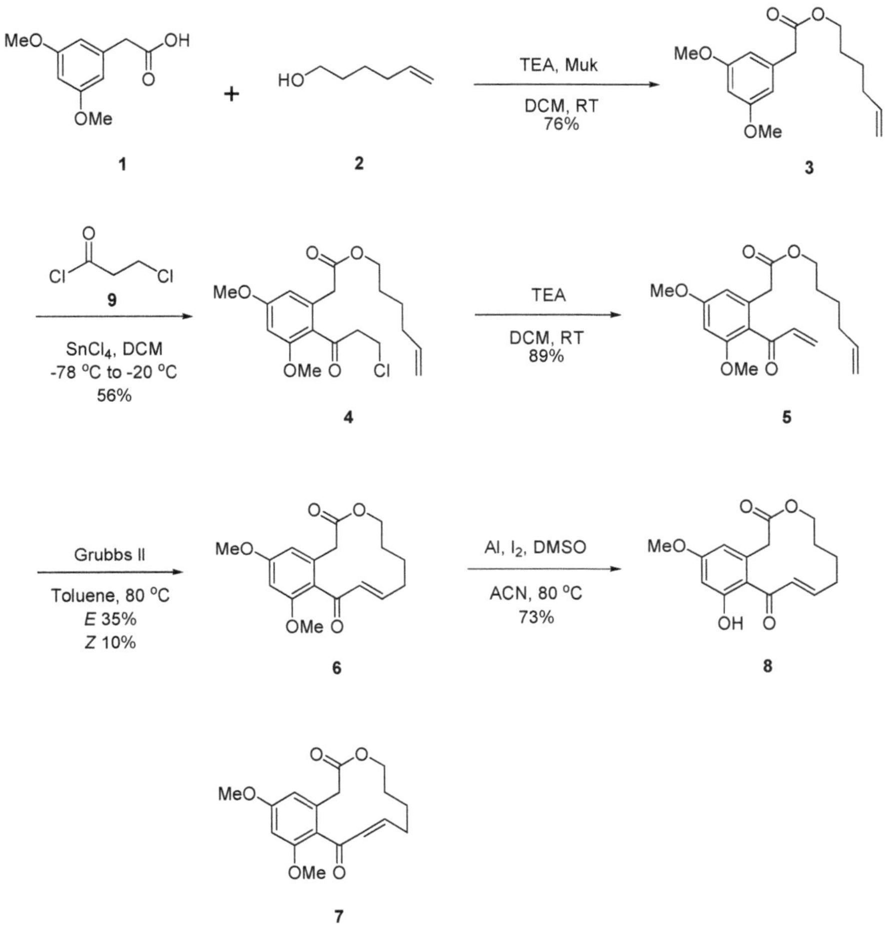

Scheme 1. Schematic outlining the preparation of Cur-5H-OMe2-F1 (**6**), Cur-5H-OMe2-F2 (**7**), and Cur-5H-OMe1 (**8**). The Supplementary Materials section describes the synthesis and properties of the compounds.

The ester intermediates were subsequently subjected to Friedel–Crafts acylation with 3-chloropropanoyl chloride (**9**). For this transformation, $SnCl_4$ was specifically selected as a mild Lewis acid catalyst; unlike stronger Lewis acids (such as $AlCl_3$), $SnCl_4$ efficiently promotes the acylation of the highly electron-rich 3,5-dimethoxyphenyl ring while preventing unwanted side reactions, such as the cleavage of the methyl ether groups (demethylation) [19,20]. This optimized acylation yielded the acylated compounds **4** (56%), **12** (78%), and **19** (72%). Following this acylation, dehydrohalogenation was performed by treating these compounds with TEA, which successfully yielded the key diene intermediates **5** (89%), **13** (88%), and **20** (63%) ready for cyclization.

The RCM step was extensively optimized using diene **20** as a model. Notably, rigorous degassing (argon bubbling for 2 h) and highly diluted conditions (4 mM in toluene) at 80 °C with 20 mol% of Grubbs II catalyst were critical to prevent intermolecular oligomerization and overcome the entropic penalties of medium-to-large macrocycle formation. Under these optimized conditions, the RCM of the dienes provided the corresponding macrolactones as mixtures of E/Z diastereomers: the 12-membered rings **6** (E, 35%) and **7** (Z, 10%); the

11-membered rings **14** (E, 40%) and **15** (Z, 13%); and the 12-membered rings **21** (E, 41%) and **22** (Z, 16%).

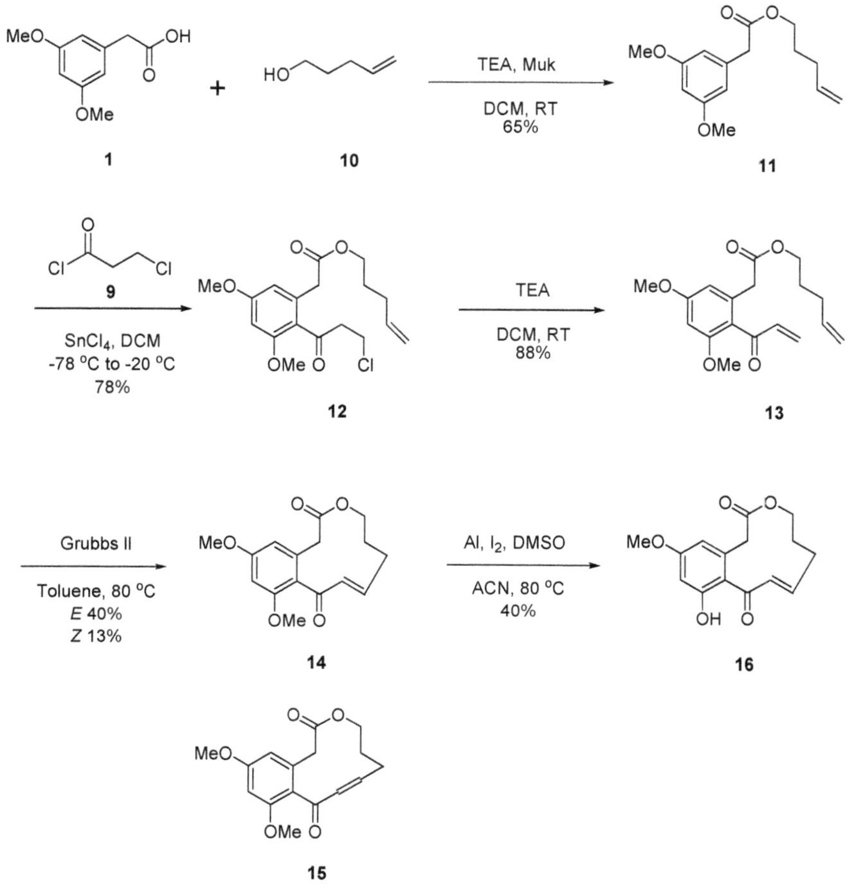

Scheme 2. Schematic outlining the preparation of Cur-4P-OMe2-F1 (**14**), Cur-4P-OMe2-F2 (**15**), and Cur-4P-OMe1 (**16**). The Supplementary Materials section describes the synthesis and properties of the compounds.

Although these conditions were heavily optimized, the overall yields remained moderate (45–57%). This can be attributed to the difficulty in the formation of medium-sized (11- to 12-membered) rings, which inherently suffer from high transannular ring strain and severe entropic penalties [21]. Furthermore, despite the highly diluted environment, unavoidable competitive intermolecular oligomerization, along with potential catalyst deactivation caused by the chelation of Lewis basic oxygen atoms (such as ester, ketone, and methoxy groups) present in the diene substrates, likely limited the final RCM efficiency [22,23].

Regarding stereoselectivity, the RCM reactions predominantly yielded the thermodynamically more stable E-isomers across all series. Interestingly, during structural characterization, we observed an isomeric phase transition for the purified Z-isomers (**7**, **15**, and **22**). When dissolved in nuclear magnetic resonance (NMR) solvents, these Z-isomers slowly and spontaneously isomerized to their respective E-isomers over time, further confirming the strong thermodynamic preference for the E-configuration in these curvularin scaffolds.

https://doi.org/10.3390/molecules31061061

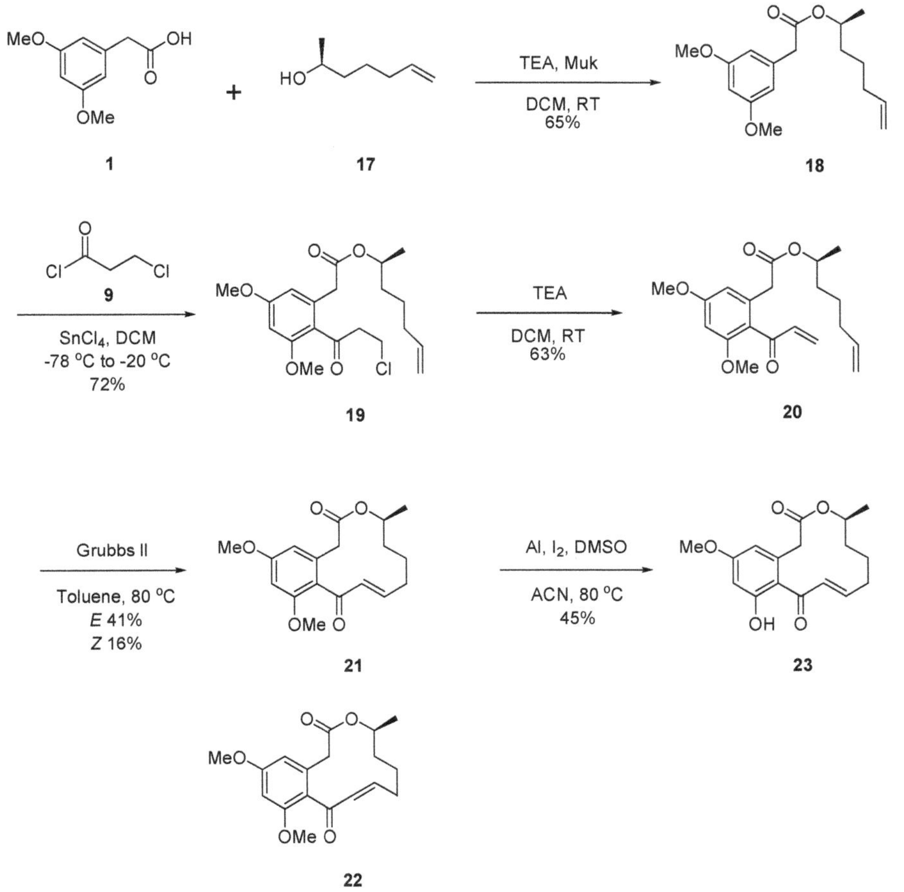

Scheme 3. Schematic outlining the preparation of Cur-OMe2-F1 (**21**), Cur-OMe2-F2 (**22**), and Cur-OMe1 (**23**). The Supplementary Materials section describes the synthesis and properties of the compounds.

Finally, to generate the mono-methoxy target derivatives, the E-isomer macrolactones (**6**, **14**, and **21**) were subjected to selective demethylation using aluminum powder, iodine, and dimethyl sulfoxide (DMSO) in acetonitrile [19]. This final step successfully afforded the newly synthesized derivatives **8** (73%), **16** (40%), and **23** (45%).

The Cur-5H-OMe, Cur-4P-OMe, and Cur-OMe analogues were designed to retain the 12-membered macrolactone skeleton, the principal physiologically active framework of curvularin, while modulating electronic and steric properties by varying substituent position and sizes. Curvularin derivatives exhibit a range of biological activities, including antibacterial [24], antifungal [25], anti-inflammatory [16,26,27], anticancer [28], and cytoprotective activity [29]. The results of this synthesis are significant because they demonstrate the successful preparation of various curvularin analogues through a unified reaction, building on previous studies [19,30,31]. Although the process is sensitive to the conditions of each step, it provides high yield and selectivity and establishes a synthetic paradigm useful for constructing future bioactivity screening libraries. This synthetic route provides a stable platform for expanding the activity range of curvularin derivatives and conducting precise SAR studies.

2.2. *Anti-Inflammatory Activity*

Inhibition of nitric oxide (NO) production in LPS-induced RAW 264.7 cells was measured to determine the nitric oxide activity of nine derivatives (compounds **6–8**, **14–16**, and **21–23**). The concentrations of the nine derivatives were 0, 0.025, 0.05, 0.1, 2.5, and 5 µg/mL. Nitrite concentrations were measured using the Griess reaction and were found to be significantly increased in LPS-stimulated RAW 264.7 cells. However, NO production was inhibited in a concentration-dependent manner by the compounds synthesized in this study, except for compound **8** (Figure 1A). Although the synthesized compounds reduced NO production at high concentrations, a corresponding decrease in cell viability was observed, indicating that the NO reduction at these doses was likely due to cytotoxicity (Figure 1B). Compound **7** exhibited an anti-inflammatory effect, inhibiting NO production by more than 70% while maintaining cell survival by more than 50% even at high concentrations.

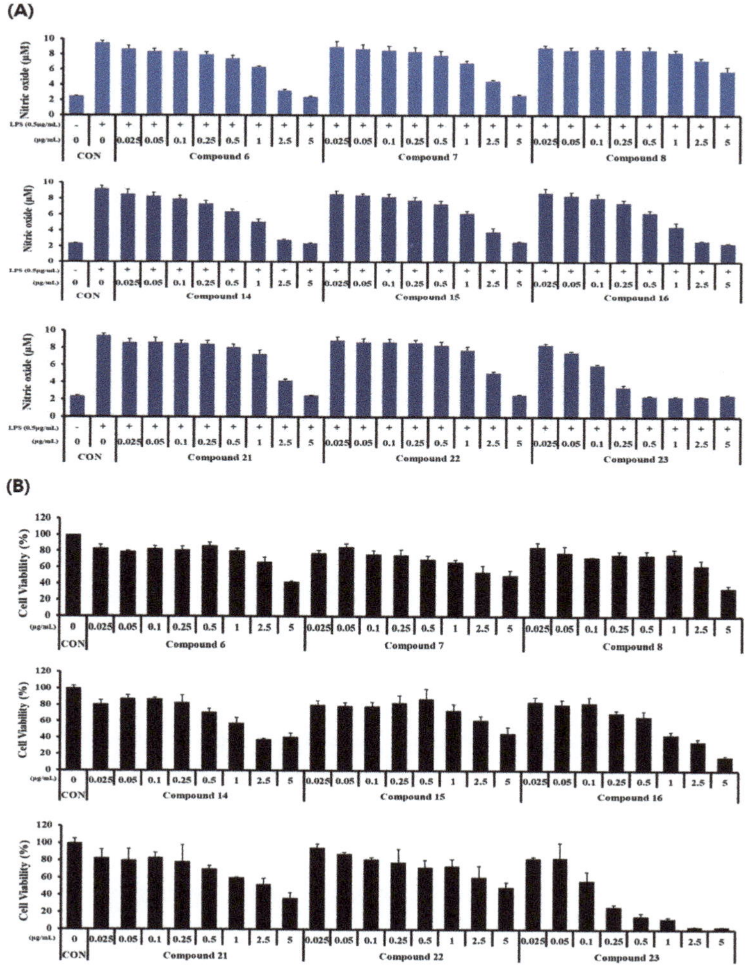

Figure 1. Effects of curvularin derivatives on (**A**) LPS-induced nitric oxide (NO) production and (**B**) cell viability in RAW 264.7 macrophages. NO levels were quantified using the Griess assay, and viability was measured using the MTT assay. Data are presented as mean ± SD (n = 3).

https://doi.org/10.3390/molecules3106106

The anti-inflammatory potential of curvularin derivatives has been previously explored [16,26,29,32]. In the present study, most derivatives inhibited NO production in LPS-induced RAW 264.7 cells. However, to distinguish true anti-inflammatory activity from cytotoxicity mediated NO reduction, a concern raised by the correlation between NO inhibition and cell death at high concentrations, we calculated the selectivity index (SI) for each compound (Table 1).

Table 1. Anti-inflammatory activity and selectivity index (SI) of derivatives.

Compounds	NO Inhibition	Cytotoxicity	SI (Selectivity Index) (Cytotoxicity/ NO Inhibition)
	IC_{50} Value (μg/mL)		
6	1.83	4.15	2.27
7	2.32	5.1	2.20
8	>5.0	3.63	<1.0
14	1.31	1.35	1.03
15	2.01	4.28	2.13
16	0.95	0.85	0.89
21	2.21	2.89	1.31
22	2.84	4.91	1.73
23	0.18	0.13	0.72

While some derivatives (e.g., compounds **16** and **23**) exhibited low SI values (<1.0), indicating that their apparent inhibitory effects were primarily due to cytotoxicity, compounds **6**, **7**, and **15** demonstrated a favourable safety profile with SI values greater than 2.0. Notably, compound **6** (SI = 2.27) and compound **7** (SI = 2.20) inhibited NO production with half-maximal inhibitory concentration (IC_{50} values) of 1.83 μg/mL and 2.32 μg/mL, respectively, which were significantly lower than their cytotoxic concentrations (IC_{50} > 4.0 μg/mL). These quantitative results suggest that compounds **6** and **7** exhibit genuine anti-inflammatory potential and warrant further development as therapeutic agents.

Notably, from an SAR perspective, compound **7** (12-membered macrocycle with a Z-alkene configuration and di-methoxy substituents) exhibited the most favourable anti-inflammatory profile (high NO inhibition with low cytotoxicity). This suggests that the increased conformational flexibility of the 12-membered ring, combined with the specific spatial arrangement dictated by the Z-geometry, may facilitate selective binding to inflammatory targets such as iNOS [33,34]. Furthermore, retaining the di-methoxy groups (avoiding free phenols) appears crucial for mitigating generalized cytotoxicity, thereby widening the therapeutic window for anti-inflammatory applications [35].

2.3. Anticancer Activity

The cytotoxicity of the synthesized curvularin derivatives was investigated on normal human keratinocytes (HaCaT) and three human cancer cell lines (HCT116, HeLa, and A375). Cells were incubated with the nine derivatives at concentrations ranging from 0 to 5 μg/mL for 24 h. The dose–response profiles are presented in Figure 2, and the corresponding half-maximal inhibitory concentration (IC_{50}) values are summarized in Table 2.

Most derivatives exhibited weak cytotoxicity against the tested cancer cell lines (Figure 2). This observation is confirmed by the quantitative analysis in Table 2, which indicates that the majority of compounds display IC_{50} values greater than 5.0 μg/mL. While compounds **16** and **23** displayed potent inhibitory effects on cancer cell growth (Figure 2), a comparison with normal HaCaT cells revealed a lack of selectivity.

https://doi.org/10.3390/molecules31061061

Specifically, compound **23** exhibited strong cytotoxicity against all cancer cell lines, with IC_{50} values ranging from 0.35 to 0.79 μg/mL (Table 2). However, it was significantly more toxic to normal HaCaT cells (IC_{50} = 0.09 μg/mL). Similarly, compound **16** displayed moderate activity against cancer cells (IC_{50} = 2.68–3.25 μg/mL) but demonstrated higher toxicity toward HaCaT cells (IC_{50} = 0.89 μg/mL).

These findings align with those of previous studies indicating that curvularin derivatives often exhibit a broad spectrum of cytotoxicity, as observed in HT1080, T46D, A2780S [36], A549, HeLa, MCF-7 [25], and other human tumour cell lines [37]. Although potent cytotoxicity has been reported repeatedly for curvularin analogues, selective activity against specific cancer cell lines remains limited. Therefore, our data suggest that while compounds **16** and **23** are potent cytotoxic agents, further target-directed design is required to improve their selectivity for cancer cells over normal cells.

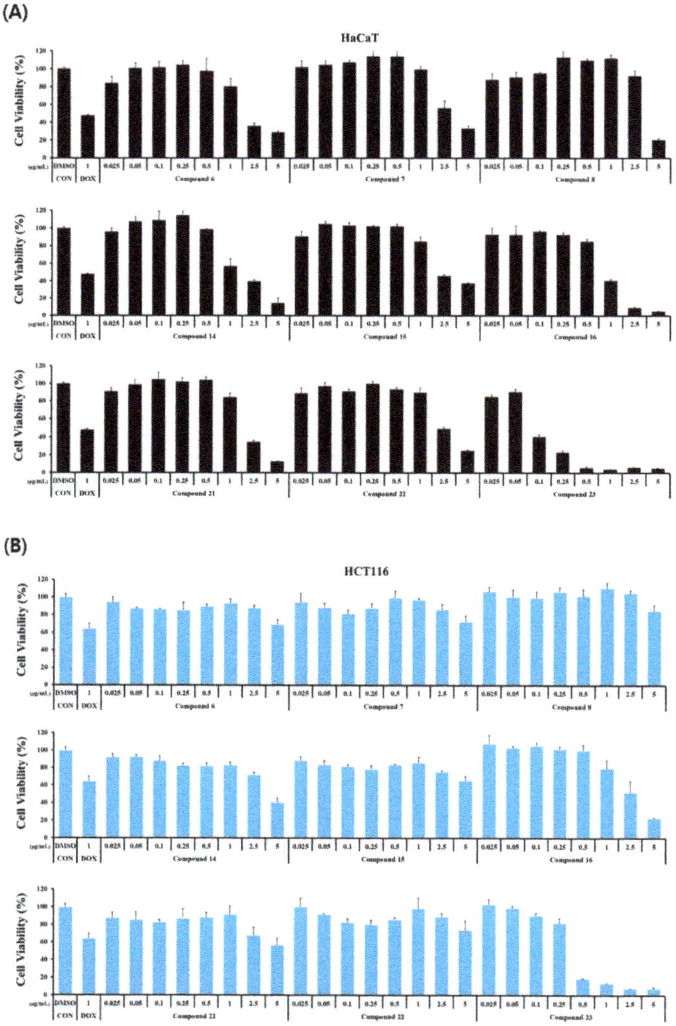

Figure 2. *Cont.*

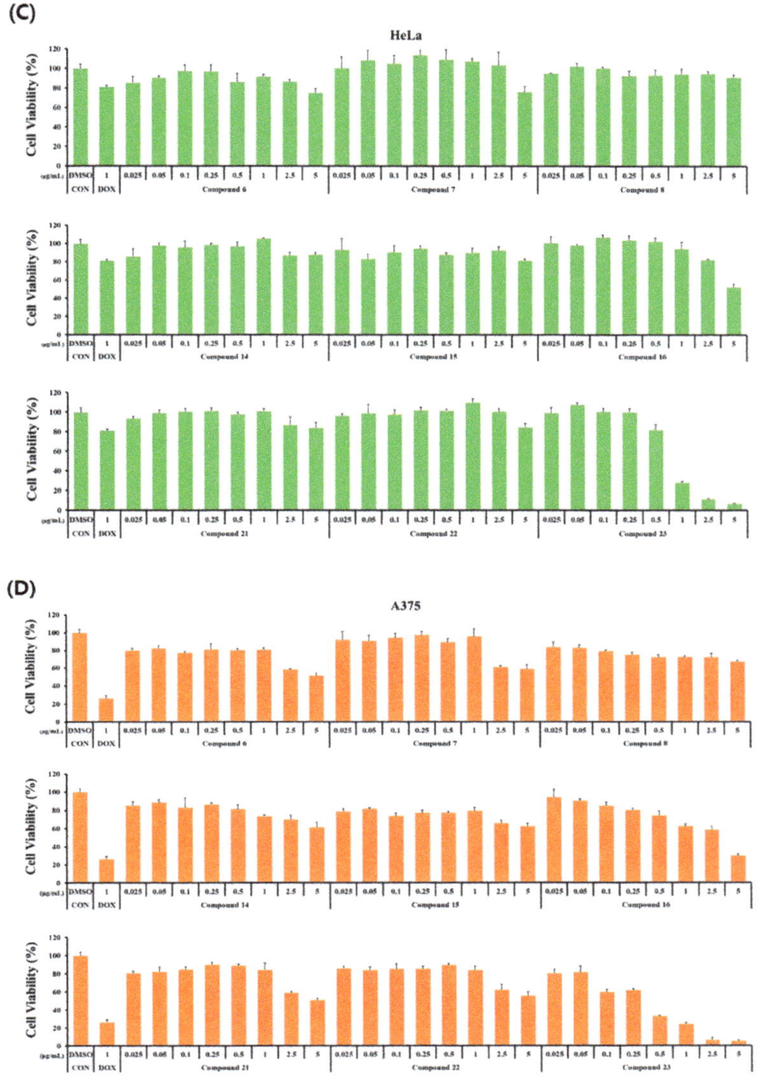

Figure 2. Cytotoxic effects of curvularin derivatives on four human cell lines. Cells were treated for 24 h with compounds (**6–8**, **14–16**, **21–23**) at concentrations of 0.025, 0.05, 0.1, 0.25, 0.5, 1, 2.5, and 5 μg/mL. Cell viability was assessed after treatment; the data represent the mean ± SD values of three independent experiments: (**A**) HaCaT keratinocytes, (**B**) HCT116 colorectal carcinoma cells, (**C**) HeLa cervical carcinoma cells, and (**D**) A375 melanoma cells.

The SAR analysis of the anticancer results reveals a clear integration of chemical modification and pharmacological outcome. The notable increase in cytotoxicity observed for compounds **16** and **23** highlights the critical role of the selective demethylation step in our synthetic strategy. The unmasking of a free phenolic hydroxyl group on the benzene ring serves as a structural "on-switch" for cytotoxicity [38]. While the completely methylated precursors were largely inactive against cancer cells, the free phenol likely enhances non-specific interactions with cell membranes, acts as a potent hydrogen bond donor to protein targets, or participates in redox cycling to generate cytotoxic reactive oxygen species (ROS) [39]. However, as this leads to potent, but non-selective, toxicity (affecting normal

HaCaT cells as well), future synthetic efforts must focus on modifying the macrocyclic backbone of these mono-hydroxy derivatives to impart target-specific selectivity.

Table 2. Anticancer activity of curvularin derivatives.

Compounds	IC$_{50}$ Value (μg/mL)			
	HaCaT	HCT116	HeLa	A375
6	2.02	>5.0	>5.0	>5.0
7	3.16	>5.0	>5.0	>5.0
8	3.97	>5.0	>5.0	>5.0
14	1.59	4.25	>5.0	>5.0
15	2.34	>5.0	>5.0	>5.0
16	0.89	2.68	>5.0	3.25
21	2.03	>5.0	>5.0	>5.0
22	2.47	>5.0	>5.0	>5.0
23	0.09	0.38	0.79	0.35

2.4. Antibacterial and Antifungal Activity

Cell inhibition against *S. aureus* and *E. coli* was measured at concentrations of 0, 2.5, 25, 50, 100, and 200 μg/mL to determine the antibacterial activity of nine derivatives (compounds **6–8**, **14–16**, and **21–23**). The IC$_{50}$ values revealed no activity against *E. coli*; however, seven derivatives (compounds **7**, **8**, **14**, **16**, **21**, **22**, and **23**) exhibited activity against *S. aureus*. Four derivatives (compounds **8**, **14**, **16**, and **23**) showed minimum inhibitory concentration (MIC) values against *S. aureus* within the tested concentration range. Compound **23** exhibited an IC$_{50}$ value of 8.90 ± 2.28 μg/mL, which was stronger than that of kanamycin (14.97 ± 6.17 μg/mL), suggesting the potential for antibiotic activity (Table 3).

Table 3. Antibacterial activity of curvularin derivatives.

Compounds	*Staphylococcus aureus*		*Escherichia coli*	
	IC$_{50}$ (μg/mL)	MIC (μg/mL)	IC$_{50}$ (μg/mL)	MIC (μg/mL)
Kanamycin [a]	14.97 ± 6.17	29.18 ± 4.71	7.31 ± 2.61	26.32 ± 2.50
6	>200	-	>200	-
7	137.43 ± 27.62	-	>200	-
8	49.94 ± 3.29	120.24 ± 6.60	>200	-
14	90.56 ± 14.35	168.50 ± 42.88	>200	-
15	>200	-	>200	-
16	39.04 ± 8.97	93.46 ± 15.59	>200	-
21	192.99 ± 14.36	-	>200	-
22	192.47 ± 5.30	-	>200	-
23	8.90 ± 2.28	25.70 ± 1.34	>200	-

[a] Positive control: Kanamycin (*S. aureus*, *E. coli*).

The growth inhibition of *C. albicans* cells was measured at concentrations of 0, 2.5, 25, 50, 100, and 200 μg/mL to determine the antifungal activity of the nine derivatives. The IC$_{50}$ values confirmed the activity of six derivatives (compounds **7**, **8**, **15**, **16**, **22**, and **23**) against *C. albicans*, whereas MIC values were obtained for three derivatives (compounds **8**, **16**, **23**). Compounds **8** and **23** exhibited low IC$_{50}$ values, indicating the potential for these derivatives to exhibit potent antifungal effects (Table 4).

In tests targeting *S. aureus* and *E. coli*, no significant inhibitory effect was observed against *E. coli*, whereas seven derivatives (**7**, **8**, **14**, **16**, **21**, **22**, **23**) displayed activity against *S. aureus*. This is consistent with previous reports indicating that the curvularin skeleton was more effective at targeting Gram-positive bacteria than Gram-negative bacteria [25].

https://doi.org/10.3390/molecules3106106

Previous studies have shown that some curvularin derivatives also displayed activity against *E. coli*; therefore, a detailed investigation is needed to understand how structural differences affect efficacy [36,40,41].

Table 4. Antifungal activity of curvularin derivatives.

Compounds	*Candida albicans*	
	IC_{50} (μg/mL)	MIC (μg/mL)
Nystatin [a]	0.83 ± 0.22	1.11 ± 0.07
6	>200	-
7	116.10 ± 9.22	-
8	6.86 ± 0.59	57.36 ± 0.85
14	>200	-
15	157.02 ± 7.53	-
16	32.25 ± 5.32	139.58 ± 12.21
21	>200	-
22	133.30 ± 9.29	-
23	5.24 ± 0.02	10.48 ± 0.04

[a] Positive control: Nystatin (*C. albicans*).

Moreover, antifungal activity against *C. albicans* was confirmed for curvularin and 11-alpha-methoxycurvularin [36,42]. Among the nine compounds synthesized in this study, compounds **8**, **16**, and **23**, each comprising only one O-methyl group on the benzene ring, exhibited relatively higher antibacterial and antifungal activities than the other compounds. This can be attributed to the fact that methylation often reduces the polarity of drug molecules, which can change the binding affinity for specific receptors and reduce toxicity [35]. The presence of two O-methyl groups on the benzene ring appears to reduce toxicity, resulting in lower antibacterial and antifungal activity in the other compounds.

Curvularin derivatives were effective against *B. megaterium*, *M. violaceum*, *S. tritici*, *C. fusca* [43], *H. pylori* [44], *P. capsici* zoospores [45], *M. marinum* [46], *V. alginolyticus*, *V. harveyi* [24], and *M. tuberculosis* [47]. The compounds synthesized in this study also need to be further studied for their efficacy against different types of pathogenic microorganisms. Compound **23**, which has antibacterial and antifungal activity, is considered a promising starting material that can be developed into an antibiotic candidate through structural optimization.

2.5. Quantitative Comparison of Activities of Curvularin and Its Derivatives

The physiological activities of curvularin, dehydrocurvularin, and the derivative (i.e., compound **23**) synthesized in this study were quantitatively compared (Table 5). Compound **23** exhibited anti-inflammatory activity with an IC_{50} value of 0.59 μM, which was higher than that of dehydrocurvularin (0.44 μM) but significantly lower than that of curvularin (18.1 μM). Although a direct comparison of anti-cancer activity was difficult, compound **23** exhibited lower IC_{50} values than that previously reported. The compounds synthesized in this study did not show MIC against *E. coli* within the tested concentrations, whereas previous studies have shown that dehydrocurvularin exhibited activity, and compound **23** exhibited better activity against *S. aureus*. Compound **23** exhibited lower MIC against *C. albicans* than that of dehydrocurvularin.

https://doi.org/10.3390/molecules31061061

Table 5. Comparison of the physiological activities of curvularin and its derivatives.

Compound	Activity				Reference
	Anti-Inflamatory (IC_{50})	Anti-Cancer (IC_{50})	Anti-Bacteria (MIC)	Anfi-Fungal (MIC)	
Curvularin	18.1 μM				[16]
Dehydrocurvularin	0.44 μM				[26]
		10.56–29.17 μM (against HT1080, T46D, and A2780S)	40 μg/mL (against *S. aureus*)	40 (μg/mL) (against *C. albicans*)	[36]
		10.3 μg/mL (against HeLa)	375 μg/mL (against *S. aureus*)	-	[25]
		1.25–30.06 μM (against human tumour cell lines)			[37]
			1 μg/mL (against *E. coli*)		[40]
		7.9 μM (against HeLa)	2.5 μg/mL (against *E. coli*)		[41]
11-α-methoxycurvularin			50 μg/mL (against *S. aureus*) 50 μg/mL (against *E. coli*)		[42]
Compound **23** (*S*,*E*)-11-hydroxy-13-methoxy-4-methyl-4,5,6,7-tetrahydro-2*H*-benzo[*d*][1]oxacyclododecine-2,10(1*H*)-dione	0.59 μM	0.38 μM (HCT116), 0.79 μM (HeLa), 0.35 μM (A375)	25.7 μg/mL (against *S. aureus*)	10.48 μg/mL (against *C. albicans*)	This study

3. Materials and Methods

3.1. General Experimental Information

Anhydrous solvents and reagents used in this study were purchased from Sigma-Aldrich (St. Louis, MO, USA) and TCI (Tokyo, Japan) and used as directed without further purification. All glassware was completely dried in a drying oven at 80 °C and cooled under dry argon immediately before use. All reactions were performed under argon gas, and solvents and liquid reagents were transferred using syringes. The organic extract was dried over $MgSO_4$ and concentrated under reduced pressure using a rotary evaporator. The residual solvent was removed under high vacuum (Vacuubrand RZ 2.5, Wertheim, Germany, 1×10^{-2} mbar). TLC was performed using 0.25 mm silica gel plates (Merck, Darmstadt, Germany) and observed under 254 nm ultraviolet light (UV CABINET CN-15.LC, VILBER, Vilber Lourmat, France). Flash column chromatography used silica gel 60 (230–400 mesh, Merck, Darmstadt, Germany) as the stationary phase and a mixture of *n*-hexane and ethyl acetate as the mobile phase. Reversed-phase flash column chromatography was performed using an AI-580S system (Yamazen Science, Inc., Burlingame, CA, USA). C18 resin (ODS-SM, 50 μm, 120 Å) was used as the stationary phase, eluted with a gradient of water and methanol.

The NMR spectra were measured using a Jeol JNM ECP-400 spectrometer (Jeol Ltd., Tokyo, Japan) and a Bruker Avance III HD 600 MHz spectrometer (Bruker Daltonics, Leipzig, Germany), using methanol-d4 (CD_3OD) or chloroform-d4 ($CDCl_3$) as the solvent. Chemical shifts were set based on internal standards or residual solvent signals (MeOH-d_4: δ_H 3.31/δ_C 49.1; $CDCl_3$: δ_H 7.27/δ_C 77.0; tetramethylsilane). Peak splitting patterns were

https://doi.org/10.3390/molecules31061061

abbreviated as m (multiplets), s (singlets), d (doublets), t (triplets), dd (doublet of doublets), dt (doublet of triplets), td (triplet of doublets), and br (broad). Accurate mass was measured using an AB Sciex Triple TOF 4600 instrument (AB Sciex, Framingham, MA, USA) in direct injection mode.

3.2. *General Procedures for the Synthesis of Curvularin Derivatives*

3.2.1. General Procedure A: Esterification (Step 1)

In a flame-dried round-bottom flask, 3,5-dimethoxyphenylacetic acid (1.0 equiv), the corresponding primary alcohol (e.g., 5-hexen-1-ol, 4-propan-1-ol, or (S)-hept-6-en-2-ol; 1.2 equiv), and 2-chloro-1-methylpyridinium iodide (Mukaiyama reagent, 1.2 equiv) were dissolved in anhydrous DCM. The mixture was stirred at room temperature for 15 min. Triethylamine (TEA, 3.0 equiv) was then added dropwise, and the reaction mixture was stirred at room temperature for 20 h. Subsequently, the mixture was filtered and concentrated under reduced pressure. The crude product was purified by flash column chromatography (hexane/EtOAc) to afford the corresponding ester intermediate.

3.2.2. General Procedure B: Friedel–Crafts Acylation (Step 2)

A flame-dried flask was charged with anhydrous DCM and cooled to −78 °C before adding 3-chloropropanoyl chloride (1.3 equiv). A 1 M solution of $SnCl_4$ in DCM (1.3 equiv) was added slowly, and the mixture was stirred for 15 min. The ester intermediate prepared in Step 1 (1.0 equiv) was dissolved in DCM and added dropwise to the reaction mixture at −78 °C. The temperature was gradually raised to −20 °C, and the mixture was stirred for 1.5–2.5 h. The reaction was quenched by pouring the mixture into a separatory funnel containing ice and water. The organic layer was separated, washed successively with saturated aqueous $NaHCO_3$ and water, dried over $MgSO_4$, filtered, and concentrated. The crude product was purified by flash column chromatography to obtain the acylated intermediate.

3.2.3. General Procedure C: Dehydrohalogenation (Step 3)

The acylated intermediate (1.0 equiv) was dissolved in anhydrous DCM, and TEA (2.0 equiv) was added. The mixture was stirred overnight at room temperature to promote dehydrohalogenation and intramolecular ring-forming rearrangement. The reaction was quenched with saturated aqueous NH_4Cl and extracted with DCM. The combined organic layers were dried over $MgSO_4$, filtered, and concentrated. The concentrate was purified using flash column chromatography (hexane/EtOAc, 8:2) to afford the diene intermediate.

3.2.4. General Procedure D: Ring-Closing Metathesis (Step 4)

In a flame-dried round-bottom flask, the diene compound prepared in Step 3 (1.0 equiv) was dissolved in anhydrous toluene to reach a highly dilute concentration (approx. 4 mM). The solution was thoroughly degassed by bubbling argon gas for 2 h. A solution of Grubbs second-generation catalyst (0.2 equiv, 20 mol%) in a small amount of toluene was added, and the reaction mixture was heated to 80 °C and stirred for 1 h. Upon completion, the mixture was concentrated, pre-adsorbed onto silica gel for dry loading, and purified using an open silica-gel column (hexane/EtOAc eluents) to yield the corresponding metathesis products (*E*/*Z* mixtures).

3.2.5. General Procedure E: Selective Demethylation (Step 5)

The metathesis product (1.0 equiv) was dissolved in anhydrous ACN, followed by the addition of aluminum powder (10.0 equiv) and DMSO (5.0 equiv). Iodine (3.3 equiv) was added in one portion, and the reaction mixture was heated to 80 °C for 30 min. After cooling to room temperature, the mixture was diluted with 2 M aqueous HCl and extracted with ethyl acetate. The combined organic layers were dried over $MgSO_4$, filtered, and

https://doi.org/10.3390/molecules31061061

concentrated. The final crude product was purified via flash column chromatography (hexane/EtOAc, 7:3) to yield the target curvularin analogues.

3.2.6. Synthesis of Hex-5-en-1-yl 2-(3,5-dimethoxyphenyl)acetate (**3**)

General procedure A. TLC R_f = 0.57 (hexane/EtOAc, 7:3). ^{1}H NMR (400 MHz, CD_3OD) δ 6.43 (d, *J* = 2.3 Hz, 1H, H-4; 1H, H-8), 6.37 (t, *J* = 2.3 Hz, 1H, H-6), 5.77 (m, 1H, H-5′), 4.98 (m, 2H, H-6′), 4.93 (m, 2H, H-6′), 4.09 (t, *J* = 6.4 Hz, 2H, H-1′), 3.75 (s, 3H, H-5 OMe; 3H, H-7 OMe), 3.52 (s, 2H, H-2), 2.04 (m, 2H, H-4′), 1.62 (m, 2H, H-2′), 1.40 (m, 2H, H-3′); 13C NMR (100 MHz, CD3OD) δ 173.4, 162.4, 162.4, 139.5, 137.7, 115.2, 108.3, 108.3, 100.0, 65.8, 55.7, 55.7, 42.3, 34.3, 29.1, 26.3; HRESIMS *m/z* 279.1596 [M + H]$^+$ (calcd for $C_{16}H_{23}O_4$, 279.1596).

3.2.7. Synthesis of Hex-5-en-1-yl 2-(2-(3-chloropropanoyl)-3,5-dimethoxyphenyl)acetate (**4**)

General procedure B. TLC R_f = 0.42 (hexane/EtOAc, 7:3). ^{1}H NMR (400 MHz, $CDCl_3$) δ 6.41 (d, *J* = 2.3 Hz, 1H, H-6), 6.37 (d, *J* = 2.3 Hz, 1H, H-4), 5.78 (m, 1H, H-5′), 5.00 (m, 2H, H-6′), 4.95 (m, 2H, H-6′), 4.08 (t, *J* = 6.9 Hz, 2H, H-1′), 3.84 (s, 3H, H-5 OMe), 3.82 (s, 3H, H-7 OMe), 3.82 (m, 2H, H-11), 3.67 (s, 2H, H-2), 3.35 (t, *J* = 7.1 Hz, 2H, H-10), 2.05 (m, 2H, H-4′), 1.64 (m, 2H, H-2′), 1.43 (m, 2H, H-3′); ^{13}C NMR (100 MHz, $CDCl_3$) δ 202.5, 171.5, 162.0, 159.5, 138.5, 135.8, 122.8, 114.9, 108.6, 97.6, 65.0, 55.8, 55.6, 47.1, 39.5, 39.2, 33.4, 28.1, 25.5; HRESIMS *m/z* 369.1476 [M + H]$^+$ (calcd for $C_{19}H_{26}ClO_5$, 369.1469).

3.2.8. Synthesis of Hex-5-en-1-yl 2-(2-acryloyl-3,5-dimethoxyphenyl)acetate (**5**)

General procedure C. TLC R_f = 0.42 (hexane/EtOAc, 7:3). ^{1}H NMR (400 MHz, $CDCl_3$) δ 6.68 (dd, *J* = 17.4, 10.5 Hz, 1H, H-10), 6.42 (d, *J* = 2.3 Hz, 1H, H-4), 6.40 (d, *J* = 2.3 Hz, 1H, H-6), 6.09 (dd, *J* = 17.4, 1.4 Hz, 2H, H-11), 5.79 (m, 2H, H-11), 5.76 (m, 1H, H-5′), 4.99 (m, 2H, H-6′), 4.94 (m, 2H, H-6′), 4.05 (t, *J* = 6.9 Hz, 2H, H-1′), 3.82 (s, 3H, H-5 OMe), 3.77 (s, 3H, H-7 OMe), 3.61 (s, 2H, H-2), 2.04 (m, 2H, H-4′), 1.40 (m, 2H, H-3′); ^{13}C NMR (100 MHz, $CDCl_3$) δ 195.8, 171.2, 161.8, 159.3, 138.5, 138.2, 135.5, 128.5, 122.3, 114.9, 107.9, 97.7, 64.9, 55.8, 38.9, 33.3, 28.1, 25.2; HRESIMS *m/z* 333.1632 [M + H]$^+$ (calcd for $C_{19}H_{25}O_5$, 333.1702).

3.2.9. Synthesis of (E)-11,13-Dimethoxy-4,5,6,7-tetrahydro-2H-benzo[d][1] oxacyclododecine-2,10(1H)-dione (**6**)

General procedure D. TLC R_f = 0.38 (hexane/EtOAc, 6:4). ^{1}H NMR (400 MHz, $CDCl_3$) δ 6.55 (d, *J* = 1.8, 1H, H-4), 6.53 (d, *J* = 2.3, 1H, H-6), 6.46 (dd, *J* = 15.8, 7.8, 1H, H-11), 6.21 (dt, *J* = 15.6, 1.2 Hz, 1H, H-10), 4.06 (t, *J* = 4.6, 2H, H-15), 3.83 (s, 3H, H-5 OMe), 3.75 (s, 3H, H-7 OMe), 3.40 (s, 2H, H-2), 2.31 (m, 2H, H-14), 1.71 (m, 2H, H-12; 2H, H-13); ^{13}C NMR (100 MHz, $CDCl_3$) δ 201.3, 172.2, 162.8, 159.4, 159.0, 133.9, 133.7, 125.6, 108.3, 98.5, 68.1, 56.2, 56.0, 40.1, 34.8, 28.3, 27.2; HRESIMS *m/z* 305.1378 [M + H]$^+$ (calcd for $C_{17}H_{21}O_5$, 305.1389).

3.2.10. Synthesis of 11,13-Dimethoxy-4,5,6,7-tetrahydro-2H-benzo[d][1] oxacyclododecine-2,10(1H)-dione (**7**)

General procedure D. TLC R_f = 0.48 (hexane/EtOAc, 6:4). ^{1}H NMR (400 MHz, $CDCl_3$) δ 6.47 (m, 1H, H-10), 6.41 (m, 1H, H-4), 6.37 (m, 1H, H-6), 5.90 (dt, *J* = 11.9, 8.2 Hz, 1H, H-11), 4.01 (m, 2H, H-15), 3.83 (s, 3H, H-5 OMe), 3.74 (s, 3H, H-7 OMe), 3.41 (s, 2H, H-2), 2.29 (m, 2H, H-12), 1.60 (m, 2H, H-13), 1.60 (m, 2H, H-14); ^{13}C NMR (100 MHz, $CDCl_3$) δ 197.6, 170.7, 161.7, 159.3, 144.1, 134.4, 133.1, 124.5, 108.0, 97.9, 67.6, 56.0, 55.6, 39.3, 28.2, 27.6, 24.8; HRESIMS *m/z* 305.1378 [M + H]$^+$ (calcd for $C_{17}H_{21}O_5$, 305.1389).

3.2.11. Synthesis of (E)-11-Hydroxy-13-methoxy-4,5,6,7-tetrahydro-2H-benzo[d][1] oxacyclododecine-2,10(1H)-dione (**8**)

General procedure E. TLC R_f = 0.53 (hexane/EtOAc, 1:1). ^{1}H NMR (600 MHz, $CDCl_3$) δ 6.70 (m, 1H, H-10), 6.60 (m, 1H, H-11), 6.42 (d, *J* = 2.6 Hz, 1H, H-6), 6.33 (t, *J* = 2.9 Hz,

1H, H-4), 4.17 (dd, *J* = 6.2, 2.5 Hz, 2H, H-15), 3.83 (s, 3H, OMe), 3.82 (s, 2H, H-2), 2.45 (m, 2H, H-14), 1.83 (m, 2H, H-12; 2H, H-13); ^{13}C NMR (150 MHz, $CDCl_3$) δ 196.0, 171.3, 167.0, 164.3, 148.1, 137.4, 131.7, 114.0, 113.6, 100.4, 66.3, 55.5, 43.9, 32.8, 27.1, 26.7; HRESIMS *m*/*z* 291.1223 $[M + H]^+$ (calcd for $C_{16}H_{19}O_5$, 291.1232).

3.2.12. Synthesis of Pent-4-en-1-yl 2-(3,5-dimethoxyphenyl)acetate (**11**)

General procedure A. TLC R_f = 0.60 (hexane/EtOAc, 7:3). ^{1}H NMR (400 MHz, $CDCl_3$) δ 6.44 (d, *J* = 2.3 Hz, 1H, H-4; 1H, H-8), 6.37 (t, *J* = 2.3 Hz, 1H, H-6), 5.77 (m, 1H, H-4′), 5.00 (m, 2H, H-5′), 4.97 (m, 2H, H-5′), 4.10 (t, *J* = 6.6 Hz, 2H, H-1′), 3.78 (s, 3H, H-5 OMe; 3H, H-7 OMe), 3.55 (s, 2H, H-2′), 2.09 (m, 2H, H-3′), 1.72 (m, 2H, H-2′); ^{13}C NMR (100 MHz, $CDCl_3$) δ 171.5, 161.0, 161.0, 137.5, 136.3, 115.4, 107.4, 107.4, 99.3, 64.4, 55.4, 55.4, 41.8, 30.1, 27.9; HRESIMS *m*/*z* 265.1437 $[M + H]^+$ (calcd for $C_{15}H_{21}O_4$, 265.1440).

3.2.13. Synthesis of Pent-4-en-1-yl 2-(2-(3-chloropropanoyl)-3,5-dimethoxyphenyl) acetate (**12**)

General procedure B. TLC R_f = 0.5 (hexane/EtOAc, 7:3). ^{1}H NMR (400 MHz, $CDCl_3$) δ 6.41 (d, *J* = 2.3 Hz, 1H, H-6), 6.37 (d, *J* = 2.3 Hz, 1H, H-4), 5.78 (m, 1H, H-4′), 5.02 (m, 2H, H-5′), 4.97 (m, 2H, H-5′), 4.09 (t, *J* = 6.6 Hz, 2H, H-1′), 3.84 (s, 3H, H-5 OMe), 3.82 (s, 3H, H-7 OMe), 3.81 (m, 2H, H-11), 3.67 (s, 2H, H-2), 3.35 (t, *J* = 7.1 Hz, 2H, H-10), 2.10 (m, 2H, H-3′), 1.72 (m, 2H, H-2′); ^{13}C NMR (100 MHz, $CDCl_3$) δ 202.5, 171.5, 162.1, 159.5, 137.6, 135.8, 122.8, 115.4, 108.6, 97.6, 64.5, 55.8, 55.6, 47.1, 39.5, 39.2, 30.1, 27.9; HRESIMS *m*/*z* 355.1286 $[M + H]^+$ (calcd for $C_{18}H_{24}ClO_5$, 355.1312).

3.2.14. Synthesis of Pent-4-en-1-yl 2-(2-acryloyl-3,5-dimethoxyphenyl)acetate (**13**)

General procedure C. TLC R_f = 0.78 (hexane/EtOAc, 6:4). ^{1}H NMR (400 MHz, $CDCl_3$) δ 6.69 (dd, *J* = 17.4, 10.5 Hz, 1H, H-10), 6.42 (d, *J* = 2.3 Hz, 1H, H-4), 6.41 (d, *J* = 2.3 Hz, 1H, H-6), 6.09 (dd, *J* = 17.4, 1.4 Hz, 2H, H-11), 5.79 (m, 2H, H-11), 5.76 (m, 1H, H-4′), 4.97 (m, 2H, H-5′), 4.05 (t, *J* = 6.6 Hz, 2H, H-1′), 3.82 (s, 3H, H-5 OMe), 3.77 (s, 3H, H-7 OMe), 3.61 (s, 2H, H-2), 2.07 (m, 2H, H-3′), 1.69 (m, 2H, H-2′); ^{13}C NMR (100 MHz, $CDCl_3$) δ 195.8, 171.2, 161.8, 159.3, 138.2, 137.6, 135.5, 128.5, 122.3, 115.3, 107.9, 97.7, 64.4, 55.8, 55.5, 38.9, 30.1, 27.8; HRESIMS *m*/*z* 319.1517 $[M + H]^+$ (calcd for $C_{18}H_{23}O_5$, 319.1545).

3.2.15. Synthesis of (E)-10,12-Dimethoxy-5,6-dihydrobenzo[d][1]oxacycloundecine-2,9(1H,4H)-dione (**14**)

General procedure D. TLC R_f = 0.44 (hexane/EtOAc, 6:4). ^{1}H NMR (400 MHz, $CDCl_3$) δ 6.43 (s, 1H, H-4; 1H, H-6), 6.38 (m, 1H, H-11), 6.25 (td, *J* = 16.0, 1.4 Hz, 1H, H-10), 4.29 (m, 2H, H-14), 3.84 (s, 3H, H-5 OMe), 3.76 (s, 3H, H-7 OMe), 3.36 (s, 2H, H-2), 2.45 (m, 2H, H-12), 2.36 (m, 2H, H-12), 2.02 (m, 2H, H-13), 1.78 (m, 2H, H-13); ^{13}C NMR (100 MHz, $CDCl_3$) δ 197.9, 171.1, 161.3, 157.8, 155.8, 133.2, 132.5, 122.9, 108.4, 97.9, 65.1, 56.1, 55.6, 40.3, 31.1, 26.9; HRESIMS *m*/*z* 291.1132 $[M + H]^+$ (calcd for $C_{16}H_{19}O_5$, 291.1232).

3.2.16. Synthesis of 10,12-Dimethoxy-5,6-dihydrobenzo[d][1]oxacycloundecine-2,9(1H,4H)-dione (**15**)

General procedure D. TLC R_f = 0.58 (hexane/EtOAc, 6:4). ^{1}H NMR (400 MHz, $CDCl_3$) δ 6.49 (s, 1H, H-4; 1H, H-6), 6.38 (m, 1H, H-11), 5.93 (m, 1H, H-10), 4.29 (m, 2H, H-14), 3.84 (s, 3H, H-5 OMe), 3.76 (s, 3H, H-7 OMe), 3.36 (s, 2H, H-2), 2.45 (m, 2H, H-12), 2.36 (m, 2H, H-12), 2.02 (m, 2H, H-13), 1.80 (m, 2H, H-13); ^{13}C NMR (100 MHz, $CDCl_3$) δ 197.9, 171.1, 161.3, 157.8, 140.4, 133.2, 132.5, 122.9, 108.4, 97.9, 65.1, 56.1, 55.6, 40.3, 31.1, 26.9; HRESIMS *m*/*z* 291.1132 $[M + H]^+$ (calcd for $C_{16}H_{19}O_5$, 291.1232).

https://doi.org/10.3390/molecules31061061

3.2.17. Synthesis of (E)-10-Hydroxy-12-methoxy-5,6-dihydrobenzo[d][1]oxacycloundecine-2,9(1H,4H)-dione (**16**)

General procedure E. TLC R_f = 0.55 (hexane/EtOAc, 1:1). ^{1}H NMR (600 MHz, $CDCl_3$) δ 6.45 (d, *J* = 2.5 Hz, 1H, H-4), 6.40 (d, *J* = 2.5 Hz, 1H, H-6), 6.36 (dt, *J* = 16.3, 1.5 Hz, 1H, H-10), 5.91 (dt, *J* = 16.2, 7.0 Hz, 1H, H-11), 4.31 (t, *J* = 6.0 Hz, 2H, H-14), 3.83 (s, 3H, OMe), 3.46 (s, 2H, H-2), 2.40 (m, 2H, H-12), 1.97 (m, 2H, H-13); ^{13}C NMR (150 MHz, $CDCl_3$) δ 200.6, 171.7, 164.4, 163.6, 147.0, 137.3, 131.7, 116.1, 113.5, 100.3, 64.8, 55.7, 42.2, 30.5, 24.6; HRESIMS *m*/*z* 277.0906 [M + H]$^+$ (calcd for $C_{15}H_{17}O_5$, 277.1076).

3.2.18. Synthesis of (S)-Hept-6-en-2-yl 2-(3,5-dimethoxyphenyl)acetate (**18**)

General procedure A. TLC R_f = 0.67 (hexane/EtOAc, 7:3). ^{1}H NMR (400 MHz, CD_3OD) δ 6.43 (d, *J* = 2.3 Hz, 1H, H-4; 1H, H-8), 6.37 (t, *J* = 2.3 Hz, 1H, H-6), 5.74 (m, 1H, H-5′), 4.96 (m, 2H, H-6′), 4.93 (m, 1H, H-1′), 4.90 (m, 2H, H-6′), 3.75 (s, 3H, H-5 OMe; 3H, H-7 OMe), 3.52 (s, 2H, H-2), 1.99 (m, 2H, H-4′), 1.54 (m, 2H, H-2′), 1.34 (m, 2H, H-3′), 1.20 (d, *J* = 6.4 Hz, 3H, H-7′); ^{13}C NMR (100 MHz, CD_3OD) δ 173.1, 162.4, 162.4, 139.6, 137.8, 115.1, 108.3, 108.3, 100.0, 72.7, 55.7, 55.7, 42.8, 36.4, 34.5, 25.7, 20.2; HRESIMS *m*/*z* 293.1761 [M + H]$^+$ (calcd for $C_{17}H_{25}O_4$, 293.1753).

3.2.19. Synthesis of (S)-Hept-6-en-2-yl 2-(2-(3-chloropropanoyl)-3,5-dimethoxyphenyl) acetate (**19**)

General procedure B. TLC R_f = 0.5 (hexane/EtOAc, 7:3). ^{1}H NMR (600 MHz, $CDCl_3$) δ 6.40 (d, *J* = 2.3 Hz, 1H, H-6), 6.37 (d, *J* = 2.3 Hz, 1H, H-4), 5.78 (m, 1H, H-5′), 5.00 (m, 2H, H-6′), 4.95 (m, 2H, H-6′), 4.89 (m, 2H, H-1′), 3.83 (s, 3H, H-5 OMe), 3.82 (m, 2H, H-11), 3.81 (s, 3H, H-7 OMe), 3.67 (s, 2H, H-2), 3.35 (t, *J* = 7.1 Hz, 2H, H-10), 2.03 (m, 2H, H-4′), 1.58 (m, 2H, H-2′), 1.49 (m, 2H, H-2′), 1.38 (m, 2H, H-3′), 1.21 (d, *J* = 6.4 Hz, 3H, H-7′); ^{13}C NMR (150 MHz, $CDCl_3$) δ 202.3, 170.9, 161.8, 159.2, 138.4, 135.6, 122.6, 114.6, 108.2, 97.3, 71.4, 55.6, 55.3, 46.9, 39.3, 39.2, 35.2, 33.4, 24.5, 19.8; HRESIMS *m*/*z* 383.1616 [M + H]$^+$ (calcd for $C_{20}H_{28}ClO_5$, 383.1625).

3.2.20. Synthesis of (S)-Hept-6-en-2-yl 2-(2-acryloyl-3,5-dimethoxyphenyl)acetate (**20**)

General procedure C. TLC R_f = 0.60 (hexane/EtOAc, 6:4). ^{1}H NMR (600 MHz, $CDCl_3$) δ 6.68 (dd, *J* = 17.4, 10.5 Hz, 1H, H-10), 6.43 (d, *J* = 2.3 Hz, 1H, H-4), 6.41 (d, *J* = 2.3 Hz, 1H, H-6), 6.09 (dd, *J* = 17.4, 1.4 Hz, 2H, H-11), 5.79 (m, 2H, H-11), 5.76 (m, 1H, H-5′), 4.96 (m, 2H, H-6′), 4.94 (m, 2H, H-6′), 4.88 (m, 2H, H-1′), 3.83 (s, 3H, H-5 OMe), 3.78 (s, 3H, H-7 OMe), 3.60 (s, 2H, H-2), 2.03 (m, 2H, H-4′), 1.57 (m, 2H, H-2′), 1.36 (m, 2H, H-3′), 1.18 (d, *J* = 6.4 Hz, 3H, H-7′); ^{13}C NMR (150 MHz, $CDCl_3$) δ 195.8, 170.7, 161.6, 159.1, 138.5, 138.1, 135.4, 128.5, 122.2, 114.6, 107.6, 97.5, 71.4, 55.6, 55.4, 38.9, 35.2, 33.4, 24.5, 19.9; HRESIMS *m*/*z* 347.1882 [M + H]$^+$ (calcd for $C_{20}H_{27}O_5$, 347.1858).

3.2.21. Synthesis of (S,E)-11,13-Dimethoxy-4-methyl-4,5,6,7-tetrahydro-2H-benzo[d][1]oxacyclododecine-2,10(1H)-dione (**21**)

General procedure D. TLC R_f = 0.38 (hexane/EtOAc, 6:4). ^{1}H NMR (400 MHz, CD_3OD) δ 6.54 (d, *J* = 2.3 Hz, 1H, H-4), 6.53 (d, *J* = 2.3 Hz, 1H, H-6), 6.48 (dd, *J* = 15.6, 7.8 Hz, 1H, H-11), 6.21 (dt, *J* = 15.6, 1.2 Hz, 1H, H-10), 4.89 (m, 1H, H-15), 3.83 (s, 3H, H-5 OMe), 3.75 (s, 3H, H-7 OMe), 3.37 (s, 2H, H-2), 2.33 (m, 2H, H-12), 2.20 (m, 2H, H-12), 1.90 (m, 2H, H-13), 1.80 (m, 2H, H-14), 1.49 (m, 2H, H-14), 1.42 (m, 2H, H-13), 1.15 (d, *J* = 6.4 Hz, 3H, H-16); ^{13}C NMR (100 MHz, CD_3OD) δ 201.2, 172.0, 162.8, 159.4, 159.0, 134.1, 133.8, 123.4, 108.3, 98.5, 74.5, 56.3, 56.0, 40.2, 35.1, 35.0, 25.5, 20.5; HRESIMS *m*/*z* 319.1466 [M + H]$^+$ (calcd for $C_{18}H_{23}O_5$, 319.1545).

https://doi.org/10.3390/molecules31061061

3.2.22. (S)-11,13-Dimethoxy-4-methyl-4,5,6,7-tetrahydro-2H-benzo[d][1]oxacyclododecine-2,10(1H)-dione (**22**)

General procedure D. TLC R_f = 0.45 (hexane/EtOAc, 6:4). ^{1}H NMR (400 MHz, CD_3OD) δ 6.53 (d, *J* = 1.8 Hz, 1H, H-4), 6.51 (d, *J* = 2.3 Hz, 1H, H-6), 6.37 (dt, *J* = 12.4, 1.6 Hz, 1H, H-10), 5.99 (dt, *J* = 12.4, 8.2 Hz, 1H, H-11), 4.80 (m, 1H, H-15), 3.83 (s, 3H, H-5 OMe), 3.78 (s, 3H, H-7 OMe), 3.37 (s, 2H, H-2), 1.40–2.25 (m, 2H, H-12; 2H, H-13; 2H, H-14), 1.13 (d, *J* = 6.4 Hz, 3H, H-16); ^{13}C NMR (100 MHz, CD_3OD) δ 199.9, 171.8, 163.6, 160.6, 146.3, 135.2, 133.8, 125.4, 109.4, 98.5, 75.2, 56.5, 56.0, 40.1, 31.9, 28.4, 26.0, 19.8; HRESIMS *m/z* 319.1466 $[M + H]^+$ (calcd for $C_{18}H_{23}O_5$, 319.1545).

3.2.23. Synthesis of (S,E)-11-Hydroxy-13-methoxy-4-methyl-4,5,6,7-tetrahydro-2H-benzo[d][1]oxacyclododecine-2,10(1H)-dione (**23**)

General procedure E. TLC R_f = 0.64 (hexane/EtOAc, 1:1). ^{1}H NMR (400 MHz, $CDCl_3$) δ 6.67 (dd, *J* = 16.5, 1.0 Hz, 1H, H-10), 6.62 (ddd, *J* = 15.1, 8.2, 3.6 Hz, 1H, H-11), 6.41 (d, *J* = 2.8 Hz, 1H, H-6), 6.32 (d, *J* = 2.8 Hz, 1H, H-4), 4.85 (m, 1H, H-15), 4.04 (d, *J* = 17.9 Hz, 2H, H-2), 3.82 (s, 3H, OMe), 3.54 (d, *J* = 17.9 Hz, 2H, H-2), 2.47 (m, 2H, H-12), 2.32 (m, 2H, H-12), 1.97 (m, 2H, H-14), 1.89 (m, 2H, H-13), 1.65 (m, 2H, H-14), 1.64 (m, 2H, H-13), 1.24 (d, *J* = 6.4 Hz, 3H, H-16); ^{13}C NMR (100 MHz, $CDCl_3$) δ 196.4, 171.3, 166.9, 164.4, 148.3, 137.6, 131.5, 114.3, 113.6, 100.6, 72.7, 55.6, 44.2, 34.1, 32.8, 24.2, 20.1; HRESIMS *m/z* 305.1411 $[M + H]^+$ (calcd for $C_{17}H_{21}O_5$, 305.1389).

3.3. Measurement of NO Production

Nitrite concentrations were measured in culture supernatants using the Griess reagent (1% sulfanilamide, 0.1% *N*-(1-naphthyl)-ethylenediamine dihydrochloride, 5% phosphoric acid). Cells were seeded in 96-well plates at a density of 5×10^5 cells/mL and treated with specific concentrations of compounds **6**, **7**, **8**, **14**, **15**, **16**, **21**, **22**, and **23** at 37 °C for 1 h. Subsequently, cells were stimulated with 0.5 µg/mL LPS (Sigma-Aldrich, Carlsbad, CA, USA) and incubated for 24 h at a final volume of 200 µL. Thereafter, 100 µL of the cell culture supernatant was mixed with an equal volume of the Griess reagent. The mixture was incubated at room temperature for 5 min. A standard curve was generated using sodium nitrite, and the absorbance was measured at 540 nm to calculate the nitrite concentration. All experiments were performed in triplicate.

3.4. Cytotoxicity Measurement

Cytotoxicity was determined using the 3-(4,5-dimethyl-2-thiazolyl)-2,5-diphenyl-2H-tetrazolium bromide (MTT, Amresco, Solon, OH, USA) assay. RAW 264.7 cells were seeded in 96-well plates at a concentration of 1×10^5 cells/mL and incubated at 37 °C for 24 h with various concentrations of compounds **6**, **7**, **8**, **14**, **15**, **16**, **21**, **22**, and **23**. Thereafter, 5 µL of a 5 mg/mL MTT solution was added to the cells and incubated at 37 °C for 4 h. To dissolve the crystals, 100 µL of fresh DMSO was added and incubated for 10 min. The absorbance was measured at 570 nm using a microplate reader (Thermo Scientific Inc., San Diego, CA, USA). Relative cell viability was calculated based on the absorbance of the untreated control. All experiments were performed in triplicate.

3.5. Measurement of Antibiotic and Antibacterial Activity

Antibacterial and antifungal activity assays were performed using 96-well plates with *Staphylococcus aureus* KCTC 3881 (bacterium), *Escherichia coli* DH5α (bacterium), and *Candida albicans* KCTC 27242 (yeast) (Korean Collection for Type Cultures, Daejeon, Republic of Korea). Cell cultures were diluted with a sterile medium to a 0.5 McFarland Standard, and the *C. albicans* culture was further diluted 100-fold before use. After dispensing 95 µL of culture medium into each well, compounds dissolved in DMSO were added to achieve final concentrations of 12.5, 25, 50, 100, and 200 μg/mL, and the final volume in each well was adjusted to 100 μL. The plates were incubated at 37 °C for 24 h, and cell inhibition was measured at 600 nm for *S. aureus* and *E. coli* and at 530 nm for *C. albicans* using a Multiskan GO Microplate Spectrophotometer (Thermo Scientific, Waltham, MA, USA). IC_{50} values were calculated using an exponential trend line in Excel software (Microsoft, Redmond, WA, USA), and values were expressed as the mean ± SD of triplicate measurements. Kanamycin was used as a positive control for bacteria (*S. aureus* and *E. coli*) and nystatin was used as the positive control for fungi (*C. albicans*).

4. Conclusions

In this study, we synthesized derivatives based on the structure of curvularin isolated from *Penicillium* sp. derived from Antarctic marine sponges, and systematically evaluated their anti-inflammatory, anticancer, and antibacterial activities. Among the synthesized derivatives, compounds **6**, **7**, **8**, **14**, **15**, **16**, and **23** were newly synthesized. Some exhibited high cancer cytotoxicity and inhibitory effects on inflammatory mediators, confirming their potential as anti-inflammatory and anticancer candidates. Furthermore, antibacterial activity tests demonstrated that the novel derivatives exhibited excellent growth inhibition against *S. aureus* and *C. albicans*, suggesting their potential as novel antibacterial agents. This study confirms the possibility of enhancing physiological activity through structural changes in curvularin derivatives, thereby suggesting an important direction for future research on natural product-based drug development.

Supplementary Materials: The following supporting information can be downloaded at: https://www.mdpi.com/article/10.3390/molecules31061061/s1, Synthetic procedures and characterization data for the compounds **3–8**, **11–16**, **18–23**.

Author Contributions: Conceptualization, J.H.Y., I.-C.K. and S.J.H.; methodology, formal analysis, and investigation, K.H.K. and T.K.K.; validation and data curation, K.H.K., J.-M.H., J.A.K. and M.J.K.; writing—original draft preparation, K.H.K. and S.J.H.; writing—review and editing, T.K.K. and S.J.H.; supervision, J.H.Y.; project administration, I.-C.K.; funding acquisition, J.-H.K. All authors have read and agreed to the published version of the manuscript.

Funding: This work was supported by the Korea Polar Research Institute (KOPRI) grant funded by the Ministry of Oceans and Fisheries (KOPRI PE26160).

Institutional Review Board Statement: Not applicable.

Informed Consent Statement: Not applicable.

Data Availability Statement: The original contributions presented in this study are included in the article. Further inquiries can be directed to the corresponding author.

Conflicts of Interest: Author Tai Kyoung Kim was employed by CRYOTECH Inc. The remaining authors declare that the research was conducted in the absence of any commercial or financial relationships that could be construed as a potential conflict of interest.

Abbreviations

The following abbreviations are used in this manuscript:

NO	Nitric oxide
RCM	Ring-closing metathesis
SAR	Structure–activity relationship
SI	Selectivity index
DMSO	Dimethyl sulfoxide

https://doi.org/10.3390/molecules3106106

NMR	Nuclear magnetic resonance
TLC	Thin-layer chromatography
MTT	3-(4,5-dimethyl-2-thiazolyl)-2,5-diphenyl-2H-tetrazolium bromide

eferences

Lund-Hansen, L.C.; Gradinger, R.; Hassett, B.; Jayasinghe, S.; Kennedy, F.; Martin, A.; McMinn, A.; Søgaard, D.H.; Sorrell, B.K. Sea ice as habitat for microalgae, bacteria, virus, fungi, meio- and macrofauna: A review of an extreme environment. *Polar Biol.* **2024**, *47*, 1275–1306. [CrossRef]

Cao, S.; Zhang, W.; Ding, W.; Wang, M.; Fan, S.; Yang, B.; McMinn, A.; Wang, M.; Xie, B.B.; Qin, Q.L.; et al. Structure and function of the Arctic and Antarctic marine microbiota as revealed by metagenomics. *Microbiome* **2020**, *8*, 47. [CrossRef] [PubMed]

Dickinson, I.; Goodall-Copestake, W.; Thorne, M.A.; Schlitt, T.; Avila-Jimenez, M.L.; Pearce, D.A. Extremophiles in an Antarctic marine ecosystem. *Microorganisms* **2016**, *4*, 8. [CrossRef] [PubMed]

Fu, Z.; Gong, X.; Hu, Z.; Wei, B.; Zhang, H. Unveiling biosynthetic potential of an Arctic marine-derived strain *Aspergillus sydowii* MNP-2. *BMC Genom.* **2024**, *25*, 603. [CrossRef]

Song, Y.P.; Ji, N.Y. Chemistry and biology of marine-derived *Trichoderma* metabolites. *Nat. Prod. Bioprospect.* **2024**, *14*, 14. [CrossRef]

Arrieche, D.; Cabrera-Pardo, J.R.; San-Martin, A.; Carrasco, H.; Taborga, L. Natural products from Chilean and Antarctic marine fungi and their biomedical relevance. *Mar. Drugs* **2023**, *21*, 00098. [CrossRef]

Cavalcante, S.B.; Dos Santos Biscaino, C.; Kreusch, M.G.; da Silva, A.F.; Duarte, R.T.D.; Robl, D. The hidden rainbow: The extensive biotechnological potential of Antarctic fungi pigments. *Braz. J. Microbiol.* **2023**, *54*, 1675–1687. [CrossRef]

Jodłowska, I.; Białkowska, A.M. Cold-adapted fungi: Goldmine of biomolecules applicable in industry. *Appl. Sci.* **2024**, *14*, 11950. [CrossRef]

Zhao, H.; Cai, C.; Liu, X.; Jiao, B.; Chen, B.; Cai, M.; He, P. Secondary metabolites of Antarctic fungi antagonistic to aquatic pathogenic bacteria. *Open Life Sci.* **2018**, *13*, 11–21. [CrossRef]

Camacho, K.F.; de Melo Carlos, L.; Bernal, S.P.F.; de Oliveira, V.M.; Ruiz, J.L.M.; Ottoni, J.R.; Vieira, R.; Neto, A.; Rosa, L.H.; Passarini, M.R.Z. Antarctic marine sediment as a source of filamentous fungi-derived antimicrobial and antitumor compounds of pharmaceutical interest. *Extremophiles* **2024**, *28*, 21. [CrossRef]

Ha, T.M.; Kim, D.C.; Sohn, J.H.; Yim, J.H.; Oh, H. Anti-inflammatory and protein tyrosine phosphatase 1B inhibitory metabolites from the Antarctic marine-derived fungal strain *Penicillium glabrum* SF-7123. *Mar. Drugs* **2020**, *18*, 247. [CrossRef]

Hang, S.; Lu, H.; Jiang, Y. Marine-derived metabolites act as promising antifungal agents. *Mar. Drugs* **2024**, *22*, 180. [CrossRef] [PubMed]

Deng, W.; Du, H.; Liu, D.; Ma, Z. Editorial: The role of natural products in chronic inflammation. *Front. Pharmacol.* **2022**, *13*, 901538. [CrossRef] [PubMed]

Newman, D.J. Natural products and drug discovery. *Natl. Sci. Rev.* **2022**, *9*, nwac206. [CrossRef] [PubMed]

Ahmad, M.; Tahir, M.; Hong, Z.; Zia, M.A.; Rafeeq, H.; Ahmad, M.S.; Rehman, S.U.; Sun, J. Plant and marine-derived natural products: Sustainable pathways for future drug discovery and therapeutic development. *Front. Pharmacol.* **2024**, *15*, 1497668. [CrossRef]

Ha, T.M.; Ko, W.; Lee, S.J.; Kim, Y.C.; Son, J.Y.; Sohn, J.H.; Yim, J.H.; Oh, H. Anti-inflammatory effects of curvularin-type metabolites from a marine-derived fungal strain *Penicillium* sp. SF-5859 in lipopolysaccharide-induced RAW264.7 macrophages. *Mar. Drugs* **2017**, *15*, 282. [CrossRef]

Valeur, E.; Bradley, M. Amide bond formation: Beyond the myth of coupling reagents. *Chem. Soc. Rev.* **2009**, *38*, 606–631. [CrossRef]

Mukaiyama, T.; Usui, M.; Shimada, E.; Saigo, K. A convenient method for the synthesis of carboxylic esters. *Chem. Lett.* **1975**, *4*, 1045–1048. [CrossRef]

Allu, S.R.; Banne, S.; Jiang, J.; Qi, N.; Guo, J.; He, Y. A unified synthetic approach to optically pure curvularin-type metabolites. *J. Org. Chem.* **2019**, *84*, 7227–7237. [CrossRef]

Punch, K.A.; Piggott, M.J. Total synthesis of monosporascone and dihydromonosporascone. *Org. Biomol. Chem.* **2014**, *12*, 2801–2810. [CrossRef]

Michaut, A.; Rodriguez, J. Selective construction of carbocyclic eight-membered rings by ring-closing metathesis of acyclic precursors. *Angew. Chem. Int. Ed. Engl.* **2006**, *45*, 5740–5750. [CrossRef]

Deiters, A.; Martin, S.F. Synthesis of oxygen- and nitrogen-containing heterocycles by ring-closing metathesis. *Chem. Rev.* **2004**, *104*, 2199–2238. [CrossRef] [PubMed]

https://doi.org/10.3390/molecules31061061

23. Vougioukalakis, G.C.; Grubbs, R.H. Ruthenium-based heterocyclic carben-coordinated olefin metathesis catalyst. *Chem. Rev.* **201** *110*, 1746–1787. [CrossRef] [PubMed]
24. Wang, Y.R.; Dong, Y.L.; Li, X.M.; Shi, X.S.; Li, H.L.; Meng, L.H.; Xu, R.; Wang, B.G. Curvularin derivatives from the marin mangrove derived fungus *Penicillium sumatrense* MA-325. *Phytochemistry* **2024**, *220*, 114000. [CrossRef] [PubMed]
25. Xie, L.W.; Ouyang, Y.C.; Zou, K.; Wang, G.H.; Chen, M.J.; Sun, H.M.; Dai, S.K.; Li, X. Isolation and difference in anti-*Staphylococcu aureus* bioactivity of curvularin derivates from fungus *Eupenicillium* sp. *Appl. Biochem. Biotechnol.* **2009**, *159*, 284–293. [CrossRe
26. Yu, C.; Xia, Z.; Xu, Z.; Tang, X.; Ding, W.; Wei, J.; Tian, D.; Wu, B.; Tang, J. Curvularin derivatives from hydrothermal ver sediment fungus *Penicillium* sp. HL-50 guided by molecular networking and their anti-inflammatory activity. *Chin. J. Nat. Me* **2025**, *23*, 119–128. [CrossRef]
27. Yang, J.; Yao, F.H.; Xu, S.F.; Shi, J.Y.; Li, X.Y.; Yi, X.X.; Gao, C.H. Mauritone A, a new polyketide from a fungal-bacterial symbior *Aspergillus spelaeus* GXIMD 04541/*Sphingomonas echinoides* GXIMD 04532. *Nat. Prod. Res.* **2025**, *39*, 6259–6264. [CrossRef]
28. Santagata, S.; Xu, Y.M.; Wijeratne, E.M.; Kontnik, R.; Rooney, C.; Perley, C.C.; Kwon, H.; Clardy, J.; Kesari, S.; Whitesell, L.; et a Using the heat-shock response to discover anticancer compounds that target protein homeostasis. *ACS Chem. Biol.* **2012**, 340–349. [CrossRef]
29. Banala, R.R.; Vemuri, S.K.; Ev, S.; Av, G.R.; Gpv, S. The anti-inflammatory and cytoprotective efficiency of curvularin, a funga macrolactone against lipopolysaccharide-induced inflammatory response in nucleus pulposus cells: An in vitro study. *Asia Spine J.* **2021**, *15*, 143–154. [CrossRef]
30. Miyagi, T.; Kuwahara, S. A concise synthetic approach to β,γ-dehydrocurvularin: Synthesis of (±)-di-O-methyl-β,γ dehydrocurvularin. *Biosci. Biotechnol. Biochem.* **2007**, *71*, 1592–1594. [CrossRef]
31. Krishna, M.R.; Sridha, G.; Syed, T.; Jayaprakash, H.V. Stereoselective total synthesis of (-)-curvularin. *Synth. Commun.* **2022**, 5 37–42. [CrossRef]
32. Zhao, Q.; Feng, M.; Jin, S.; Liu, X.; Li, S.; Guo, J.; Cheng, X.; Zhou, G.; Yu, X. 10,11-Dehydrocurvularin attenuates inflammation b suppressing NLRP3 inflammasome activation. *Chin. J. Nat. Med.* **2023**, *21*, 163–171. [CrossRef] [PubMed]
33. Elzner, S.; Schmidt, D.; Schollmeyer, D.; Erkel, G.; Anke, T.; Kleinert, H.; Forstermann, U.; Kunz, H. Inhibitors of inducible NC synthase expression: Total synthesis of (S)-curvularin and its ring homologues. *ChemMedChem* **2008**, *3*, 924–939. [CrossRef [PubMed]
34. Mallinson, J.; Collins, I. Macrocycles in new drug discovery. *Future Med. Chem.* **2012**, *4*, 1409–1438. [CrossRef] [PubMed]
35. Sun, S.; Fu, J. Methyl-containing pharmaceuticals: Methylation in drug design. *Bioorg. Med. Chem. Lett.* **2018**, *28*, 3283–328 [CrossRef]
36. Wang, C.; Zaman, K.; Sarotti, A.M.; Wu, X.; Zheng, S.L.; Cao, S. NF-κB inhibitory, antimicrobial and antiproliferative potentials o compounds from Hawaiian fungus *Aspergillus polyporicola* FS910. *3 Biotech* **2021**, *11*, 391. [CrossRef]
37. Greve, H.; Schupp, P.J.; Eguereva, E.; Kehraus, S.; Kelter, G.; Maier, A.; Fiebig, H.H.; Konig, G.M. Apralactone A and a nev stereochemical class of curvularins from the marine fungus *Curvularia* sp. *Eur. J. Org. Chem.* **2008**, *2008*, 5085–5092. [CrossRef]
38. Chang, C.F.; Ke, C.Y.; Wu, Y.C.; Chuang, T.H. Structure-Activity Relationship of Synthetic 2-Phenylnaphthalenes with Hydroxy Groups that Inhibit Proliferation and Induce Apoptosis of MCF-7 Cancer Cells. *PLoS ONE* **2015**, *10*, e0141184. [CrossRef]
39. Selassie, C.D.; Kapur, S.; Verma, R.P.; Rosario, M. Cellular apoptosis and cytotoxicity of phenolic compounds: A quantita-tiv structure-activity relationship study. *J. Med. Chem.* **2005**, *48*, 7234–7242. [CrossRef]
40. Zhang, A.Y.; Shi, X.S.; Wang, B.G.; Meng, L.H. New curvularin and norlignanolide derivatives from cocultures of marin mangrove endophytic fungus *Penicillium brocae* MA-231 and phytopathogen *Curvularia spicifera* QA-26. *Chem. Biodivers.* **2025**, 2 e202500761. [CrossRef]
41. Zhou, J.T.; Wu, Q.; Zhao, J.X.; Wu, L.L.; He, X.H.; Liang, L.Q.; Zhang, G.H.; Li, J.; Xu, W.F.; Yang, R.Y. Sucurchalasins A and E sulfur-containing heterodimers of a cytochalasan and a macrolide from the endophytic fungus *Aspergillus spelaeus* GDGJ-286. *Nat. Prod.* **2024**, *87*, 2327–2334. [CrossRef]
42. Busi, S.; Peddikotla, P.; Upadyayula, S.M.; Yenamandra, V. Secondary metabolites of Curvularia oryzae MTCC 2605. *Rec. Na Prod.* **2009**, *3*, 204–208.
43. Dai, J.; Krohn, K.; Flörke, U.; Pescitelli, G.; Kerti, G.; Papp, T.; Kövér, K.E.; Bényei, A.C.; Draeger, S.; Schulz, B.; et al. Curvularin type metabolites from the fungus *Curvularia* sp. isolated from a marine alga. *Eur. J. Org. Chem.* **2010**, *2010*, 6928–693 [CrossRef]
44. Huo, R.Y.; Zhang, J.X.; Jia, J.; Bi, H.K.; Liu, L. Alternarialone A, a new curvularin-type metabolite from the mangrove-derivec fungus *Alternaria longipes*. *J. Asian Nat. Prod. Res.* **2023**, *25*, 610–616. [CrossRef]
45. Mondol, M.A.; Farthouse, J.; Islam, M.T.; Schuffler, A.; Laatsch, H. Metabolites from the endophytic fungus *Curvularia* sp. M12 ac as motility inhibitors against *Phytophthora capsici* zoospores. *J. Nat. Prod.* **2017**, *80*, 347–355. [CrossRef]

https://doi.org/10.3390/molecules3106106

Cadelis, M.M.; Li, S.A.; van de Pas, S.J.; Grey, A.; Mulholland, D.; Weir, B.S.; Copp, B.R.; Wiles, S. Antimicrobial natural products from plant pathogenic fungi. *Molecules* **2023**, *28*, 1142. [CrossRef]
de Souza, A.O.; Galetti, F.; Silva, C.L.; Bicalho, B.; Parma, M.M.; Fonseca, S.F.; Marsaioli, A.J.; Trindade, A.C.L.B.; Freitas Gil, R.P.; Bezerra, F.S.; et al. Antimycobacterial and cytotoxicity activity of synthetic and natural compounds. *Química Nova* **2007**, *30*, 1563–1566. [CrossRef]

https://doi.org/10.3390/molecules31061061

molecules

MDPI

Article

Metabolomics, Molecular Networking and Phytochemical Investigation of *Psiadia dentata* (Cass.) DC., Endemic to Reunion Island: Discovery of Novel Bioactive Molecules

Lantomalala Elsa Razafindrabenja [1,2], Keshika Mahadeo [1], Gaëtan Herbette [3,4], Lúcia Mamede [5], Michel Frederich Carole Di Giorgio [6], Béatrice Baghdikian [7], Patricia Clerc [1], Hippolyte Kodja [8], Isabelle Grondin [1] and Anne Gauvin-Bialecki [1,*]

1 Laboratoire de Chimie et de Biotechnologie des Produits Naturels (ChemBioPro), Université de La Réunion, 15 Avenue René Cassin, CS 92 003, 97744 Saint-Denis Cedex 9, La Réunion, France; lantomalala.razafindrabenja@univ-reunion.fr (L.E.R.); keshika.mahadeo@univ-reunion.fr (K.M.); isabelle.grondin@univ-reunion.fr (I.G.)
2 SPO, University of Montpellier, INRAE, Institut Agro, 34060 Montpellier, France
3 Aix-Marseille Université, CNRS, Centrale Méditerranée, FSCM, Spectropole, Campus de St Jérôme-Service 511, 13397 Marseille, France; gaetan.herbette@univ-amu.fr or gaetan.herbette@cnrs.fr
4 Université Paris-Saclay, CNRS, Institut de Chimie des Substances Naturelles, UPR 2301, 91198 Gif-sur-Yvette, France
5 Pharmacognosy Laboratory, Department of Pharmacy, Centre Interfacultaire de Recherche sur le Médicament (CIRM), University of Liège, Campus du Sart-Tilman, Quartier Hôpital, Avenue Hippocrate, 15, B36, 4000 Liège, Belgium; lucia.c.mamede@outlook.pt (L.M.); m.frederich@uliege.be (M.F.)
6 IMBE, Aix Marseille Université, Avignon Université, CNRS 7263, IRD 237, 27 Bd Jean Moulin, Service of Environmental Mutagenesis, Faculty of Pharmacy, 13385 Marseille, France; carole.di-giorgio@univ-amu.fr
7 IMBE, Aix Marseille Université, Avignon Université, CNRS 7263, IRD 237, 27 Bd Jean Moulin, Service of Pharmacognosy, Faculty of Pharmacy, 13385 Marseille, France; beatrice.baghdikian@univ-amu.fr
8 Qualisud, University Montpellier, CIRAD, Institut Agro, Avignon Université, Université de La Réunion, 34093 Montpellier, France; hippolyte.kodja@univ-reunion.fr
* Correspondence: anne.bialecki@univ-reunion.fr; Tel.: +33-262-(0)2-6293-8197

Abstract

The genus *Psiadia* (Asteraceae), widely distributed in Madagascar and the Mascarene Islands (Mauritius, La Réunion, Rodrigues), is traditionally used to treat bronchitis, asthma, colds, abdominal pain, and other inflammatory disorders. However, few studies have scientifically validated these traditional medicinal uses. To assess *P. dentata* as a valuable source of bioactive natural products, a combined ^{1}H NMR-based metabolomic, molecular networking, and phytochemical study was conducted. Multivariate analysis (PLS-DA) of crude extracts from *Psiadia* species collected on Reunion Island enabled rapid discrimination of active extracts from *P. dentata* and revealed two methoxylated flavonoids and one coumarin as metabolites correlated with its antiplasmodial and anti-inflammatory activities. Additionally, UHPLC-DAD-ESI-QTOF-MS/MS molecular networking approach enabled detailed chemical profiling of this species, allowing the annotation of 25 compounds (**1–25**) in this species. Subsequent phytochemical investigation of *P. dentata* leaves led to the isolation and identification of 25 metabolites, including nine new diterpenes (**26–34**), one new coumarin (**35**), and 15 known compounds (**1–8**, **11**, **18**, **19** and **36–39**) from the diterpenoid, flavonoid, and coumarin families. The structures of the new compounds were elucidated using spectroscopic methods, including extensive 1D and 2D NMR and HRESIMS analyses. Biological evaluation of the isolated compounds showed that compounds **1**, **7**, **26** and **27** showed antiplasmodial activity against *Plasmodium falciparum* (3D7 strain, IC_{50} = 7.25–13.46 μM). Compounds **7**, **26**, **27**, **31** and **32** inhibited nitric oxide production (IC_{50} = 0.87–27.71 μM), indicating potential anti-inflammatory effects. Only compound **1** displayed moderate cytotoxicity against HepG2 and HT29 cancer cell lines (IC_{50} = 25.67 and 18.35 μM, respectively).

Keywords: *Psiadia dentata*; metabolomics; PLS-DA; molecular networking; diterpenes; coumarin; antiplasmodial; anti-inflammatory; cytotoxic activities

1. Introduction

The importance of plant-specialized metabolites in medicine has encouraged many researchers to investigate the phytochemistry of diverse plants. Indeed, medicines derived from natural sources constitute a significant proportion of our therapeutic drugs. Over the past four decades, more than 400 naturally occurring products and their structural derivatives have been approved as therapeutic agents [1]. Their structural diversity and biological relevance have led to major advances in the treatment of cancer, infectious diseases, and inflammatory disorders.

Advances in phytochemistry and analytical chemistry have significantly enhanced our ability to explore plant composition in depth and to identify bioactive constituents through high-throughput, integrative approaches, such as metabolomics or molecular networking. These advanced strategies allow rapid profiling of complex mixtures and offer an improved expedited route for drug discovery.

Metabolomics is the comprehensive, large-scale study of primary and specialized metabolites in an organism at a given time. The development of analytical tools for metabolomics, such as proton nuclear magnetic resonance (^{1}H NMR) spectroscopy and mass spectrometry (MS), has led to rapid progress in this field. ^{1}H NMR analysis is currently used for the simultaneous detection and quantification of multiple metabolites in a single, non-destructive measurement with minimal sample preparation [2]. This approach has been widely applied in clinical research to elucidate disease mechanisms or identify biomarkers for disease diagnosis [3]. ^{1}H NMR-based metabolomic fingerprinting, combined with statistical analysis methods, has received considerable attention in the quality control of medicinal plants used in herbal preparations, particularly for detecting adulteration and identifying adulterants [4–7]. Recently, metabolomics has emerged as a rapid, alternative screening method to correlate chemical and biological data of natural products. Several studies have demonstrated the ability of NMR spectroscopy, when coupled with multivariate analysis, to predict the bioactivity of plant extracts. For instance, Heyman et al. showed that metabolomics could be used to discriminate bioactive fractions and identify bioactive compounds in crude extracts without requiring bioassay-guided fractionation and purification steps [8–10].

Another powerful and emerging approach to the dereplication of complex natural products is liquid chromatography-tandem mass spectrometry (LC-MS/MS)-based molecular networking. This approach organizes and visualizes MS/MS data through a spectral similarity map, revealing clusters of compounds that share similar fragmentation patterns [11]. Using this tool, comprehensive metabolic profiling can be achieved through a single analysis of the plant extract. The introduction of Global Natural Product Social molecular networking (GNPS) web platform (http://gnps.ucsd.edu; accessed on 22 May 2020) has enabled automatic spectral mining [12,13].

In the present work, the contributions of these approaches to identify molecular families and putative bioactive candidate molecules in natural product research are explored. As part of our ongoing efforts to discover bioactive natural products within the genus *Psiadia*, and to apply this integrated metabolomic and phytochemical strategy, *Psiadia dentata* (Cass.)

https://doi.org/10.3390/molecules31060973

DC. (Figure 1), an endemic species of Reunion Island with recognized ethnobotanical and pharmacological relevance, was selected for this study.

Figure 1. Exterior view of *P. dentata* (Cass.) DC. (Photo by the author, 2015).

Species of the genus *Psiadia*, belonging to the (Asteraceae) family, are native to both Madagascar and the Mascarene Islands (Reunion, Mauritius and Rodrigues). This genus comprises approximately 28 species in Madagascar and 26 species in the Mascarene Islands [14]. The leaves of *Psiadia* plants have long been used in ethnomedicine to treat a variety of ailments such as abdominal pains, fever, bronchitis, and asthma [15]. *Psiadia dentata* (Cass.) DC. commonly known as *Ti-mangue*, *Bois collant*, *Bois de marron*, *Bois de reinette blanc*, or *Herbe trois jours* [14], is traditionally used in the form of cataplasm to treat abscesses and skin infections caused by fungi or insect bites [16,17]. In previous studies, *P. dentata* has demonstrated a range of interesting biological activities, including antiviral [16–20], anti-inflammatory [21], antiplasmodial [21,22] and cytotoxic effects [21].

Despite this therapeutic promise and the preliminary phytochemical and bioactivity studies reported for *P. dentata*, the chemical diversity and pharmacological potential of this species remain largely underexplored. Only a limited number of metabolites have been identified to date, including five flavonoids [16,23] and one coumarin [17,23]. Furthermore, few studies have scientifically validated its traditional medicinal uses, leaving most of its metabolome uncharacterized and structure–activity relationships entirely unexplored [15].

In this context, the present work was undertaken to address this gap through an integrative investigation of the leaf metabolites of *P. dentata* using a combination of NMR metabolomics, molecular networking (MN), and phytochemical analysis. To the best of our knowledge, no comprehensive study combining these complementary approaches has yet been conducted on this species. The integration of NMR metabolomics and molecular networking provides a more holistic view of the metabolome. The MN approach, implemented using LC–MS/MS, facilitates the identification of structurally related molecules and the visualization of their relationships. In parallel, NMR-based metabolomics provides complementary relative quantitative information on metabolite abundance within a sample and reveals signatures linking metabolite profiles to biological activity. Finally, the phytochemical investigation, involving fractionation, isolation, and structure elucidation, confirmed the results obtained from the metabolomic study and the molecular networking analysis.

As a first step, anti-inflammatory and antiplasmodial activities of 11 species endemics to Reunion Island were evaluated. In the second step, a UHPLC-DAD-ESI-QTOF-MS/MS molecular networking approach was applied to *P. dentata* in order to explore its chemical diversity. The relationships between the metabolites and the biological activity of each species extracts were also evaluated separately within the NMR datasets, coupled with multivariate analysis. Following these analyses, eight new diterpenes, one new coumarin,

https://doi.org/10.3390/molecules31060973

and 15 known compounds were isolated and identified from the EtOAc (ethyl acetate) extract of *P. dentata* leaves. These compounds were tested for their *in vitro* antiplasmodial, anti-inflammatory and cytotoxicity properties. All the obtained results are discussed below.

2. Results and Discussion

2.1. Biological Activities of Psiadia Species

As part of our ongoing search for bioactive natural products within the genus *Psiadia*, eleven endemic species from Reunion Island were selected. To assess the geographical variability of their chemical profiles, several collection sites were analyzed for each species during the summer season. All ethyl acetate crude extracts were tested *in vitro* for their antiplasmodial activity against the chloroquine-sensitive (3D7) strain of *Plasmodium falciparum* and for their anti-inflammatory properties using the horseradish peroxidase (HRP) enzyme assay (Figure 2).

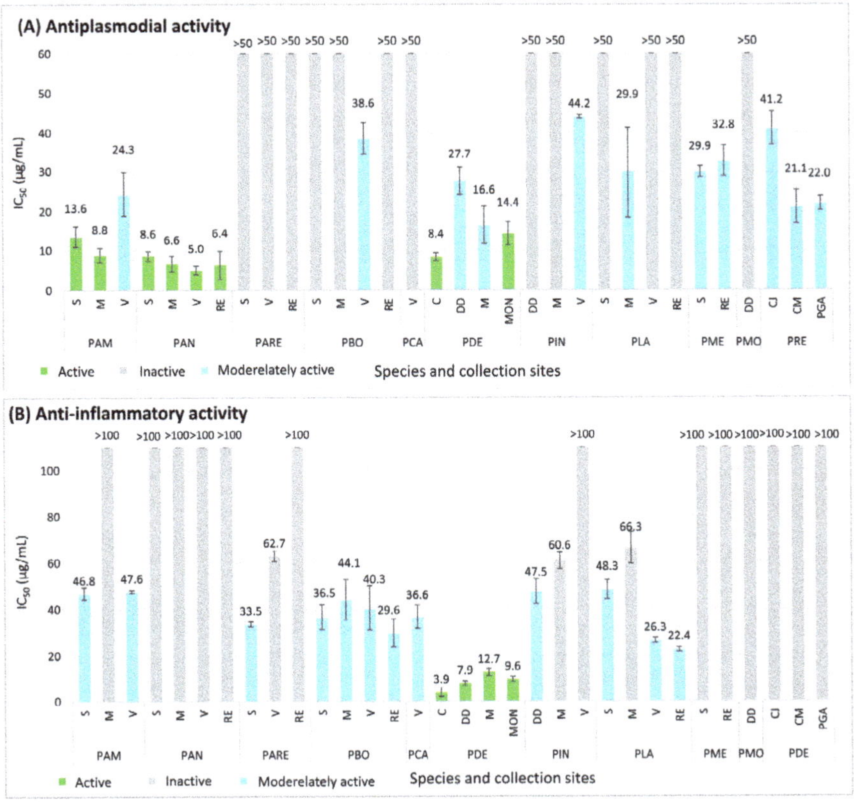

Figure 2. (**A**) *In vitro* antiplasmodial activity (3D7 strain) of *Psiadia* species crude extracts (IC_{50} in µg/mL). Positive control: artemisinin (IC_{50} = 0.004 ± 0.001 µg/mL). Inactive extract ($IC_{50} \geq 50$ µg/mL). (**B**) *In vitro* anti-inflammatory activity (HRP enzyme) of *Psiadia* species crude extracts (IC_{50} in µg/mL). Positive control: quercetin (IC_{50} = 10.8 ± 1.9 µg/mL). Inactive extract ($IC_{50} \geq 100$ µg/mL). Plant codes: PAM (*P. amygdalina*), PAN (*P. anchusifolia*), PARE (*P. argentea*), PBO (*P. boivinii*), PCA (*P. callocephala*), PDE (*P. dentata*), PIN (*P. insignis*), PLA (*P. laurifolia*), PME (*P. melastomatoides*), PMO (*P. montana*), PRE (*P. retusa*). Location codes: S (Salazie), M (Maïdo), V (Volcan), RE (Roche-Ecrite), C (Colorado), DD (Dos d'Ane), MON (Montauban), CJ (Cap Jaune), CM (Cap Méchant), PGA (Piton Grande Anse). Active compound: <15 µg/mL, moderately active compound: 15–50 µg/mL, inactive compound ($IC_{50} \geq 50$ µg/mL) [24,25].

Among the tested extracts, three species exhibited significant *in vitro* antiplasmodial activity (IC_{50} < 15 μg/mL, according to the classification criteria of [24,25]): *P. amygdalina* (PAM), *P. anchusifolia* (PAN) and *P. dentata* (PDE). The activity of *P. amygdalina* and *P. dentata* extracts varied depending on the collection site, indicating chemical profile. In contrast, the antiplasmodial activity of *P. anchusifolia* crude extracts was only slightly affected by geographical location. Although the IC_{50} values of all tested crude extracts were higher than that of the positive control, artemisinin (IC_{50} = 0.004 ± 0.001 μg/mL), these extracts nevertheless remain promising for the discovery of antiplasmodial compounds.

With regard to anti-inflammatory activity, *P. dentata* was the only species to display significant *in vitro* activity against the HRP enzyme (IC_{50} < 15 μg/mL, according to [24,25]), with minimal variation across collection sites. Moreover, the anti-inflammatory activity of *Psiadia dentata* crude extracts, while lower than that of the positive control quercetin (IC_{50} = 10.8 ± 1.9 μg/mL), suggests the presence of highly active compounds in these extracts.

Crude extracts are complex mixtures containing hundreds of molecules, some of which may exhibit high biological activity when isolated. Based on the biological activity results, *Psiadia dentata* was selected as the most promising species due to its potent anti-inflammatory and antiplasmodial properties. Consequently, to identify the compound families or specific molecules that are responsible for these activities, a metabolomic analysis was conducted on the plant extracts using ^{1}H NMR spectroscopy.

2.2. Molecular Networking

To obtain the first molecular fingerprint of *P. dentata* and gain insights about the chemodiversity of the species, by highlighting distinct metabolite features, a MN approach was performed. Isohexane and methanolic fractions of the crude extract were analyzed by UHPLC-HRMS-ESI-QTOF in positive mode. The resulting data were subsequently processed on the GNPS platform, where tandem mass spectrometry (MS/MS) spectra were compared, organized and visualized through spectral similarity mapping, revealing clusters of structurally related metabolites [26]. The generated MS/MS molecular network is shown in Figure 3. It consists of 154 nodes, forming 8 clusters (connected nodes), and 48 self-loops (unlinked nodes), using a cosine score threshold of 0.62. Dereplication was performed on GNPS using a large, community-acquired spectral library. Additional in silico databases, such as MetFrag [27] and CFM-ID [28], were employed to support the compound annotation complemented by manual inspection of the MS/MS.

Annotation of molecular networks revealed a rich chemical diversity, with 25 compounds distributed among four major molecular families: flavonoids, diterpenoids, coumarins, and alkaloids (Table 1).

The first family, flavonoids (Figure 3A–D) included 10 nodes (green) corresponding to known flavonoids previously reported in several *Psiadia* species [15], such as ermanin (**1**), kaempferol 3,7,4′-trimethylether (**1**), patchypodol (**3**), retusin (**4**), penduletin (**5**), quercetin 3-methylether (**6**), isokaempferide (**7**), quercetin 3,3′-dimethylether (**8**), kaempferol (**9**) and chrysosplenol D (**10**). Most of these compounds have been previously isolated from the leaves of *P. trinervia* [29], *P. terebinthina* [30], and *P. dentata* [16,23].

https://doi.org/10.3390/molecules3106097

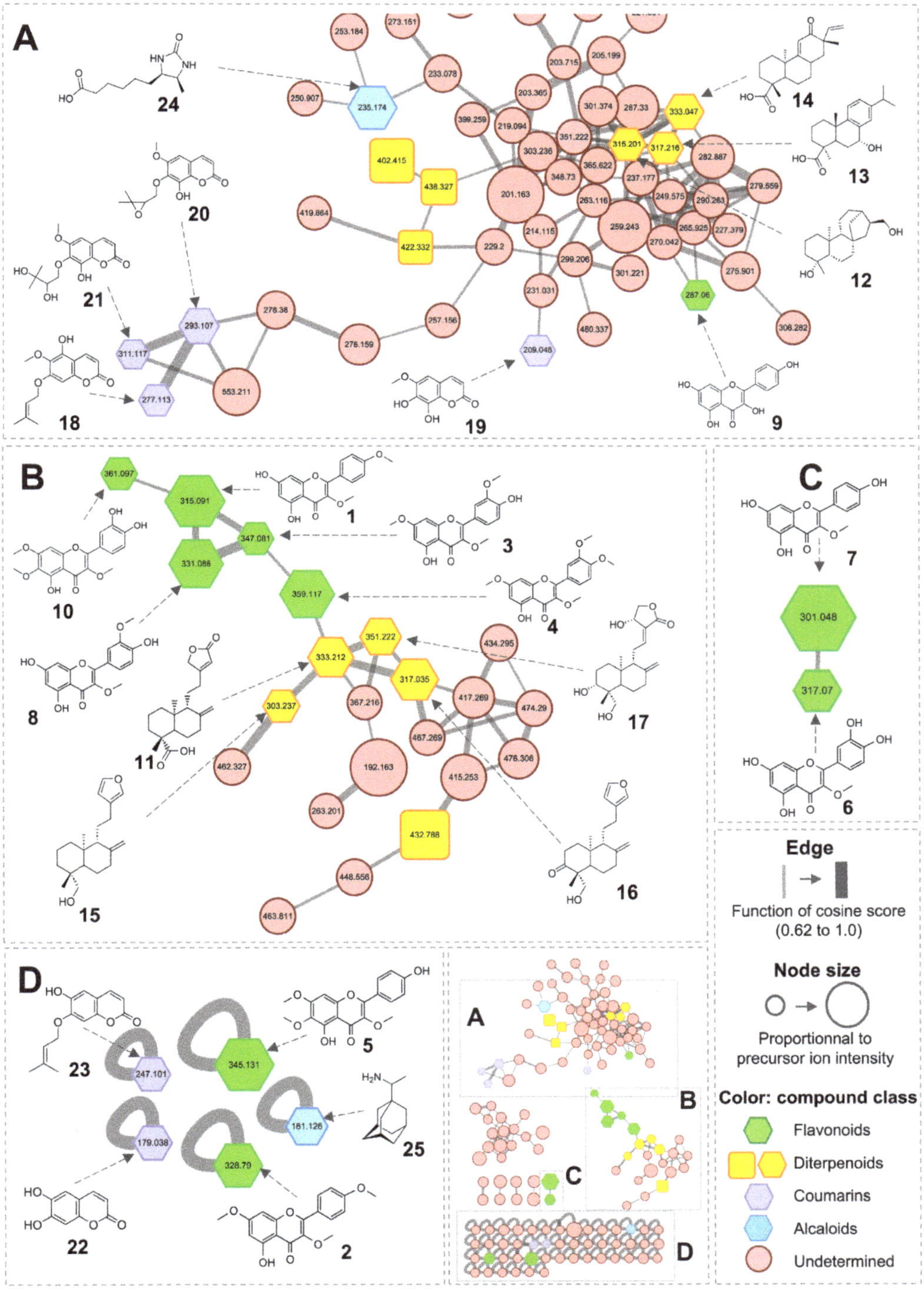

Figure 3. Annotation of the molecular network of isohexane and methanolic fractions derived from *P. dentata*.

https://doi.org/10.3390/molecules31060973

Table 1. Summary table of molecular network annotations for *Psiadia dentata* leaf extract.

Natural Product Family	Compound	*m*/*z*	Precursor Ion	Raw Formula	Structural Analogue
Flavonoids	**1**	315.091	$[M + H]^+$	$C_{17}H_{14}O_6$	ermanin
	2	328.79	$[M + H]^+$	$C_{18}H_{16}O_6$	kaempferol 3,7,4′-trimethylether
	3	347.081	$[M + H]^+$	$C_{18}H_{16}O_7$	patchypodol
	4	359.117	$[M + H]^+$	$C_{19}H_{18}O_7$	retusin
	5	345.131	$[M + H]^+$	$C_{18}H_{16}O_7$	penduletin
	6	317.070	$[M + H]^+$	$C_{16}H_{12}O_7$	quercetin 3-methylether
	7	301.048	$[M + H]^+$	$C_{16}H_{12}O_6$	isokaempferide
	8	331.086	$[M + H]^+$	$C_{17}H_{14}O_7$	quercetin 3,3′-dimethylether
	9	287.060	$[M + H]^+$	$C_{15}H_{10}O_6$	kaempferol
	10	361.097	$[M + H]^+$	$C_{18}H_{16}O_8$	chrysosplenol D
Diterpenoids	**11**	333.212	$[M + H]^+$	$C_{20}H_{28}O_4$	*ent*-labda-8(17),13-dien-15,16-olide-19-oic acid
	12	315.201	$[M + H]^+$	$C_{19}H_{32}O_2$	annosquamosine C
	13	317.216	$[M + H]^+$	$C_{20}H_{38}O_3$	7-hydroxycallitrisic acid
	14	333.047	$[M + NH_4]^+$	$C_{20}H_{28}O_3$	methyl-*ent*-12-oxopimara-9(11),15-dien-19-oic acid
	15	303.237	$[M + H]^+$	$C_{20}H_{30}O_2$	6-deoxypsiadiol
	16	317.035	$[M + H]^+$	$C_{20}H_{28}O_3$	3-oxo-19-hydroxy-13-furyl-ent-labda-8(17)-ene
	17	351.222	$[M + H]^+$	$C_{20}H_{30}O_5$	andrographolide
Coumarins	**18**	277.113	$[M + H]^+$	$C_{15}H_{16}O_5$	isoobtusitin
	19	209.048	$[M + H]^+$	$C_{10}H_8O_5$	fraxetin
	20	293.107	$[M + H]^+$	$C_{15}H_{16}O_6$	7-(2′,3′-epoxy-3′-methylbutoxy)-8-hydroxy-6-methoxycoumarin
	21	311.117	$[M + H]^+$	$C_{15}H_{18}O_7$	7-(2′,3′-dihydroxy-3′-methylbutoxy)-8-hydroxy-6-methoxycoumarin
	22	179.038	$[M + H]^+$	$C_9H_6O_4$	esculetin
	23	247.101	$[M + H]^+$	$C_{14}H_{14}O_4$	prenyletin
Alkaloids	**24**	235.174	$[M + Na]^+$	$C_{10}H_{18}N_2O_3$	desthiobiotin
	25	181.126	$[M + H]^+$	$C_{12}H_{21}N$	rimantadine

The second family observed consists of diterpenoids (yellow nodes) (Figure 3A,B). A total of 7 diterpenes were revealed, including *ent*-labda-8(17),13-dien-15,16-olide-19-oic acid (**11**), annosquamosine C (**12**), 7-hydroxycallitrisic acid (**13**), methyl-*ent*-12-oxopimara-9(11),15-dien-19-oic acid (**14**), 6-deoxypsiadiol (**15**), 3-oxo-19-hydroxy-13-furyl-ent-labda-8(17)-ene (**16**), and andrographolide (**17**). All of these compounds share a labdane-type structure, with bicyclic carbon skeletons and various oxygenated functional groups. Several labdane- and kaurane-type diterpenes in *Psiadia* species such as psiadiol and 6-deoxypsiadiol (**15**) [31], as well as *ent*-16β,17-dihydroxykauran-20-oic acid, 2-oxotrachyloban-18,19-diol, and trachyloban-2β,6β,19-triol [32] have previously been reported. Their occurrence in the MN is therefore consistent with these previous findings. In addition to the 7 annotated diterpenes, the MN also revealed seven other diterpenes that could not be annotated. However, based on their connectivity within the network and their fragmentation patterns, these compounds are also likely to share a labdane-type skeleton.

Another class of molecular compounds in MN was annotated as coumarins (purple nodes) (Figure 3A,D). Six coumarins were suggested: isoobtusitin (**18**), and fraxetin (**19**), 7-(2′,3′-epoxy-3′-methylbutoxy)-8-hydroxy-6-methoxycoumarin (**20**), 7-(2′,3′-dihydroxy-

3′-methylbutoxy)-8-hydroxy-6-methoxycoumarin (**21**), esculetin (**22**) and prenyletin (**23**). Isoobtusitin has also been previously reported from *P. dentata* [17], confirming its consistent occurrence, whereas esculetin and prenyletin have been documented in several Asteraceae species [33,34].

Finally, two nodes potentially corresponded to alkaloids (light blue nodes) identified as desthiobiotin (**24**) and rimantadine (**25**) (Figure 3A,D). To date, no alkaloid has been isolated from species of the genus *Psiadia*, although many Asteraceae, especially *Senecio* species, are known sources of pyrrolizidine alkaloids [35].

Nodes that could not be annotated suggest the presence of compounds with a certain structural originality.

2.3. ^1H NMR Metabolomic Analysis

2.3.1. PLS-DA of ^1H NMR Data of *P. dentata* Crude Extracts and Correlation with the Antiplasmodial Activity

The ^1H NMR spectra of all crude extracts were analyzed using a partial least square (PLS) regression (Figure 4A) to evaluate the correlation between the antiplasmodial activity and the detected metabolites. The model was generated using a single quantitative Y variable corresponding to the IC_{50} values of each crude extract. The resulting PLS model exhibited the following statistical parameters: R^2 (X) = 0.39, R^2 (Y) = 0.84 and Q^2 = 0.71. The R^2, Q^2 values, and the permutation test confirmed the model's validity. From a general point of view, the samples tended to cluster into two distinct groups along the first principal component, which accounted for 27.1% of the total variation. On the PLS score plot, the distribution of samples within each cluster suggested distinct metabolic profiles. Interestingly, some moderately active extracts (IC_{50}: $22 \leq IC_{50} \leq 44$ µg/mL) were grouped with either active or inactive extracts, which may reflect a shared chemical composition and the presence of low levels of active compounds in the moderately active samples. In contrast, the moderately active extracts grouped with the active ones displayed lower IC_{50} ($16 \leq IC_{50} \leq 22$ µg/mL).

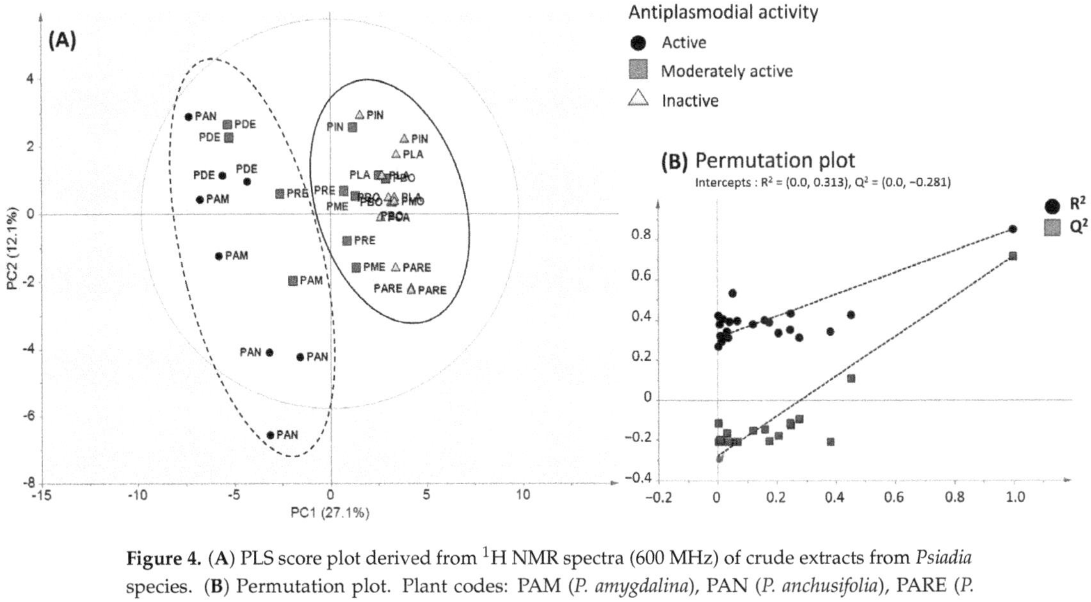

Figure 4. (**A**) PLS score plot derived from ^1H NMR spectra (600 MHz) of crude extracts from *Psiadia* species. (**B**) Permutation plot. Plant codes: PAM (*P. amygdalina*), PAN (*P. anchusifolia*), PARE (*P. argentea*), PBO (*P. boivinii*), PCA (*P. callocephala*), PDE (*P. dentata*), PIN (*P. insignis*), PLA (*P. laurifolia*), PMO (*P. montana*), PME (*P. melastomatoides*), PRE (*P. retusa*).

Therefore, to highlight molecules or compound families responsible for the antiplasmodial activity of *P. dentata* crude extracts, multivariate data analysis comparing the active crude extracts *versus* all inactive extracts was performed. Since the four *P. dentata* extracts (both active and moderately active extracts) are grouped, they seem to have a similar composition.

Therefore, a contribution plot was generated, with the comparison of all inactive extracts from *Psiadia* species with those to all *P. dentata* extracts (Figure 5A). A similar analysis was performed by comparing all moderately active and inactive extracts to *P. dentata* extracts (Figure S1).

Examination of the contribution plot highlighted distinct signals in the following regions of the ^{1}H NMR spectrum: the aliphatic region (0.69–1.93 ppm), the sugar-based/N-containing region (3.83–4.65 ppm), the olefinic region (5.49 ppm), and the aromatic region (6.4–8.0 ppm) of the ^{1}H NMR spectrum. The ^{1}H NMR spectra of the active ethyl acetate crude extracts of *P. dentata* were analyzed, and the chemical shifts of interest were compared with all compounds previously identified within the genus *Psiadia*, based on our earlier review of this genus [15]. This comparison enabled the assignment of specific spectral variables (1.77, 1.81, 3.93, 4.65, 5.49, and 6.45 ppm) to the chemical shifts of the compound isoobtusitin (**18**) (Figure 5B,C). All the chemical shifts corresponding to isoobtusitin were identified in the ^{1}H NMR spectrum of the most active extract of *P. dentata*.

Further analysis of the ^{1}H NMR spectrum revealed additional signals at δ_H 6.29 (d, J = 2.1 Hz), δ_H 6.41 (d, J = 2.1 Hz), δ_H 6.93 (d, J = 9.0 Hz) and δ_H 8.00 (d, J = 9.0 Hz) which may correspond to flavonols. In addition, the latter signals at δ_H 3.85 (s) and 3.90 (s) suggested the presence of two specific flavonols in the active crude extract: ermanin (**1**) and isokaempferide (**7**). These three molecules had previously been isolated from *P. dentata* crude extract [16,17]. Their higher abundance in the most active extract correlates with the observed antiplasmodial activity, suggesting a potential contribution, although the participation of other minor or as-yet unidentified compounds cannot be ruled out. Moreover, these molecules were also detected in the moderately active extract, at lower concentrations. Indeed, the concentrations of ermanin and isokaempferide are 5 times lower in the moderately active extract (collected at Dos d'Ane, IC_{50} = 27.7 μg/mL), while the concentration of isoobtusitin is 8 times lower (Figure S2). Several studies have reported that flavonoids and coumarins exhibit antimalarial activity [36–39]. Furthermore, signals observed in the aliphatic region also appeared to play a significant role in the discrimination between active and inactive extracts. However, precise identification of these was not possible due to overlapping peaks in this spectral region, suggesting that additional, yet unidentified, compounds may contribute to the observed bioactivities.

2.3.2. PLS-DA of ^{1}H NMR Data of *P. dentata* Crude Extracts and Correlation with the Anti-Inflammatory Activity

The correlation between the anti-inflammatory activity of *P. dentata* crude extracts and their chemical composition was analyzed using an OPLS-DA model. The model exhibited the following statistical parameters: R^2 (X) = 0.39, R^2 (Y) = 0.56, and Q^2 = 0.38. The model revealed two distinct groups: the first group consisted of active crude extracts of *P. dentata*, while the second group included all inactive and moderately active extracts (Figure 6). Therefore, a contribution plot was generated by comparing the spectral variables of all active extracts (group 1) with those of inactive and moderately active extracts (group 2). Examination of the contribution plot highlighted signals in the following regions of the ^{1}H NMR spectrum: the aliphatic region (0.69–1.93 ppm), the sugar-based/N-containing (3.83–4.65), and the aromatic region (6.4–8.0 ppm). Comparison of the chemical shifts with all compounds previously identified within the genus *Psiadia* allowed the attribution of spectral variables at 1.81, 3.84, 6.28, 6.40, 7.02, and 8.06 ppm to the compound

https://doi.org/10.3390/molecules3106097

isokaempferide (7). However, other spectral variables highlighted by the contribution plot could not be attributed to known compounds detected in *Psiadia* species. A more comprehensive study on the chemical composition of *P. dentata* active crude extract is required to identify the other potential anti-inflammatory compounds.

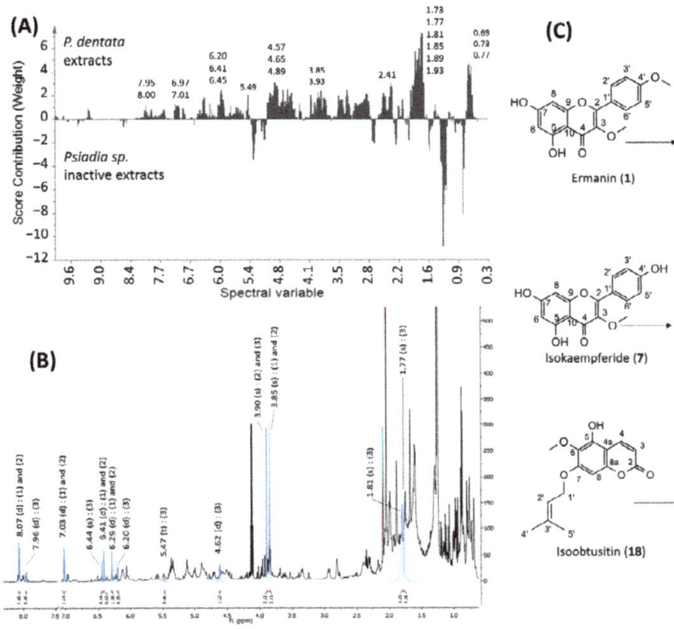

Position	δ_{1H} in ppm (multiplicity, J en Hz) Experimental (CD$_3$OD, 600 MHz)	Literature (DMSO-d_6, 700 MHz)
2	-	-
3	-	-
4	-	-
5	-	-
6	6.21 (d, 2.1)	6.21 (d, 1.9)
7	-	
8	6.43 (d, 2.1)	6.46 (d, 1.7)
9	-	-
10	-	-
1′	-	-
2′,6′	8.09 (d, 9.1)	8.03 (d, 9.1)
3′,5′	7.09 (d, 9.1)	7.14 (d, 9.1)
4′	-	-
CH$_3$O-3	3.79 (s)	3.80 (s)
CH$_3$O-4′	3.90 (s)	3.87 (s)

Position	δ_{1H} in ppm (multiplicity, J in Hz) Experimental (CDCl$_3$, 600 MHz)	Literature (DMSO-d_6, 400 MHz)
2	-	-
3	-	-
4	-	-
5	-	-
6	6.20 (d, 2.1)	6.19 (d, 2.0)
7	-	-
8	6.40 (d, 2.2)	6.43 (d, 2.1)
9	-	-
10	-	-
1′	-	-
2′, 6′	8.00 (d, 9.0)	7.93 (d, 8.9)
3′, 5′	6.93 (d, 9.0)	6.94 (d, 8.9)
4′	-	-
CH$_3$O-3	3.77 (s)	3.77 (s)

Position	δ_{1H} in ppm (multiplicity, J en Hz) Experimental (CD$_3$OD, 600 MHz)	Literature (CDCl$_3$, 500 MHz)
2	-	-
3	6.15 (d, 9.6)	6.20 (d, 9.7)
4	8.12 (d, 9.6)	7.96 (d, 9.7)
4a	-	-
5	-	-
6	-	-
7		
8	6.52 (sl)	6.44 (s)
8a	-	-
1′	4.65 (dl, 6.8)	4.62 (dl, 6.7)
2′	5.52 (tspt, 1.4 ; 6.8)	5.47 (t, 6.7)
3′	-	-
4′	1.82 (dl, 1.4)	1.81 (s)
5′	1.80 (dl, 1.4)	1.79 (s)
CH$_3$O-6	3.78 (s)	3.90 (s)

Figure 5. (**A**) Contribution plot of the PLS model comparing *P. dentata* active crude extracts with inactive extracts from other *Psiadia* species; (**B**) ^{1}H NMR spectrum (CDCl$_3$, 600 MHz) of an active crude extract of *P. dentata*, highlighting the assignments of ermanin (**1**), isokaempferide (**7**), and isoobtusitin (**18**); (**C**) molecular structures of ermanin (**1**), isokaempferide (**7**), and isoobtusitin (**18**), along with their experimental and literature ^{1}H NMR data [17,40,41].

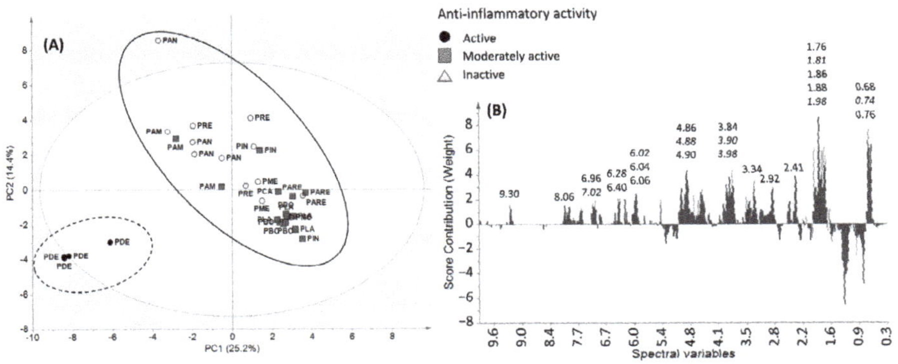

Figure 6. (**A**) Score plot of PLS-DA model generated from the ^{1}H NMR spectra (600 MHz) of *Psiadia* species crude extracts for the anti-inflammatory activity. (**B**) Contribution plot of the PLS-DA model of *P. dentata* active crude extracts against *Psiadia* sp. inactive and moderately active extracts. Plant codes: PAM (*P. amygdalina*), PAN (*P. anchusifolia*), PARE (*P. argentea*), PBO (*P. boivinii*), PCA (*P. callocephala*), PDE (*P. dentata*), PIN (*P. insignis*), PLA (*P. laurifolia*), PMO (*P. montana*), PME (*P. melastomatoides*), PRE (*P. retusa*).

https://doi.org/10.3390/molecules31060973

2.4. Phytochemical Investigation

Among the four EtOAc extracts of *P. dentata* collected in summer and evaluated for their biological activities (Section 2.1), two extracts from Colorado and Montauban exhibited the highest antiplasmodial (IC_{50} = 8.4 and 14.4 µg/mL respectively) and anti-inflammatory activities (IC_{50} = 3.9 and 9.6 µg/mL respectively). Comparative chemical profiling using HPLC-CAD (high-performance liquid chromatography coupled with charged aerosol detection) revealed that both extracts shared broadly similar chromatographic profiles, suggesting the presence of related compounds. However, Montauban extract exhibited a slightly richer chromatographic profile (Figure 7) and was selected for subsequent fractionation and purification.

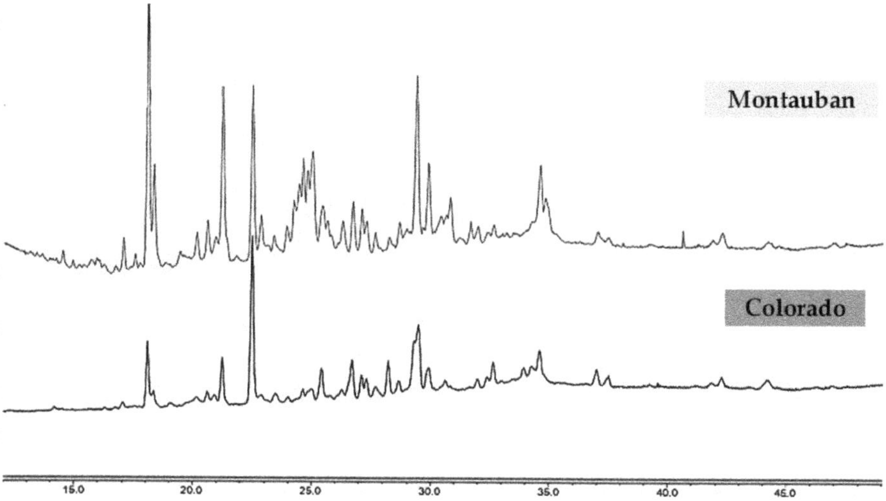

Figure 7. HPLC-CAD chromatograms of two bioactive extracts of *P. dentata*.

The selected EtOAc leaf extract of *P. dentata* was subjected to liquid–liquid partitioning using methanol (MeOH) and isohexane, yielding two fractions: a hydrocarbon fraction and an oxygenated fraction. These two fractions were then evaluated for their biological activities, including anti-inflammatory and cytotoxic properties. Both fractions exhibited anti-inflammatory effects, with IC_{50} values for NO release of 0.04 and 0.05 µg/mL, respectively. They also showed cytotoxic effects against HepG2 hepatocarcinoma cells (IC_{50} = 18.65 and 15.94 µg/mL, respectively) and HT-29 colorectal cancer cells (IC_{50} = 18.63 and 10.38 µg/mL, respectively). Interestingly, the crude EtOAc extract was more cytotoxic than either fraction, with IC_{50} values of 8.64 and 6.34 µg/mL, against HepG2 and HT-29 cells, respectively. This difference may be attributed to synergistic interactions between compounds in the crude extract that are lost after fractionation.

Both fractions were further subjected to purification of their compounds using a combination of normal-phase medium-pressure liquid chromatography (MPLC) and repeated reversed-phase semi-preparative HPLC. This process led to the isolation of nine new diterpenes (**26–34**), one coumarin (**35**), and fifteen known compounds (**1–8**, **11**, **18**, **19** and **36–39**) (Figure 8). The latter were identified by comparison with previously reported spectroscopic data as ermanin (**1**) [41], kaempferol 3,7,4′-trimethylether (**2**) [42], patchypodol (**3**) [43], retusin (**4**) [44], penduletin (**5**) [45], quercetin 3-methylether (**6**) [46], isokaempferide (**7**) [40], quercetin 3,3′-dimethylether (**8**) [47], *ent*-8(17),13-labdadien-15,16-olid-18-oic acid (**11**) [48], isoobtusitin (**18**) [17], fratexin (**19**) [49], kumatakilin (**36**) [50],

https://doi.org/10.3390/molecules3106097

penduletin 4-methylether (**37**) [51], chrysosplenetin (**38**) [52] and casticin (**39**) [53]. The structure of the new compounds **26–35** was established by the analysis of their UV, IR, NMR (^{1}H and ^{13}C NMR, COSY, HSQC, HMBC and NOESY) spectroscopic data, together with HRESIMS spectrometric data.

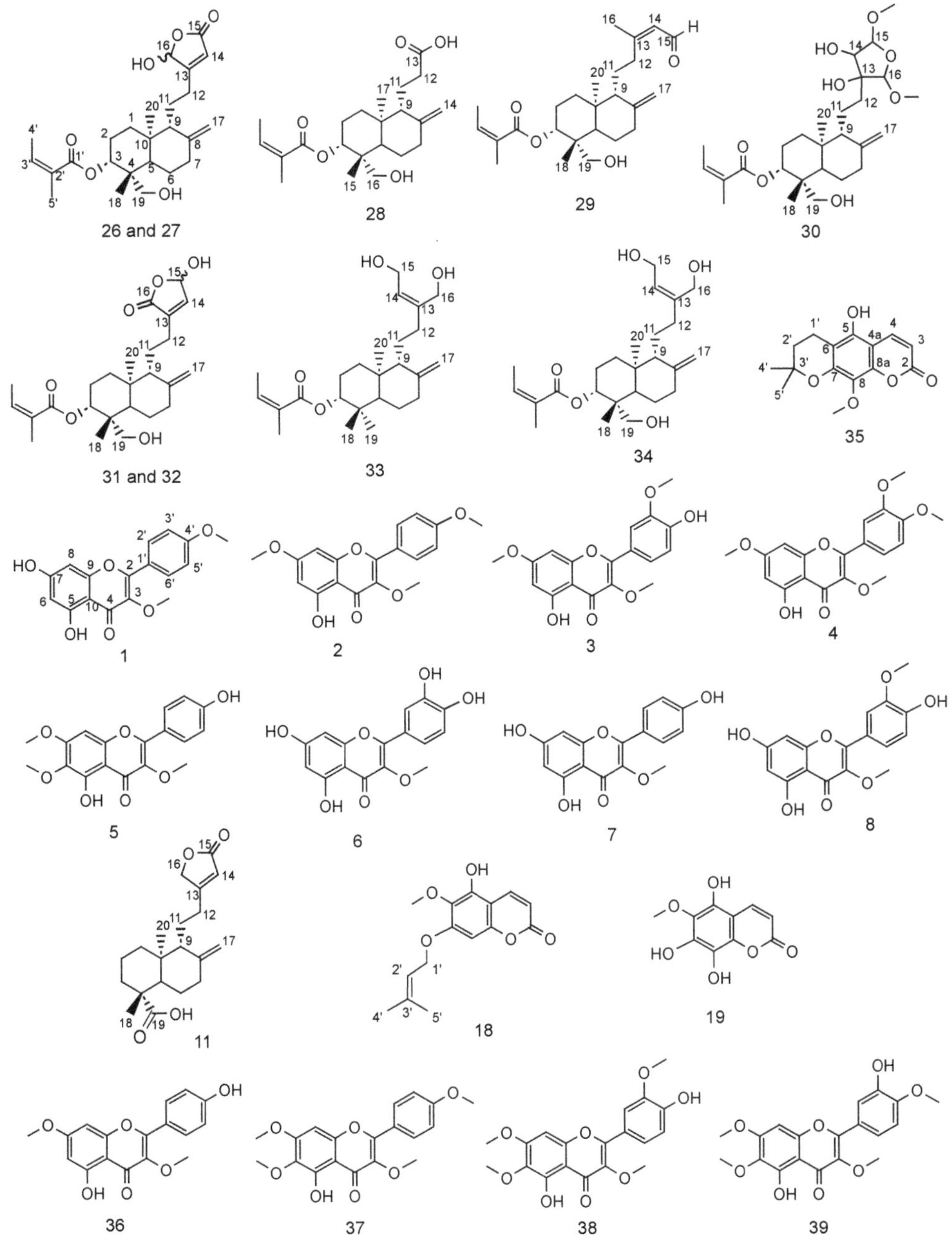

Figure 8. Structures of new compounds **26–35** and known compounds **1–8**, **11**, **18**, **19** and **36–39** isolated from *P. dentata*.

https://doi.org/10.3390/molecules31060973

Interestingly, the isolated metabolites are consistent with the molecular network analysis, which had already highlighted the presence of flavonoids, coumarins, and diterpenoids in this species. Particularly, the identification of known diterpene *ent*-8(17),13-labdadien-15,16-olid-18-oic acid (**11**), together with the coumarins isoobtusitin (**18**), and fratexin (**19**), as well as several flavonoids including ermanin (**1**), kaempferol 3,7,4′-trimethylether (**2**), patchypodol (**3**), retusin (**4**), penduletin (**5**), quercetin 3-methylether (**6**), isokaempferide (**7**) and quercetin 3,3′-dimethylether (**8**), provides experimental confirmation of the chemical diversity predicted by the MN approach.

2.4.1. Structural Elucidation of Compounds **26–35**

Compounds **26** and **27** were isolated as yellow powders in 1:1 mixture. Their molecular formula $C_{25}H_{36}O_6$ was determined by HRESIMS, which showed a $[M + H]^+$ ion at m/z 433.2587 (calcd. for $C_{25}H_{37}O_6$ $[M + H]^+$, 433.2585), corresponding to 8 degrees of unsaturation. The UV spectrum displayed an absorption band typical of conjugated double bonds at λ_{max} 225 nm. In the IR spectrum, characteristic vibration bands were observed at 3371 cm^{-1} (O–H stretching) 2925 cm^{-1} and 2853 cm^{-1} (CH2 asymmetric and symmetric stretching); 1757 cm^{-1} (conjugated C=O), 1646 cm^{-1} (conjugated C=C) and 1260 cm^{-1} (C–C stretching) (Figures S3–S5). Analysis of the ^{13}C NMR spectrum (Figure S7) revealed 10 characteristic carbon signals of a labdane-type diterpenoid: C-1 (δc 36.8), C-2 (δ_C 24.4), C-3 (δ_C 74.3), C-4 (δ_C 42.9), C-5 (δ_C 46.0 and 46.1), C-6 (δ_C 23.3), C-7 (δ_C 37.6), C-8 (δ_C 147.2 and 147.3), C-9 (δ_C 55.9 and 56.1) and C-10 (δ_C 39.2). The duplication of several carbon signals confirmed that the sample contained two diastereoisomers in a 1:1 ratio, as further supported by their relative peak intensities. The ^1H NMR data (Table 2 and Figure S6) corroborated this structure framework, showing characteristic resonances for H-1 (δ_H 1.29 and 1.82), H-2 (δ_H 1.73 and 1.91), H-3 (δ_H 4.97), H-5 (δ_H 1.74), H-6 (δ_H 1.37 and 1.75), H-7 (δ_H 2.08 and 2.39) and H-9 (δ_H 1.75). The deshielded signal of H-3 (δ_H 4.97) indicated its attachment to an oxygen atom. The diterpene was methyl-substituted at C-10 and disubstituted at C-4 [54]. Moreover, broad singlets at δ_H 4.51 and 4.89 (H-17) indicated a methylidene group. Additional resonances at δ_H 1.98 (H-4′), 1.87 (H-5′), and 6.12 (H-3′) supported the presence of an angeloyl group [55,56]. COSY and HMBC experiments confirmed this assignment (Figures S8 and S10): COSY correlations were observed between H-4′ and H-3′, while HMBC correlations of H-5′ with C-1′ (δ_C 169.3), C-2′ (δ_C 127.7), and C-3′ (δ_C 139.5), and similarly for H-4′ indicated that the angeloyl group was esterified at C-3, by the presence of HMBC correlations of H-3 with C-1′ (δ_C 169.3). Further HMBC correlations of H-3 with C-18 (δ_C 13.0) and C-19 (δ_C 64.5), and of H-18 (δ_H 0.68) and H-19 (δ_H 2.91, 3.36) with C-4 (δ_C 42.9) indicated their spatial proximity. The deshielding of C-19 (δ_C 64.5) and its attached protons (δ_H 2.91; 3.36) suggested the presence of a neighbouring oxygen atom (Figure 9). HMBC correlations of H-17 with C-7 (δ_C 37.6), C-8 (δ_C 147.3), and C-9 (δ_C 56.1) confirmed the attachment of the methylidene group at C-8. HMBC correlations between H-14 (δ_H 5.86) and C-15 (δ_C 170.2)/C-16 (δ_C 99.4), as well as between H-16 (δ_H 5.97) and C-14 (δ_C 117.3)/C-15 (δ_C 170.2), established the presence of a *β*-substituted butenolide ring. The chemical shift of C-16 further supported the presence of a hydroxyl group at this position. The ^1H NMR spectrum also showed methylene signals at δ_H 1.68–1.80 (H-11) and δ_H 2.34–2.52 (H-12). COSY correlations between H-9 (δ_H 1.75) and H-11, as well as between H-11 and H-12, indicated the sequential connectivity of C-9, C-11, and C-12. Moreover, HMBC correlations between H-12 and C-14 (δ_C 117.3) confirmed the hydroxybutenolide ring was attached to C-12 (Figure 9). The duplicated ^{13}C signals were mainly associated with the butenolide moiety, suggesting that compounds **26** and **27** are diastereoisomers differing only at the chiral centre C-16 (*R* or *S* hydroxyl) (Figure S7). Altogether, these data allowed the identification of compounds **26** and **27** as 3α-angeloyloxy-19,16(*R*/*S*)-dihydroxy-*ent*-

https://doi.org/10.3390/molecules31060973

labda-8(17),13-dien-15,16-olide. Their ^{1}H and ^{13}C NMR data closely resembled those of 3α-angeloyloxy-19-hydroxy-*ent*-labda-8(17),13-dien-15,16-olide, previously isolated from *Gutierrezia solbrigii* (*Asteraceae*) [57]. However, the main difference lies at the C-16 position, where compounds **26** and **27** bear a hydroxyl group, whereas the reported diterpene contains a methylene group. The stereochemistry of these compounds and of the remaining eight new diterpenes, is discussed further in Section 2.4.2.

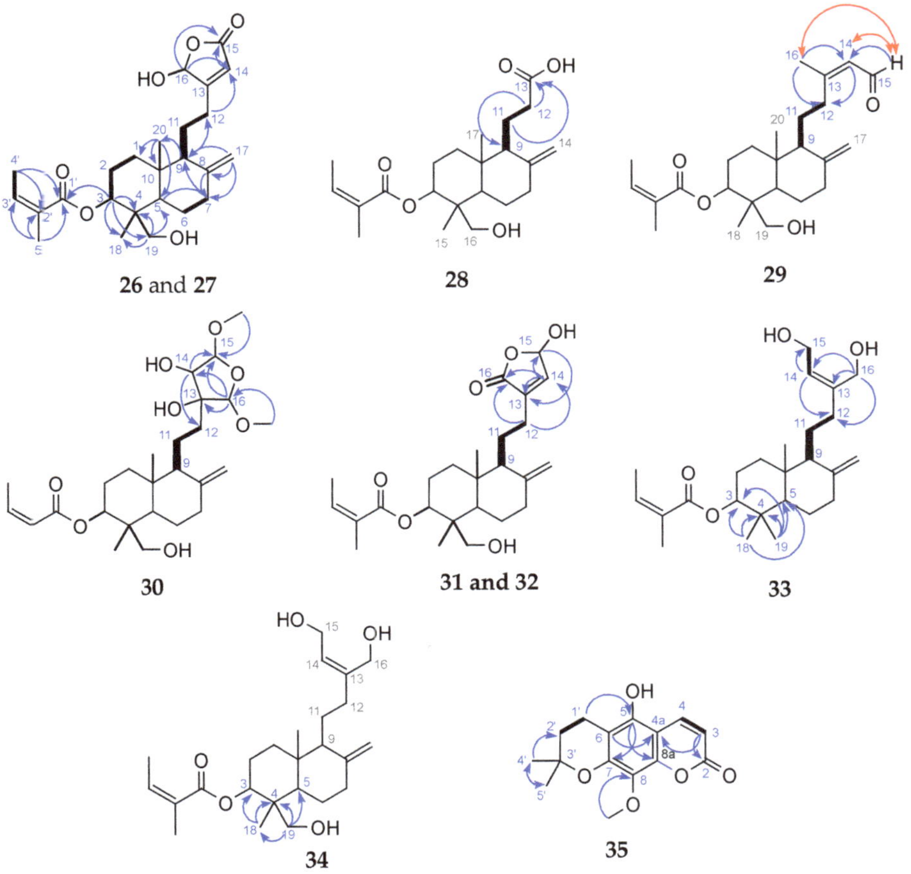

Figure 9. Key COSY (**bold** chain indicates vicinal proton correlations only), NOE (red arrows) and HMBC (blue arrows, H → C) correlations for compounds **26–35**.

Compound **28** was obtained as a white paste. Its molecular formula $C_{22}H_{34}O_5$ was determined by HRESIMS, which showed a $[M + H]^+$ ion at m/z 379.2486 (calcd. for $C_{22}H_{35}O_5^+$, 379.2400), corresponding to 6 degrees of unsaturation. The UV spectrum displayed an absorption band at λ_{max} 199 nm, attributable to a carbonyl group. The IR spectrum showed absorption bands at 3305 cm^{-1} (O–H stretching), 2963 cm^{-1} (CH_3 asymmetric vibrations), 2923 cm^{-1} and 2853 cm^{-1} (CH2 asymmetric and symmetric vibrations), 1715 cm^{-1} (C=O), and 1261 cm^{-1} (C–C stretching) (Figures S12–S14). The ^{1}H, COSY, HSQC and HMBC NMR data (Figures S15–S18) of **28** were highly similar to those of compounds **26** and **27** (Tables 2 and 4), indicating a labdane skeleton substituted at C-3 by an angeloyl group, at C-4 (δ_C 43.5) by a methyl and a hydroxymethyl group, at C-8 (δ_C 147.4) by an exocyclic methylene, and at C-10 (δ_C 39.2) by a methyl group. The main difference between compound **28** and the two diastereoisomers **26** and **27** was located at C-12. Specifically, the chemical shift of C-13 (δ_C 177.2) in compound **28** indicated the presence of a carbonyl function. In the HMBC spectrum, correlations of H-11 (δ_H 1.68, 1.90) and H-12 (δ_H 2.32, 2.53) with C-13 supported the attachment of a carboxylic acid group at C-12, as depicted in Figure 9. Accordingly, compound **28** was identified as 3α-angeloyloxy-16-hydroxy-*ent*-labda-8(17)-en-13-oic acid.

Compound **29** was isolated as a white oil with HRESIMS analysis indicating an $[M + NH_4]^+$ peak at m/z 420.3108 (calcd. for $C_{25}H_{42}O_4N^+$, 420.3109), corresponding to 7 degrees of unsaturation. The UV spectrum of compound **29** revealed the same absorption maximum at λ_{max} 227 nm as observed for compounds **26** and **27**. Its IR spectrum showed close similarities to those of **26** and **27**: absorption bands were observed at 3394 cm^{-1} (O–H stretching), 2961.09 cm^{-1} (CH_3 asymmetric vibrations), 2924 cm^{-1} and 2854 cm^{-1} (CH2 asymmetric and symmetric vibrations), 1733 cm^{-1} (conjugated C=O), 1673 cm^{-1} (conjugated C=C), 1260 cm^{-1} (C–C stretching), 1022 cm^{-1} (C–O), and 798 cm^{-1} (aromatic C–H) (Figures S20–S22). The ^{1}H and ^{13}C NMR spectroscopic data determined with edited HSQC and HMBC spectra (Tables 2 and 4) were closely related to those of compounds **26** and **27**, except for the moiety attached to carbon C-12. The ^{1}H NMR spectrum (Figure S23) showed additional signals corresponding to the methyl protons H-16 (δ_H 2.22), a deshielded aldehydic proton H-15 (δ_H 9.96), and an olefinic proton H-14 (δ_H 5.83, J = 8.2 Hz) coupled to the aldehydic proton H-15. HMBC correlations of H-16 (δ_H 2.22) with C-12 (δ_C 40.5), C-14 (δ_C 128.1), and C-15 (δ_C 193.6), as well as correlations of H-14 (δ_H 5.83) with C-12 (δ_C 40.5) and C-16 (δ_C 17.6); and of H-15 (δ_H 9.96) with C-14 (δ_C 128.1) established the carbon chain motif (Figure 9). Thus, compound **29** was identified as 3α-angeloyloxy-19-hydroxy-*ent*-labda-8(17),13*E*-dien-15-al.

Compound **30** was purified as a white paste with HRESIMS analysis indicating $[M - H]^-$ ion peak at m/z 495.2049 (calcd. for $C_{27}H_{43}O_8^-$, 495.2963), accounting for 6 degrees of unsaturation. Inspection of its UV spectrum indicated an absorption band at λ_{max} 197 nm, attributable to a carbonyl group. The IR spectrum displayed absorption bands at 3467 cm^{-1} (O–H stretching), 2953 cm^{-1} (asymmetric CH_3 vibrations), 2929 cm^{-1} and 2854 cm^{-1} (asymmetric and symmetric CH_2 vibrations), 1692 cm^{-1} (conjugated C=O), 1644 cm^{-1} (conjugated C=C), 1260 cm^{-1} (C–C stretching), and 1022 cm^{-1} (C–O) (Figures S28–S30). The ^{1}H and ^{13}C NMR spectral data (Tables 2 and 4) of compound **30** closely resembled those of compounds **26–29**, consistent with a labdane skeleton substituted at C-4 by a methyl and hydroxymethyl group, at C-3 by an angeloyl group, at C-8 by an exocyclic methylene, and with a linear connectivity between C-9, C-11 and C-12. Moreover, three deshielded signals H-14 (δ_H 3.95), H-15 (δ_H 4.90), and H-16 (δ_H 4.75) suggested proximity to oxygenated functionalities. The chemical shift of C-14 (δ_C 80.4) further indicated a hydroxyl substitution. Two methoxy groups at δ_H 3.47 and δ_H 3.40 were attached to carbons C-15 and C-16, respectively. Thus, compound **30** was characterized as 3α-angeoyloxy-19-hydroxy-15,16-dimethoxy-*ent*-labda-8(17)-en-furan-13,14-diol.

Compounds **31** and **32** were isolated as white paste in 1:1 mixture with HRESIMS analysis indicating $[M + H]^+$ peak at m/z 433.2589 (calcd. for $C_{25}H_{37}O_6^+$, m/z 433.2585), accounting for 8 degrees of unsaturation. The UV spectrum displayed a maximum absorption at λ_{max} 197 nm, identical to that of **30**. The IR spectrum and NMR spectra indicated marked structural similarities with compounds **26** and **27** (Figures S37–S39), with the main differences observed in the butenolide moiety. In fact, HMBC correlations between the olefinic proton H-14 (δ_H 6.85) and carbons C-13 (δ_C 138.6), C-15 (δ_C 96.9), and C-16 (δ_C 171.8), together with the correlations between H-15 (δ_H 6.08) and carbons C-13 (δ_C 138.6) and C-16 (δ_C 171.8), supported the presence of an α,β-unsaturated γ-lactone ring [58]. In addition, HMBC correlations from protons H-12 (δ_H 2.16 and 2.46) to carbons C-14 (δ_C 143.6) and

https://doi.org/10.3390/molecules3106097

C-16 (δ_C 171.8) confirmed the attachment of this lactone to C-12 (Figure 9). The downfield chemical shift of C-15 (δ_C 96.9) further suggested a hydroxyl substitution at this position. The duplicated signals observed in the ^{13}C NMR spectrum in the lactone moiety supported the presence of two diastereoisomers, differing solely in the stereochemical configuration of C-15 (*R*/*S* hydroxyl) (Figure S41). Based on these combined spectroscopic data of compounds **31** and **32** were thus identified as 3*α*-angeloyloxy-19,15(*R*/*S*)-dihydroxy-*ent*-andrograpanin.

Compound **33** was isolated as white crystals. Its molecular formula was determined by HRESIMS, which showed a $[M + H]^+$ ion at *m*/*z* 427.2815 (calcd. for $C_{25}H_{41}O_4^+$, 427.2819), corresponding to 6 degrees of unsaturation. Inspection of its UV spectrum indicated similarities with those of **26** and **27**. Its IR spectrum exhibited similarities to that of compound **28**, with additional absorption bands at 1603 cm^{-1} (asymmetric C=C), 1456 cm^{-1} (asymmetric CH_3), and 1384 cm^{-1} (symmetric CH_3) (Figures S37–S39). The ^{1}H and ^{13}C NMR data from edited HSQC and HMBC spectra (Tables 3 and 4) of compound **33** displayed the characteristic features in the other compounds: a labdane skeleton with an angeloyl group at C-3, an exocyclic methylene group at C-8, and a linear connectivity of C-9, C-11 and C-12. However, in the ^{1}H NMR spectrum (Figure S49), several deshielded signals were observed: H-14 (δ_H 5.62), H-15 (δ_H 4.22), and H-16 (δ_H 4.16 and 4.20). The COSY correlation between H-14 and H-15 indicated that these protons are attached to vicinal carbons (Figure 9). Furthermore, the chemical shifts of C-15 (δ_C 59.1) and C-16 (δ_C 61.5) suggest that these carbons are bonded to oxygen atoms. In the HMBC spectrum (Figure S52), correlations were observed between H-14 and carbons C-12 (δ_C 35.3) and C-15 (δ_C 59.1), as well as between H-16 and carbons C-12, C-13 (δ_C 144.5), and C-14 (δ_C 126.6). Based on these data, compound **33** was identified as 3*α*-angeloyloxy-15,16-dihydroxy-*ent*-labda-8(17),13*E*-diene.

Compound **34** was isolated as white crystals. Its HRESIMS spectrum displayed a pseudomolecular ion $[M + Na]^+$ at *m*/*z* 443.2764, consistent with the molecular formula (calcd. for $C_{25}H_{40}O_5Na^+$, 443.2768), corresponding to 6 degrees of unsaturation. The UV spectrum displayed an absorption band at λ_{max} 194 nm, similar to that of compound **28**, which may be attributed to the carbonyl group (Figures S53 and S54). Its IR and ^{1}H NMR spectra (Figures S55 and S56) were very similar to those of compound **33**, except at the C-19 position. In compound **33**, H-19 appeared as a methyl singlet (δ_H 0.90), whereas in compound **34** two doublets (δ_H 2.92, 3.35) indicated the presence of a hydroxymethyl group. HMBC correlations between protons H-19 and carbons C-3 (δ_C 74.3), C-4 (δ_C 42.9), C-5 (δ_C 46.1), and C-18 (δ_C 13.1) confirmed the assignment of the hydroxymethyl group at C-4 (Figure 9). Compound **34** was therefore assigned as 3*α*-angeloyloxy-15,16,19-trihydroxy-*ent*-labda-8(17),13*E*-diene.

Compound **35** was obtained as a mixture with fraxetin (**19**) [49] as pale yellow powders. Its molecular formula was established by HRESIMS, which showed a $[M + H]^+$ ion at *m*/*z* 277.1083 (calcd. for $C_{15}H_{17}O_5^+$, 277.1071), corresponding to 8 degrees of unsaturation. Based on the relative intensities of the split signals observed in the ^{1}H NMR spectrum, the mixture ratio was estimated to be 65:35. The UV spectrum displayed an absorption band at λ_{max} 211 nm, which may be related to an aromatic or conjugated system. The IR spectrum showed characteristic absorption bands at 3293 cm^{-1} (O–H stretching), 2968 cm^{-1} (asymmetric CH_3), 2933 cm^{-1} 2857 cm^{-1} (asymmetric and symmetric CH_2 stretching), 1700 cm^{-1} (conjugated C=O), 1620 cm^{-1} (conjugated C=C), 1503 cm^{-1} (aromatic C=C), 1467 cm^{-1} (asymmetric CH_3), 1377 cm^{-1} (symmetric CH_3), 1181 cm^{-1} (isopropyl C–C), 1029–1143 cm^{-1} (C–O), and 804–927 cm^{-1} (aromatic C–H) (Figures S60–S62). The ^{1}H NMR spectrum (Figure S63) revealed two doublets, H-3 (δ_H 6.13) and H-4 (δ_H 8.13), coupled with J = 9.6 Hz, indicating a 1,2-benzopyrone nucleus substituted at C-5 by an oxygen [59]. Additional signals included a singlet at δ_H 1.41 (H-4′ and H-5′) corresponding to two methyl groups, two triplets at δ_H 1.88 (H-2′) and δ_H 2.82 (H-1′) corresponding to two

https://doi.org/10.3390/molecules31060973

methylene groups, and a singlet at δ_H 3.78 representing a methoxy group, suggesting that compound **35** is a pyranocoumarin [60]. COSY correlations between protons H-3 (δ_H 6.13) and H-4 (δ_H 8.13), and between protons H-1′ (δ_H 2.82) and H-2′ (δ_H 1.88), indicated vicinal coupling. Furthermore, HMBC correlations between H-4′/H-5′ (δ_H 1.41) and C-2′ (δ_C 32.6), as well as between the proton signal at δ_H 3.78 and C-8 (δ_C 133.6), allowed the assignment of the two methyl groups at C-3′ and the methoxy group at C-8, respectively (Figure 9). Compound **35** was thus unambiguously identified as 5-hydroxydihydroluvangetin.

Table 2. ^{1}H NMR data of compounds **26–30** (600 MHz, $CDCl_3$, 300 K).

Position	26 and 27		28		29		30	
	δ_H	(*J* in Hz)	δ_H	(*J* in Hz)	δ_H	(*J* in Hz)	δ_H	(*J* in Hz)
1	1.29; 1.82	m; m	1.86; 1.34	m; m	1.29; 1.85	m; m	1.28; 1.82	m; m
2	1.73; 1.91	m; m	1.73; 1.90	m; m	1.75; 1.80	m; m	1.70; 1.90	m; m
3	4.97	-	5.01	dd (12.2, 4.3)	4.97	dd (12.0, 4.5)	5	dd (12.1, 4.5)
4	-	-	-	-	-	-	-	-
5	1.74	m	1.76	m	1.73 (m)	m	1.74	m
6	1.37; 1.75	m; m	1.37; 1.75	m; m	1.75 (m)	m	1.37; 1.74	m; m
7	2.08; 2.39	m; m	2.07; 2.39	m; m	2.07; 2.40	m; m	2.06; 2.38	brtd (12.9, 4.6); ddd (12.9, 4.2, 2.3)
8	-	-	-	-	-	-	-	-
9	1.75	m	1.73	m	1.71	m	1.66	m
10	-	-	-	-	-	-	-	-
11	1.68; 1.80	m; m	1.68; 1.90	m; m	1.62; 1.76	m; m	1.73; 1.90	m; m
12	2.34; 2.52	m; m	2.32; 2.53	m; m	2.13; 2.41	m; m	1.46; 1.82	m; m
13	-	-	-	-	-	-	-	-
14	5.86	s	4.52; 4.88	brs; brs	5.83	dsxt (8.2, 1.2)	3.95	brd (3.6)
15	-	-	0.69	s	9.96	d (8.2)	4.9	brd (3.6)
16	5.97	s	2.92; 3.35	d (12.6); d (12.6)	2.22	d (2.2)	4.75	s
17	4.51; 4.89	brs; brs	0.77	s	4.57; 4.92	brd (1.2); brs	4.58; 4.84	brs; brs
18	0.68	s	-	-	0.76	s	0.68	s
19	2.91; 3.36	d (12.5); d (12.7)	-	-	3.13; 3.30	brd (11.6); brd (11.6)	2.92; 3.34	brd (12.5); brd (12.5)
20	0.77	s	-	-	0.81	s	0.76	s
1′	-	-	-	-	-	-	-	-
2′	-	-	-	-	-	-	-	-
3′	6.12	ddq (14.7, 7.3, 1.5)	6.12	qq (12.1, 1.5)	6.09	brqq (7.2, 1.4)	6.11	brqq (7.2, 1.4)
4′	1.98	dq (7.3, 1.4)	1.99	dq (7.0, 1.5)	1.96	brdq (7.2, 1.4)	1.99	brdq (7.2, 1.4)
5′	1.87	m	1.88	m	1.87	brqt (1.4)	1.88	brqt (1.4)
HO-13	-	-	-	-	-	-	2.44	brs
HO-14	-	-	-	-	-	-	2.56	brs
CH_3O-15	-	-	-	-	-	-	3.47	s
CH_3O-16	-	-	-	-	-	-	3.4	s

br: broad (large signal).

https://doi.org/10.3390/molecules3106097

Table 3. ^{1}H NMR data of compounds **31–35** (600 MHz, $CDCl_3$, 300 K).

Position	31 and 32		33		34		35	
	δ_H	(*J* in Hz)	δ_H	(*J* in Hz)	δ_H	(*J* in Hz)	δ_H	(*J* in Hz)
1	1.26; 1.82	m; m	1.25; 1.81	m; m	1.28; 1.84	m; m	-	-
2	1.71; 1.89	m; m	1.65; 1.80	m; m	1.72; 1.91	m; m	6.13	d (9.6)
3	4.97	dt (11.9, 4.5)	4.99	dd (11.9, 4.0)	5.00	dd (12.0, 4.4)	8.13	d (9.6)
4	-	-	-	-	-	-	-	-
5	1.74	m	1.21	m	1.75	m	-	-
6	1.36; 1.73	m; m	1.41; 1.75	dd (12.9, 4.2); m	1.35; 1.76	dd (13.1, 4.2); m	-	-
7	2.08; 2.39	m; dm (12.4)	2.02; 2.41	m; m	2.08; 2.39	m; m	-	-
8	-	-	-	-	-	-	-	-
9	1.73	m	1.61	m	1.71	m	-	-
10	-	-	-	-	-	-	-	-
11	1.64; 1.74	m; m	1.52; 1.61	m; m	1.51; 1.64	m; m	-	-
12	2.16; 2.46	m; m	1.92; 2.33	m; m	1.92; 2.32	m; m	-	-
13	-	-	-	-	-	-	-	-
14	6.85	m	5.62	t (6.8)	5.62	t (7.1)	-	-
15	6.08	dt (6.3; 1.9)	4.22	dd (6.7, 2.9)	4.22	dd (6.8, 3.2)	-	-
16	-	-	4.16; 4.20	d (12.5); d (12.3)	4.16; 4.19	d (12.3); d (12.4)	-	-
17	4.58; 4.89	d (3.5); brs	4.56; 4.87	brs; brs	4.56; 4.87	brs; brs	-	-
18	0.68	s	0.88	s	0.68	s	-	-
19	2.91; 3.36	brd (12.6); brd (12.6)	0.9	s	2.92; 3.35	d (12.7); d (12.7)	-	-
20	0.75	s	0.72	s	0.76	s	-	-
1′	-	-	-	-	-	-	2.82	t (6.8)
2′	-	-	-	-	-	-	1.88	t (6.8)
3′	6.11	ddq (14.2, 7.1, 1.3)	6.03	qq (12.0, 1.5)	6.12	qq (12.1, 1.4)	-	
4′	1.98	dq (7.4; 1.6)	1.96	brdq (7.2, 1.5)	1.96	brdq (7.3, 1.4)	1.41	s
5′	1.88	m	1.89	m	1.88	m	1.41	s
CH_3O-8	-	-	-	-	-	-	3.78	s

br: broad (large signal).

2.4.2. Relative Configuration of Diterpenes **26–34**

The absolute configurations of the new diterpenes could not be determined due to the limited quantities available. Nevertheless, their relative configurations were established using Carman's method [61]. Labdanes occur in nature as two antipodal groups of bicyclic molecules, commonly referred to as the "normal" and "*ent*" series, in which C-9 and C-10 adopt a *cis* configuration. Typically, compounds from the same series are found together, although co-occurrence of *normal*- and *ent*-labdanes has also been reported [61]. Optical rotation provides valuable insights into stereochemical assignment [62]. In labdane diterpenes, *normal*-labdanes typically show positive specific rotations [63,64], whereas *ent*-labdanes display negative values [64–66]. All newly isolated diterpenes reported herein exhibited negative rotations ($[\alpha]^{25}_D$ −145 to −30), indicating that these compounds belong to the *ent*-labdane series, consistent with the known *ent*-labdane (**11**), *ent*-8(17),13-labdadien-15,16-olide-19-oic acid [48].

Table 4. ^{13}C NMR data of compounds **26**–**35** (150 MHz, $CDCl_3$, 300 K).

Position	26 and 27		28		29		30		31 and 32		33		34		35	
	δ_C	Type	δ_C *	Type	δ_C *	Type	δ_C	Type	δ_C	Type	δ_C *	Type	δ_C *	Type	δ_C *	Type
1	36.8	CH_2	36.9	CH_2	37.6	CH_2	36.8	CH_2	36.8	CH_2	37.3	CH_2	36.1	CH_2	-	-
2	24.4	CH_2	24.3	CH_2	24.8	CH_2	24.5	CH_2	23.3	CH_2	25	CH_2	24.4	CH_2	163.9	C
3	74.3	CH	74.3	CH	75.5	CH	74.4	CH	74.4	CH	80.7	CH	74.3	CH	110.0	CH
4	42.9	C	43.5	C	43.1	C	43	C	42.9	C	38.5	C	42.9	C	141.6	CH
5	46.0; 46.1	CH	46.6	CH	47.3	CH	46.3	CH	46.1	CH	55.1	CH	46.1	CH	152.6	C
6	23.3	CH_2	23.3	CH_2	24.7	CH_2	23.4	CH_2	23.1	CH_2	24.3	CH_2	23.4	CH_2	102.1	C
7	37.6	CH_2	37.8	CH_2	38.6	CH_2	37.8	CH_2	37.6	CH_2	38.2	CH_2	37.7	CH_2	150.5	C
8	147.2; 147.3	C	147.4	C	148.9	C	147.8	C	147.2	C	148.2	C	147.2	C	133.6	C
9	55.9; 56.1	CH	55.8	CH	57.2	CH	57	CH	55.6; 55.7	CH	56.4	CH	56.1	CH	-	-
10	39.2	C	39.2	C	40.1	C	39.3	C	39.1	C	39.8	C	39.1	C	-	-
11	21.0; 21.2	CH_2	19.1	CH_2	22.6	CH_2	24.5	CH_2	21.6; 21.7	CH_2	22.4	CH_2	22.5	CH_2	-	-
12	26.7; 26.9	CH_2	32.4	CH_2	40.5	CH_2	32.8	CH_2	24.3; 24.4	CH_2	35.2	CH_2	34.8	CH_2	-	-
13	171.4	C	177.2	C	168	C	81.5	C	138.5; 138.6	C	144.5	C	144.5	C	-	-
14	117.3	CH	107.1	CH_2	128.1	CH	80.4	CH	143.6	CH	126.6	CH	126.3	CH	-	-
15	170.1; 170.2	C	12.6	CH_3	193.6	C	110.7	CH	96.9; 97.0	CH	59.1	CH_2	58.8	CH_2	-	-
16	99.0; 99.4	CH	64.5	CH_2	17.6	CH_3	109	CH	171.8; 172.0	C	61.5	CH_2	61.2	CH_2	-	-
17	107.1; 107.3	CH_2	15.2	CH_3	107.5	CH_2	107.3	CH_2	107.3	CH_2	107.3	CH_2	107	CH_2	-	-
18	13	CH_3	-	-	14	CH_3	13	CH_3	13	CH_3	17.3	CH_3	13.1	CH_3	-	-
19	64.5	CH_2	-	-	64.6	CH_2	64.9	CH_2	64.5	CH_2	28.7	CH_3	64.6	CH_2	-	-
20	15.3	CH_3	-	-	15.5	CH_3	15.3	CH_3	15.3	CH_3	15.1	CH_3	15.4	CH_3	-	-

https://doi.org/10.3390/molecules3106097

Table 4. *Cont.*

Position	26 and 27		28		29		30		31 and 32		33		34		35	
	δ_C	Type	δ_C *	Type	δ_C *	Type	δ_C	Type	δ_C	Type	δ_C *	Type	δ_C *	Type	δ_C *	Type
1′	169.3	C	169.3	C	169.1	C	169.2	C	169.3	C	168.2	C	169.1	C	17.1	CH_2
2′	127.7	C	127.7	C	129.5	C	127.9	C	127.7	C	128.5	C	127.5	C	32.6	CH_2
3′	139.5	CH	139.2	CH	138.8	CH	138.9	CH	139.3	CH	138.1	CH	139.1	CH	77.0	C
4′	16.1	CH_3	16.1	CH_3	16.1	CH_3	16	CH_3	16.1	CH_3	16.2	CH_3	16.2	CH_3	27.0	CH_3
5′	20.8	CH_3	20.9	CH_3	20.9	CH_3	20.8	CH_3	20.8	CH_3	21.1	CH_3	20.9	CH_3	27.0	CH_3
4a	-	-	-	-	-	-	-	-	-	-	-	-	-	-	103.9	C
8a	-	-	-	-	-	-	-	-	-	-	-	-	-	-	-	C
CH_3O-8	-	-	-	-	-	-	-	-	-	-	-	-	-	-	61.4	CH_3
CH_3O-15	-	-	-	-	-	-	56.5	CH_3	-	-	-	-	-	-	-	-
CH_3O-16	-	-	-	-	-	-	55.4	CH_3	-	-	-	-	-	-	-	-

* ^{13}C chemical shifts determined by 2D NMR (edited-HSQC, HMBC spectra).

https://doi.org/10.3390/molecules31060973

The relative stereochemistry of these new diterpenes was further corroborated by NOESY analysis (Figures S11, S19, S27, S36 and S45). Within the bicyclic ring system, characteristic NOE correlations between H-3/H-5, H-18/H-5, and H-18/H-6 established the orientation of the C-18 methyl group, the C-6 methylene, and the axial protons H-3 and H-5. In derivatives bearing a hydroxymethyl group at C-19, correlations between H-19 and H-20 confirmed their co-facial arrangement, further supported by the H-11/H-20 interaction. Altogether, these data unambiguously establish the *ent*-configuration for the new labdanes (Figure 10). Additionally, NOE correlations between H-3′/H-4′ and H-3′/H-5′ in the angeloyl moiety allowed assignment of a (*Z*)-configuration for the double bond. However, the configurations of chiral centres C-13, C-15, and C-13′ located on the lateral chains could not be determined.

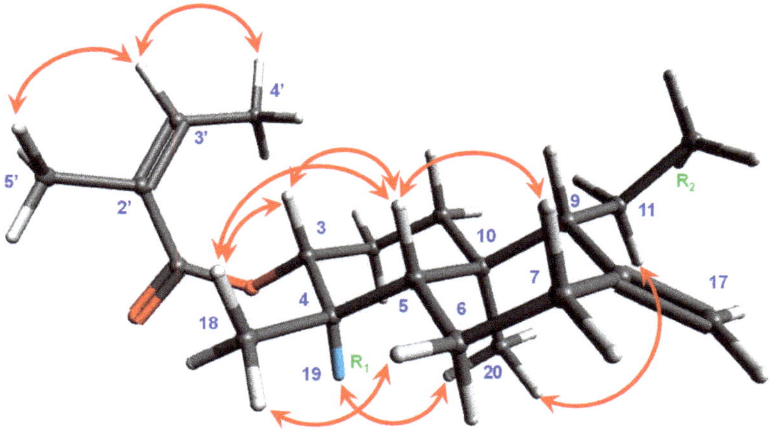

	26 and 27	28	29	30	31 and 32	33	34
R1	CH2OH	CH2OH	CH2OH	CH2OH	CH2OH	CH3	CH2OH
R2							

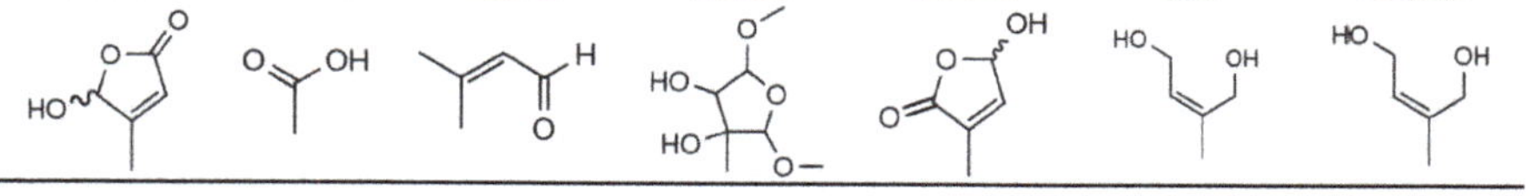

Figure 10. Key NOESY (red arrows) correlations observed for the decalin nucleus of compounds **26–34**.

2.5. Biological Activities of the Isolated Compounds

2.5.1. Antiplasmodial Activity of the Isolated Compounds

The *in vitro* antiplasmodial activity of isolated compounds was evaluated, for the first time, against the *Plasmodium falciparum* 3D7 strain. The results are presented in Figure 11, showing the concentrations required to inhibit 50% of parasite growth (IC_{50}). Three samples could not be tested due to limited availability: kumatakilin (**36**) (isolated in mixture with ermanin (**1**)), and the mixture of retusin (**4**) and penduletin 4-methylether (**37**).

Among the tested samples, according to classification [24,25] (active compound: 5–15 µg/mL), the mixture containing the newly identified compounds **26** and **27**, 3α-angeloyloxy-19,16(*R*/*S*)-dihydroxy-*ent*-labda-8(17),13-dien-15,16-olide, together with the known metabolites ermanin (**1**) and isokaempferide (**7**), displayed the highest inhibitory activity against *Plasmodium falciparum* within the series, with IC_{50} values ranging from 7.25 to 13.46 µg/mL. According to the same classification [24,25]), several other compounds (**2, 3, 5, 6, 8, 11, 18, 28, 30, 19** and **35, 29** and **37, 31** and **32**) showed moderate activity, with IC_{50} values ranging from 15.41 to 46.21 µg/mL. Notably, the antiplasmodial effects of

https://doi.org/10.3390/molecules31060973

ermanin (**1**) (IC_{50} = 13.46 μg/mL) and quercetin 3-methylether (**6**) (IC_{50} = 46.21 μg/mL), have been previously reported [67,68], consistent with the current observations.

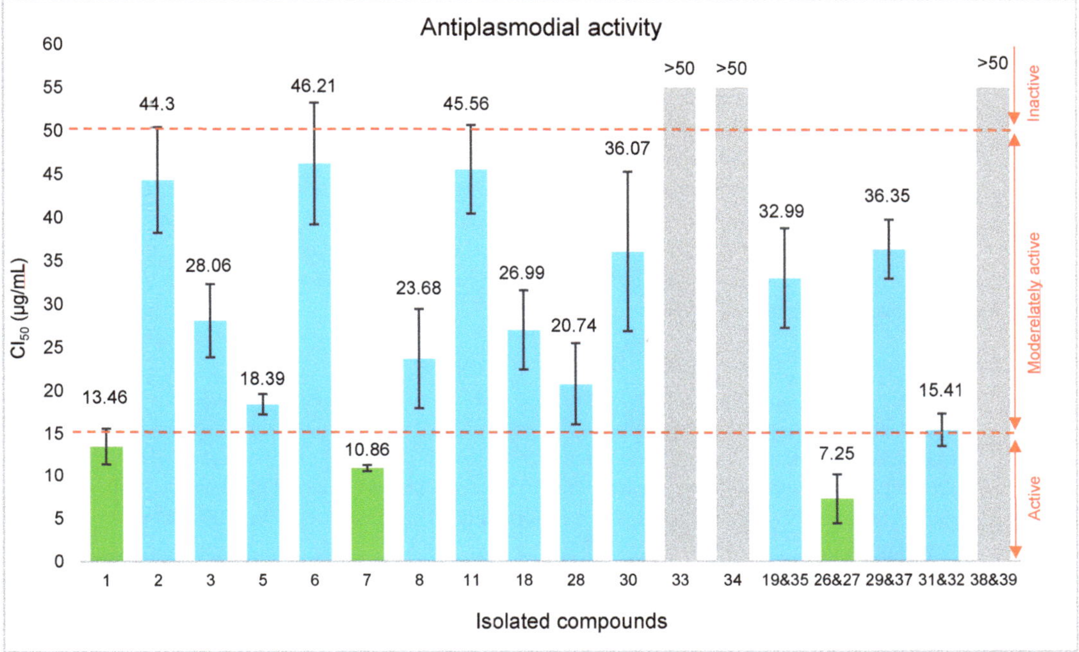

Figure 11. *In vitro* antiplasmodial activity (3D7 strain) of isolated compounds (IC_{50} in μg/mL). Positive control: artemisinin (IC_{50} = 0.003 ± 0.001 μg/mL). Active compound: 5–15 μg/mL, moderately active compound: 15–50 μg/mL, inactive compound ($IC_{50} \geq 50$ μg/mL) [24,25].

These results indicate that both flavonoid derivatives and some diterpenoid constituents may contribute to the antiplasmodial activity of *P. dentata*. However, although their IC_{50} values indicate inhibition of parasite growth, their potency remains considerably lower than that of the reference drug artemisinin (IC_{50} = 0.003 μg/mL). Therefore, they can be considered weak to moderately active and are not by themselves sufficient for high anti-malarial activity. Nevertheless, they may provide useful starting points for further optimization or structure–activity relationship studies.

Regarding the new compounds, labdane diterpenoids have been reported to exhibit moderate antiplasmodial activity *in vitro* [69–71]. The isolated diterpenes are quite similar, sharing a common *ent*-labdane skeleton featuring an angeloyl group at C-3 (except compound **11**), an exocyclic methylene group at C-8, and a linear connectivity between C-9, C-11, and C-12. They mainly differ in the functional groups extending from C-13. Taking their antiplasmodial activities into account, it is suggested that the hydroxyfuranone group in the mixtures of compounds **26** and **27** as well as **31** and **32** may improve the antiplasmodial activity, as these derivatives showed the highest activity within the tested series. These observations align with prior reports on angeloyloxylabdanes, where hydroxyfuranone substitution at C-13 enhances antiplasmodial potency [72]. In contrast, compounds **33** and **34** were inactive ($IC_{50} \geq 50$ μg/mL), likely due to the presence of identical non-conjugated allylic diol functionalities at C-13. Compound **30** showed only weak activity, further supporting the contribution of the furanone moiety to the observed activity.

https://doi.org/10.3390/molecules31060973

These results complement the metabolomic findings. Specifically, the antiplasmodial activities detected for isolated compounds such as ermanin (**1**), isokaempferide (**7**), and isoobtusitin (**18**), are consistent with the statistical correlations from metabolomic analysis, suggesting that these compounds, particularly isokaempferide (**7**), the major constituent of the extract may contribute to the antiplasmodial potential of *P. dentata*. Nevertheless, these compounds alone are unlikely to fully account for the activity of the crude extract. Minor active compounds could not be clearly identified due to signal overlap in the NMR spectra. In addition, signals detected in the aliphatic region, possibly corresponding to diterpenoids or other constituents, may also contribute to the overall antiplasmodial effect. The phytochemical investigation therefore complemented the metabolomic approach by enabling the isolation and identification of additional active constituents. Furthermore, the possible presence of chrysosplenol D (**10**), annotated in the molecular network and previously reported to possess antiplasmodial activity [73], may also play a role in the overall activity of the crude extract.

2.5.2. Anti-Inflammatory Activity of the Isolated Compounds

Isolated compounds were tested for *in vitro* anti-inflammatory activity by measuring the capacity of RAW 264.7 macrophages to generate an inflammatory response when stimulated with antigens, inducing nitric oxide (NO) release [74]. Due to limited availability, quercetin 3-methylether (**6**) and the sample with kumatakilin (**36**) were not evaluated. The results, presented in Figure 12, show the concentrations required to inhibit 50% of NO production (CI_{50}).

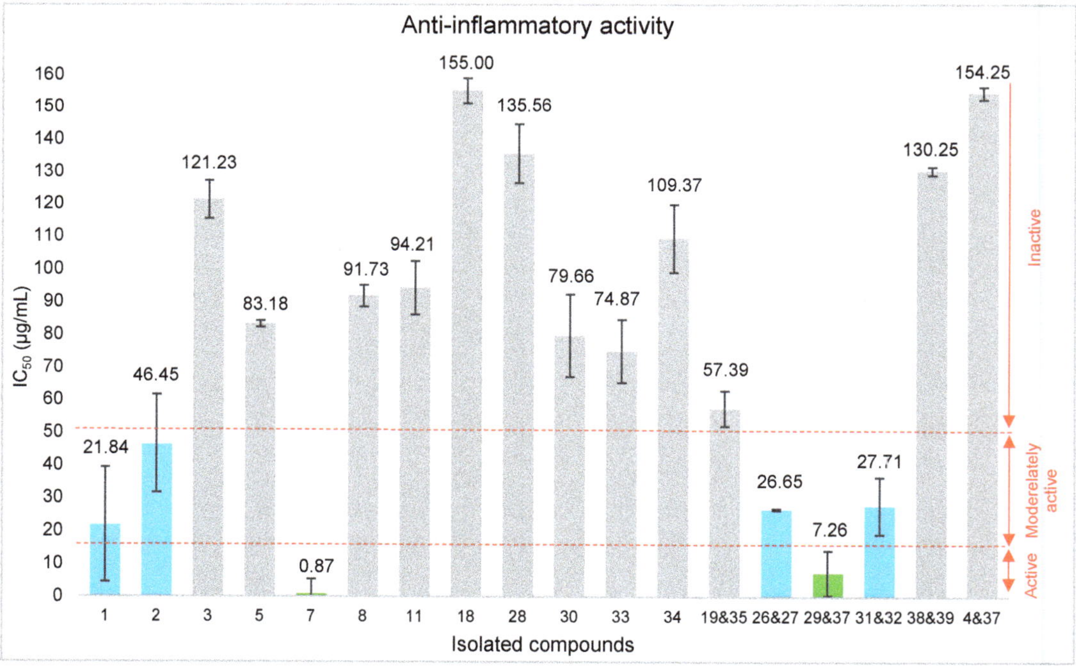

Figure 12. *In vitro* anti-inflammatory activity (RAW 264.7 cell line) of isolated compounds (IC_{50} in µg/mL). Positive control: dexamethasone (IC_{50} = 3.36 ± 1.04 µg/mL). Active compound: 5–15 µg/mL, moderately active compound: 15–50 µg/mL, inactive compound ($IC_{50} \geq 50$ µg/mL) [24,25].

https://doi.org/10.3390/molecules3106097

Three tested compounds displayed a significant anti-inflammatory activity: the mixture containing the newly identified compound 3α-angeloyloxy-19-hydroxy-*ent*-labda-8(17),13*E*-dien-15-al (**29**) and penduletin 4-methylether (**37**), together with isokaempferide (**7**), inhibited NO production, with IC_{50} values of 7.26 and 0.87 µg/mL respectively (IC_{50} < 15 µg/mL, classification according to [24,25]). Meanwhile, the mixture containing the newly identified compounds **26** and **27**, 3α-angeloyloxy-19,16(*R*/*S*)-dihydroxy-*ent*-labda-8(17),13-dien-15,16-olide, compounds **31** and **32**, 3α-angeloyloxy-19,15(*R*/*S*)-dihydroxy-*ent*-andrograpanin, as well as ermanin (**1**) and kaempferol 3,7,4′-trimethylether (**2**) displayed moderate activity, with IC_{50} values ranging from 21.84 to 46.45 µg/mL (IC_{50} < 50 µg/mL, classification according to [24,25]). Their cytotoxicity (IC_{50}) towards mouse macrophages was also evaluated. Among the active compounds, **29** and **37** showed significant cytotoxicity with IC_{50} of 22.27 µg/mL, whereas the remaining compounds exhibited IC_{50} values above 250 µg/mL, indicating low toxicity. Consequently, the mixture containing **29** and **37** cannot be considered a promising anti-inflammatory agent due to its cell cytotoxicity. The observed activities of ermanin (**11**) (IC_{50} = 21.84 µg/mL) and isokaempferide (**21**) (IC_{50} = 0.87 µg/mL) are consistent with previous reports [75,76], supporting their contribution to the anti-inflammatory potential of *P. dentata*.

Several labdane diterpenoids have been shown to inhibit NO production in RAW264.7 cells [77–79], which supports the present findings. However, the underlying mechanisms remain only partially elucidated. Analysis of the diterpenoid tested in the present study suggests that the anti-inflammatory effects of epimeric mixtures **26** and **27** as well **31** and **32** might be associated with the presence of a hydroxyfuranone moiety at C-13. This hypothesis is further supported by previous studies [77,78], in which several labdane-type diterpenoids featuring furanone derivative showed significant NO inhibition.

The metabolomic analysis revealed signals corresponding to isokaempferide that may be positively correlated with the anti-inflammatory activity of the crude extract. These results illustrate the ability of metabolomic profiling to link chemical composition with biological activity and the subsequent isolation and biological evaluation of isokaempferide provide experimental support for this prediction. However, since most isolated compounds were inactive or only moderately active, the bioactivity of the crude extract may be influenced by synergistic or matrix effects that are lost upon isolation, or by other compounds that were not isolated but may contribute to this activity.

2.5.3. Cytotoxic Activity of the Isolated Compounds

The isolated compounds were tested for cytotoxic activity against human liver carcinoma (HepG2) and human colon and colorectal adenocarcinoma (HT29) cell lines, using the red dye assay. Due to limited availability, quercetin 3-methylether (**6**) and the mixture with kumatakilin (**36**) were not tested. The results shown in Figure 13 indicate the concentrations required to inhibit 50% of cell viability (IC_{50}). Among the tested compounds, only ermanin (**1**) exhibited moderate cytotoxicity, with IC_{50} values of 25.67 and 18.35 µg/mL against the HepG2 and HT29 cell lines, respectively. This activity is consistent with previous reports, where the compound also exhibited cytotoxic effects against breast cancer (MCF-7) and oral cancer (BHY) cell lines [80,81].

https://doi.org/10.3390/molecules31060973

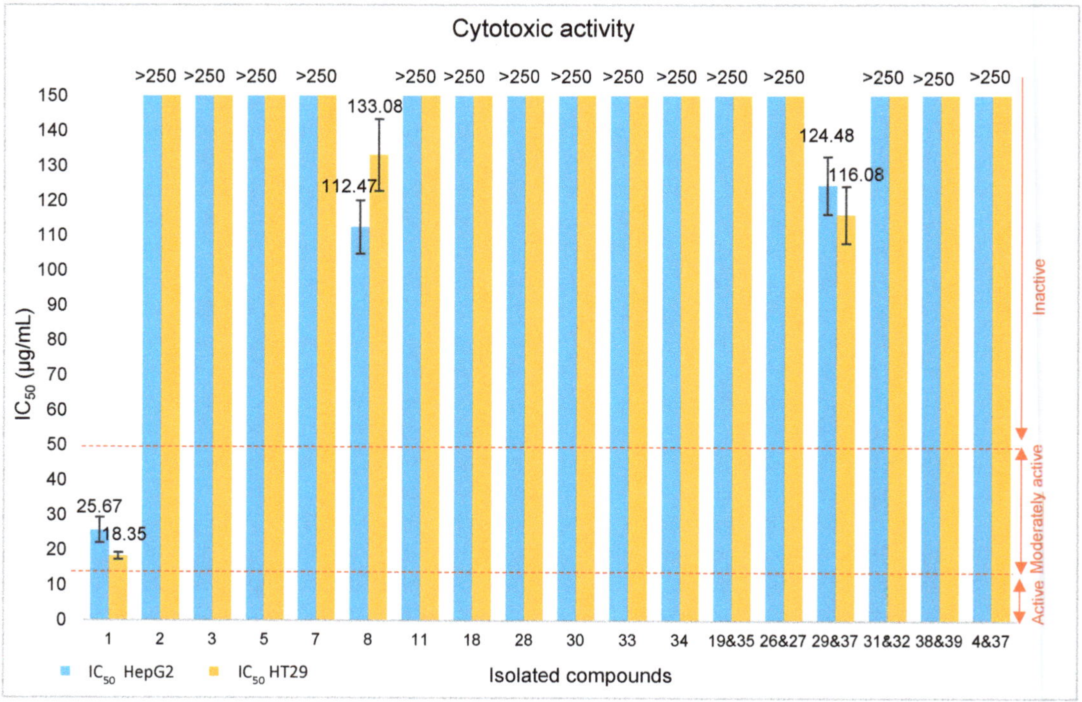

Figure 13. *In vitro* cytotoxic activity (HepG2 and HT29 cancer cell lines) of isolated compounds (IC_{50} in µg/mL). Active compound: 5–15 µg/mL, moderately active compound: 15–50 µg/mL, inactive compound ($IC_{50} \geq 50$ µg/mL) [24,25].

3. Materials and Methods

3.1. General Experimental Procedures

Optical rotations $[\alpha]_D$ were carried out on an Anton Paar MCP 200 polarimeter (Anton Paar, Graz, Austria) at 25 °C, at 589 nm wavelength, with MeOH used as solvent, *c* (g/100 mL), in a 10 × 5 mm i.d., 0.2 mL sample cell. UV spectra were acquired on a Thermo Scientific DAD spectrophotometer (Thermo Scientific, Waltham, MA, USA). IR spectra were collected on an ATR-FTIR Bruker Tensor 27 spectrometer (Bruker, Billerica, MA, USA) with OPUS 7.2 software. The ^1H NMR spectra for metabolomic study were acquired at 298 K on a Bruker Avance III 600 MHz NMR spectrometer (Bruker, Billerica, MA, USA) equipped with a TCI Cryoprobe and operated at 600.13 MHz for ^1H. The 1D (^1H and ^{13}C) and 2D (COSY, HSQC, HMBC and NOESY) NMR spectra for compound identification were acquired at 300 K on a Bruker Avance II+ 600 MHz NMR spectrometer (Bruker, Billerica, MA, USA) equipped with a TCI Cryoprobe and operated at 600.13 MHz for ^1H and 150 MHz for ^{13}C. MS/MS spectra were recorded on a UHPLC system (Dionex Ultimate 3000, Thermo Scientific, Carlsbad, CA, USA) coupled to a high-resolution mass spectrometer HRMS-QTOF Impact II equipped with an electrospray ionization source (Bruker Daltonics, Bremen, Germany) in positive and negative ionization modes (20 eV and 40 eV). The following parameters were used: dry gas (N_2): 4.00 L.min^{-1}; nebulizer (N_2) 3.5 bars; full-scan mode: m/z 50–1200; capillary voltage: 3.0 kV, Luna C18 reversed phase column, 1.6 µm, 150 × 2.1 mm i.d. (Phenomenex, Torrance, CA, USA). MPLC separations were carried out in normal phase on Buchi Sepacore flash systems C-605/C-615/C-660. A glass column (230 × 15 mm i.d.) packed with silica gel (40–63 µm). TLC

https://doi.org/10.3390/molecules3106097

was performed in normal phase on aluminum analytical plates precoated with 60 Å silica gel impregnated with a UV 254 nm fluorescence indicator. Spots were visualized on the basis of the UV absorbance at 254 nm and by heating silica gel plates sprayed with vanillin-sulphuric acid reagent. Analytical HPLC was carried out using a Gemini C18 (150 × 4.6 mm i.d., 3 µm) column (Phenomenex, Torrance, CA, USA) and was performed on a Thermo Scientific Dionex Ultimate 3000 system (Thermo Scientific, Waltham, MA, USA) equipped with a DAD photodiode array detector and a charged aerosol detector (CAD) with Chromeleon 7.2.8 software. Semi-preparative HPLC was carried out using a Gemini C18 (250 × 4.6 mm i.d., 5 µm) column (Phenomenex, Torrance, CA, USA) and was performed on a Thermo Scientific Dionex Ultimate 3000 system equipped with a DAD photodiode array detector or a Waters 2545 autopurification system process via MassLynx 4.2 software and coupled in series with a DAD photodiode array detector (190–800 nm, Waters 2998) and a DEDL light-scattering evaporative detector (Waters 2424) or a Agilent Technologies 1100 series (Agilent, Waldbronn, Germany) equipped with a DAD photodiode array detector and processed by Chemstation B.04.03 software. All solvents were analytical or HPLC grade and were used without further purification.

3.2. Plant Material

Psiadia species leaves (phylum Tracheophyta, class Magnoliopsida, order Asterales, family *Asteraceae*) were collected in 2015 in several collection sites on Reunion Island. Each species was collected in summer. A voucher specimen was deposited at the Herbarium of the University of Reunion Island for identification (identification performed by Jacques Fournel, Herbarium of the University of La Réunion). The collection sites and voucher numbers of all species are reported in Table 5.

Table 5. Area of collects and voucher number of each *Psiadia* species.

Psiadia Species	Collection Site	Voucher Number
P. amygdalina	Salazie	REU13109
	Maïdo	REU13115
	Volcan	REU13130
P. anchusifolia	Salazie	REU13111
	Maïdo	REU13117
	Volcan	REU13130
	Roche-Ecrite	REU13131
P. argentea	Salazie	REU13112
	Volcan	REU13143
	Roche-Ecrite	REU13135
P. boivinii	Salazie	REU13113
	Maïdo	REU13116
	Volcan	REU13142
	Roche-Ecrite	REU13133
P. callocephala	Volcan	REU13140
P. dentata	Maïdo	REU13119
	Colorado	REU13139
	Dos d'Ane	REU13128
	Montauban	REU13487
P. insignis	Maïdo	REU13484
	Volcan	REU13483
	Dos d'Ane	REU13125

https://doi.org/10.3390/molecules31060973

Table 5. *Cont.*

Psiadia Species	Collection Site	Voucher Number
P. laurifolia	Salazie	REU13114
	Maïdo	REU13118
	Volcan	REU13141
	Roche-Ecrite	REU13132
P. melastomatoides	Salazie	REU13110
	Roche-Ecrite	REU13134
P. montana	Dos d'Ane	REU13127
P. retusa	Cap Jaune	REU13129
	Cap Méchant	REU13486
	Piton Grande Anse	REU13485

3.3. *Extraction*

Prior to extraction, the leaves collected were dried at 40 °C. Dried and powdered leaves of each species were extracted using an accelerated solvent extractor (ASE 300 Thermo Scientific Dionex, Sunnyvale, CA, USA). It was operated at 40 °C and 100 bar, in 5 cycles with a purge time of 120 s and a pause time of 5 min. Four successive extractions were carried out with ethyl acetate (analytical grade, Carlo Erba, Val de Reuil, France) to exhaust the plant material. The extracts were pooled and evaporated under reduced pressure and kept at 4 °C until analysis.

3.4. *Molecular Networking Parameters*

Molecular networking permits the visualization of related molecules based on the comparison of their MS/MS data [12,26]. The network was created from isohexane and methanolic fractions of AcOEt extract of *Psiadia dentata* leaves using the online workflow at the GNPS web platform (http://gnps.ucsd.edu; accessed on 22 May 2020). LC-MS/MS data were acquired from UHPLC-HRMS-ESI-QTOF in positive ionization mode (20 eV and 40 eV). Analytical HPLC was performed using a Luna C_{18} column (150 mm × 2.1 mm i.d., 1.6 μm) (Phenomenex, Torrance, CA, USA) with 0.5 mL·min^{-1} gradient elution with 10% ACN-H_2O (+0.1% formic acid) to 100% ACN-H_2O (+0.1% formic acid) over 15 min. The MS/MS data were converted to the .mzXML format using MS-Convert 3.0.24 software, part of the ProteoWizard package [82] and then uploaded to the GNPS web platform for MN. The data were clustered with a parent mass tolerance of 0.02 Da and an MS/MS fragment ion tolerance of 0.02 Da to create consensus spectra. A network was then created where edges were filtered to have a cosine score above 0.62 and more than six matched peaks. The following parameters were also specified: network TopK: 10, minimum cluster size: 2, and maximum connected component size: 100. The spectral library matching was performed with cosine score above 0.7 and six matched fragment ions. The generated molecular network was visualized using Cytoscape 3.8.2 software [83].

3.5. *Metabolomic Analysis*

3.5.1. ^{1}H NMR Spectroscopy of Plant Extracts for Metabolomics

^{1}H NMR metabolomic analysis was adapted from previously described protocols [84]. For each sample, 15 ± 0.2 mg of ethyl acetate crude extract was transferred to a 2 mL Eppendorf tube. A volume of 1.5 mL of deuterated chloroform ($CDCl_3$, D 99.9 atom %, Eurisotop, Saint-Aubin, France) containing 0.03% TMS (tetramethylsilane) was added to each sample. The mixture was vortexed at room temperature for 1 min. An aliquot of 0.6 mL was transferred into 5 mm NMR tubes for NMR analysis. ^{1}H NMR spectra were recorded in $CDCl_3$ and consisted of 64 scans. $CDCl_3$ was used as the internal lock. The

https://doi.org/10.3390/molecules3106097

acquisition consisted of 64 scans and the spectra had a spectral width of 12 ppm (7211 Hz). Each ^1H-NMR spectrum was recorded with the following parameters: a standard one-pulse sequence with 30° flip angle and a relaxation delay (RD) = 1.0 s. Free induction decays (FIDs) were Fourier-transformed with line broadening (LB) = 0.3 Hz. The resulting spectra were manually phased, baseline-corrected, and calibrated to TMS at 0.00 ppm using MestReNova 10.0 (Mestrelab Research S.L.). Spectral intensities were scaled to total intensity and bucketed (bin width of 0.02 ppm) from the spectral region of 0.32–10.02 ppm. The region between δ_H 7.05 and 7.49 ppm was removed before further analysis due to the residual solvent signal.

3.5.2. Multivariate Data Analysis

Chemometric analysis was performed with SIMCA version 12.0 (Umetrics, Umea, Sweden) in the form of a supervised orthogonal partial least squares discriminant analysis. The scaling was based on the Pareto method. Rx^2 and Q^2 values described the quality of the model. Rx^2 indicated goodness of fit and is defined as the proportion of the variance in the data observed in the model. Q^2 is defined as the proportion of variance in the data that was predictable by the model.

3.6. Phytochemical Investigation and Compound Isolation

Eight grams of extract was partitioned with isohexane and MeOH, yielding 4.53 and 3.36 g, respectively of isohexane and methanolic fractions. The hydrocarbon fraction (4.25 g) was separated by MPLC on normal phase (Buchi, Singapore, 230 × 15 mm i.d.) 15 mL. min^{-1} using a gradient of isohexane and EtOAc of increasing polarity. The eluates were monitored by TLC. Fractions with the same chromatographic profile were combined. Altogether, 8 fractions, F1–F8, were obtained. HPLC analysis of these 8 fractions allowed the selection of 4 fractions, F2, F4, F7 and F8, containing the majority of prevalent compounds for further fractionation.

Fraction F2 (50 mg) was subjected to semipreparative HPLC (Gemini C_{18} prep column, 5 µm, 250 × 10 mm i.d.), 4.5 mL·min^{-1} gradient elution with 60% ACN−H_2O (+0.1% formic acid) to 75% ACN−H_2O (+0.1% formic acid) over 35 min (UV 210 nm) to give pure compounds **1** (0.3 mg), **2** (0.8 mg), **18** (0.2 mg) in mixture with another flavonoid, **26** and **27** (mixture 50:50, 0.6 mg) and **28** (0.3 mg).

Purification of fraction F4 (45 mg) was performed by semipreparative HPLC (Gemini C_{18} prep column, 5 µm, 250 × 10 mm i.d.), 4.5 mL·min^{-1} gradient elution with 50% ACN−H_2O (+0.1% formic acid) to 85% ACN−H_2O (+0.1% formic acid) over 35 min (UV 210 nm) to give compounds **1** (mixture 50:50 with compound **36**, 1.3 mg), **3** (0.8 mg), **4** and **37** (mixture 50:50, 0.9 mg), and **29** (mixture 43:39:13:5 with **37**, 0.3 mg and another non-identified derived diterpene *ent*-labdane and a flavonoid).

Fraction F7 (70 mg) was purified by semipreparative HPLC (Gemini C_{18} prep column, 5 µm, 250 × 10 mm i.d.), 4.5 mL·min^{-1} isocratic elution with 50% ACN−H_2O (+0.1% formic acid) over 15 min and elution gradient to 95% ACN−H_2O (+0.1% formic acid) over 15 min and maintained at 95% ACN−H_2O (+0.1% formic acid) over 15 min (UV 210 nm). Purification yielded compounds **5** (0.5 mg), **11** (0.6 mg), **30** (1.3 mg), **31** and **32** (mixture 50:50, 3.8 mg), **38** and **39** (mixture 67:33, 1.1 mg), and the compounds **26** and **27** (mixture 50:50, 2.1 mg), which had already been isolated from fraction F2.

Fraction F8 was first submitted to MPLC on normal phase (Buchi, 230 × 15 mm i.d), 15 mL·min^{-1} using a gradient of isohexane and EtOAc of increasing polarity. Seven sub-fractions SF1–SF7 (4.4 mg to 165 mg) were obtained based on TLC profiles. After HPLC analysis, SF4 and SF5 (83.5 g) were grouped and re-purified by MPLC on normale phase (Buchi, 100 × 15 mm i.d.) 10 mL·min^{-1} using a gradient of isohexane and EtOAc of

https://doi.org/10.3390/molecules31060973

increasing polarity to yield ten subfractions. Separation of subfraction SF4 + 5.6 (6.7 mg) was performed by semipreparative HPLC (Gemini C_{18} prep column, 5 µm, 250 × 10 mm i.d.), 4.5 mL·min^{-1} isocratic elution with 50% ACN−H_2O (+0.1% formic acid) over 38 min, then a gradient to 65% ACN−H_2O (+0.1% formic acid) over 15 min and maintained at 65% ACN−H_2O (+0.1% formic acid) over 15 min (UV 210 nm) to give pure compound **33** (0.4 mg). Similarly, SF4 + 5.8 (11.1 mg) was purified by semi-preparative HPLC (50% ACN−H_2O (+0.1% formic acid) over 25 min, elution gradient to 100% ACN−H_2O (+0.1% formic acid) over 1 min and maintained at 100% ACN−H_2O (+0.1% formic acid) over 10 min (UV 210 nm), giving pure compound **34** (0.5 mg).

The MeOH-soluble fraction (60 mg) was subjected to semipreparative HPLC (Gemini C_{18} prep column, 5 µm, 250 × 10 mm i.d.), 4.5 mL·min^{-1} gradient elution with 90% ACN−H_2O (+0.1% formic acid) to 98% ACN−H_2O (+0.1% formic acid) over 15 min, 98% ACN−H_2O (+0.1% formic acid) to 100% ACN−H_2O (+0.1% formic acid) over 15 min and isocratic elution with 100% ACN−H_2O (+0.1% formic acid) over 10 min to produce compounds **5** (0.8 mg), **6** (1.1 mg), **7** (5.0 mg), **8** (2.3 mg), **18** (0.6 mg) and compounds **19** and **35** (mixture 50:50, 2.9 mg).

3.7. NMR Sample Preparation and Spectroscopic Analysis

NMR analyses were conducted on compounds obtained following chromatographic isolation. Individual compounds were analyzed in their pure form when available, while other samples were characterized as mixtures of 2 or 3 compounds, as described in the purification section. Each sample was dissolved in an appropriate deuterated solvent according to its solubility properties, using either chloroform-d ($CDCl_3$, 99.96% D, Eurisotop, ref. D029T) or methanol-d_4 (CD_3OD, 99.95% D, Eurisotop, ref. D048T). Compounds isolated from the isohexane fraction (**1–4**, **11**, **26–34**, and **36–39**) were dissolved in $CDCl_3$, whereas those obtained from the methanolic fraction (**5–8**, **18**, **19**, and **35**) were dissolved in CD_3OD. Approximately 0.2–3.8 mg of isolated compound was dissolved in 85 µL of the selected deuterated solvent. The solution was vortexed to ensure complete dissolution and then transferred into 2.0 mm NMR tubes (final volume = 85 µL). Prior to analysis, samples were centrifuged to remove any insoluble particles. Tubes were mounted on a spinner turbine and directly introduced into the spectrometer for data acquisition. The deuterated solvent used for each compound is specified in the corresponding NMR spectrum provided in the Supplementary Materials.

The 1D (^{1}H and ^{13}C) and 2D (COSY, HSQC, HMBC and NOESY) NMR spectra were acquired at 300 K with the high-field NMR spectrometer previously described, operating at 600.13 MHz for ^{1}H and 150 MHz for ^{13}C. NMR spectra were recorded using standard Bruker pulse sequences and processed with TopSpin 4.1.1 software (Bruker Biospin, Billerica, MA, USA). Chemical shifts (δ) are expressed in parts per million (ppm), using the residual $CDCl_3$ signal (δ_H 7.26; δ_C 77.16) or CD_3OD signal (δ_H 3.31; δ_C 49.0) as internal references for ^{1}H and ^{13}C NMR experiments. Coupling constants (*J*) are reported in Hertz (Hz).

3.8. Compound Characterization

3α-angeloyloxy-19-hydroxy-*ent*-labda-8(17),13-dien-15,16-olide (**26** and **27**): yellow powders; $[\alpha]_D^{25}$: −145 (*c* 0.02, $CHCl_3$); λ_{max} 225 nm; IR (ATR) *υ* 2925, 2853, 1757, 1646, 1260 cm^{-1}; ^{1}H and ^{13}C NMR data, see Tables 2 and 4; HREIMS $[M + H]^+$ *m*/*z* 433.2587 (calcd. for $C_{25}H_{37}O_6$ $[M + H]^+$, *m*/*z* 433.2585).

3α-angeloyloxy-16-hydroxy-*ent*-labda-8(17)-en-13-oic acid (**28**): white paste; $[\alpha]_D^{25}$: −50 (*c* 0.02, MeOH); λ_{max} 199 nm; IR (ATR) *υ* 3305, 2963, 2923, 2853, 1715, 1261 cm^{-1}; ^{1}H and ^{13}C NMR data, Tables 2 and 4; HREIMS $[M + H]^+$ *m*/*z* 379.2486 (calcd. for $C_{22}H_{35}O_5^+$, *m*/*z* 379.2400).

https://doi.org/10.3390/molecules3106097

3α-angeloyloxy-19-hydroxy-*ent*-labda-8(17),13*E*-dien-15-al (**29**): white oil; $[\alpha]_D^{25}$: −30 (*c* 0.04, MeOH); λ_{max} 227 nm, IR (ATR) *υ* 2924, 2854, 1733, 1673, 1260, 1022, 798 cm^{-1}; ^1H and ^{13}C NMR data, see Tables 2 and 4; HREIMS $[M + NH_4]^+$ *m*/*z* 420.3108 (calcd. for $C_{25}H_{42}O_4N^+$, *m*/*z* 420.3109).

3α-angeoyloxy-19-hydroxy-15,16-dimethoxy-*ent*-labda-8(17)-en-furan-13,14-diol (**30**): white paste; $[\alpha]_D^{25}$: −51 (*c* 0.04, MeOH); λ_{max} 197 nm; IR (ATR) *υ* 2953, 2929, 2854, 1692, 1644, 1260, 1022 cm^{-1}; ^1H and ^{13}C NMR data, Tables 2 and 4; HREIMS $[M - H]^-$ *m*/*z* 495.2049 (calcd. for $C_{27}H_{43}O_8^-$, *m*/*z* 495.2963).

3α-angeloyloxy-19,15(*R*/*S*)-dihydroxy-*ent*-andrograpanin (**31** and **32**): white paste; $[\alpha]_D^{25}$: −145 (*c* 0.02, MeOH); λ_{max} 197 nm; IR (ATR) *υ* 3384, 2936, 2855, 1761, 1709, 1643, 1603, 1440, 1386, 1237–1261, 1041 cm^{-1}; ^1H and ^{13}C NMR data, see Tables 3 and 4; HREIMS $[M + H]^+$ *m*/*z* 433.2589 (calcd. for $C_{25}H_{37}O_6^+$, *m*/*z* 433.2585).

3α-angeloyloxy-15,16-dihydroxy-*ent*-labda-8(17),13*E*-diene (**33**): white crystals; $[\alpha]_D^{25}$: −54 (*c* 0.04, MeOH); IR (ATR) *υ* 3354, 2961, 2926, 2855, 1713, 1603, 1456, 1384, 1261, 1027–1093 cm^{-1}; ^1H and ^{13}C NMR data, see Tables 3 and 4; HRESIMS $[M + H]^+$ *m*/*z* 427.2815 (calcd. for $C_{25}H_{41}O_4^+$, *m*/*z* 427.2819).

3α-angeloyloxy-15,16,19-trihydroxy-*ent*-labda-8(17),13*E*-diene (**34**): white crystals; $[\alpha]_D^{25}$: −135 (*c* 0.02, MeOH); λ_{max} 194 nm; IR (ATR) *υ* 3297, 2965, 2931, 2827, 1700, 1621, 1466, 1357, 1270, 1028–1115 cm^{-1}; ^1H and ^{13}C NMR data, see Tables 3 and 4; HRESIMS $[M + Na]^+$ *m*/*z* 443.2764 (calcd. for $C_{25}H_{40}O_5Na^+$, *m*/*z* 443.2768).

5-hydroxydihydroluvangetin (**35**): pale yellow powders; $[\alpha]_D^{25}$: −63 (*c* 0.04, MeOH); λ_{max} 211 nm; IR (ATR) *υ* 3293, 2968, 2933, 2857, 1700, 1620, 1503, 1467, 1377, 1181, 1029–1143, 804–927 cm^{-1}; ^1H and ^{13}C NMR data, see Tables 3 and 4; HRESIMS $[M + H]^+$ *m*/*z* 277.1083 (calcd. for $C_{15}H_{17}O_5^+$, *m*/*z* 277.1071).

3.9. *Biological Assays*

3.9.1. *In Vitro* Antiplasmodial Assay

The *P. falciparum* strains utilized and details of the assay protocols have been previously reported [69,85].

3.9.2. *In Vitro* HRP Inhibition Assay

The HRP (horseradish peroxidase) anti-peroxidase activity was evaluated using chemiluminescence technique according to the method previously reported [86].

3.9.3. *In Vitro* NO Inhibition Assay

It was determined in mouse immortalized macrophages (RAW 264.7 cell line, Sigma-Aldrich) as described previously [74,87].

3.9.4. *In Vitro* Cytotoxic Assay

HepG2 human liver hepatocellular carcinoma and HT29 human colorectal adenocarcinoma cells, provided by ATCC HTB-38 were used to assess the toxicity of crude extracts, fractions and pure compounds. The protocol used was previously reported [88,89].

4. Conclusions

In this study, ^1H NMR-based metabolomics was applied to investigate the biological potential of eleven *Psiadia* species in Reunion Island. The ^1H NMR fingerprints revealed distinct chemical profiles among all studied species of the genus *Psiadia*. All *Psiadia* ethyl acetate crude extracts were evaluated for antiplasmodial and anti-inflammatory activities with three species (*P. amygdalina*, *P. anchusifolia*, and *P. dentata*) showing particularly notable activity. Among the most active species, *P. dentata*, a particularly promising candidate, was selected for subsequent studies because of its notable antiplasmodial and anti-inflammatory

https://doi.org/10.3390/molecules31060973

activities. Correlation of the ^{1}H NMR data and the IC_{50} values using PLS-DA modelling highlighted three distinct groups based on activity classes. This multivariate analysis identified the flavonoids ermanin (**1**) and isokaempferide (**7**), as well as the coumarin isoobtusitin (**18**) as key contributors to the antiplasmodial activity of *P. dentata*, while minor compounds present in trace amounts may also contribute.

The MN approach, applied to *P. dentata*, revealed 25 specialized metabolites (**1–25**) from diverse chemical families, including diterpenoids, flavonoids, coumarins, and alkaloids, highlighting the chemodiversity of this species.

Phytochemical investigation of *P. dentata* led to the isolation and structural identification of 25 metabolites. To our knowledge, compounds **26–35** are reported here for the first time, nine new diterpenes (**26–34**) and one new coumarin (**35**). Compounds **1**, **7**, **26** and **27** displayed moderate antiplasmodial activity against *Plasmodium falciparum* (IC_{50} = 7.25–13.46 μM). Five compounds, **7**, **26**, **27**, **31** and **32** (IC_{50} = 0.87–27.71 μM), showed inhibition of nitric oxide production, consistent with the traditional medicinal uses of this species. Only compound **1** showed cytotoxicity against HepG2 and HT29 cell lines (IC_{50} = 25.67 and 18.35 μM, respectively).

Overall, this study demonstrates that the combination of ^{1}H NMR-based metabolomics, molecular networking and phytochemical investigation constitutes a powerful strategy for the identification of bioactive metabolites. The chemical composition of *P. dentata*, underscores its richness in specialized metabolites, representing a promising source of therapeutic agents.

Supplementary Materials: The following supporting information can be downloaded at: https://www.mdpi.com/article/10.3390/molecules31060973/s1, including the contribution plot of the PLS model of *P. dentata* crude extracts against *Psiadia* sp. inactive and moderately active extracts for the antiplasmodial activity., ^{1}H NMR spectrum of the active crude extract and moderately active extract of *P. dentata*, and UV spectra, IR, 1D/2D NMR, HRMS data of compounds **26–35**, and chromatograms of o biologically tested compounds.

Author Contributions: Conceptualization, L.E.R., K.M., I.G. and A.G.-B.; methodology, L.E.R., K.M., C.D.G., G.H., B.B., P.C., M.F., L.M., H.K., I.G. and A.G.-B.; validation, A.G.-B., I.G., G.H., K.M., H.K. and M.F.; formal analysis, L.E.R., K.M. and G.H.; chemical investigation, L.E.R., K.M. and G.H., resources, K.M. and A.G.-B.; writing—original draft preparation, L.E.R. and K.M.; writing—review and editing, L.E.R., K.M., G.H., C.D.G., P.C., M.F., L.M., B.B., H.K., I.G. and A.G.-B.; visualization, L.E.R., K.M., G.H., I.G. and A.G.-B.; supervision, M.F., G.H., B.B., H.K., I.G. and A.G.-B.; project administration, A.G.-B. and I.G.; funding acquisition, A.G.-B. All authors have read and agreed to the published version of the manuscript.

Funding: This work was funded by the European Regional Development Funds GURDTI 2018-1828-0002370 (FEDER PHAR, EU-Région Réunion-French State national counterpart). LC-MS/MS experiments were conducted on the regional platform MALLABAR funded by the Institute of Ecology and Environment (INEE) of the French National Centre for Scientific Research (CNRS) and the French Sud PACA regional council. Razafindrabenja Lantomalala Elsa is a recipient of a fellowship from the Région Réunion (DIRED/20171403).

Institutional Review Board Statement: Not applicable.

Informed Consent Statement: Not applicable.

Data Availability Statement: NMR raw data (^{1}H, ^{13}C, gCOSY, gHSQC, gHMBC and NOESY) of compounds **26–35** are made freely available at https://doi.org/10.5281/zenodo.18294328. Raw data from this analysis data were deposited in the MassIVE Public GNPS data set (http://massive.ucsd.edu; accessed on 27 May 2020): MSV000088099. The molecular networking job on GNPS can be found at https://gnps.ucsd.edu/ProteoSAFe/status.jsp?task=12c133ae14b240a8ae5e74b60068bd92; accessed on 27 May 2020.

Acknowledgments: The authors express their gratitude to Hermann Thomas and Gabriel Deguigne, from the Parc National de La Réunion, for their invaluable assistance during the plant collections. We acknowledge Jacques Fournel, for his expertise in plant identification. Our sincere thanks Guillaume Smagghe, student from the University of Liège, for his contribution to compound isolation. We also thank Valérie Monnier, from the Mass Spectrometry Department at Spectropole, Aix-Marseille University, for performing the mass spectra of some isolated compounds. Finally, we extend our sincere appreciation to Stéphane Greff from IMBE and the Service of Chemical Ecology and Metabolomics (Aix-Marseille University) for the MS/MS data acquisition.

Conflicts of Interest: The authors declare no conflicts of interest. The funders had no role in the design of the study; in the collection, analyses, or interpretation of data; in the writing of the manuscript; or in the decision to publish the results.

eferences

Li, G.; Lou, M.; Qi, X. A brief overview of classical natural product drug synthesis and bioactivity. *Org. Chem. Front.* **2022**, *9*, 517–571. [CrossRef]

Markley, J.L.; Brüschweiler, R.; Edison, A.S.; Eghbalnia, H.R.; Powers, R.; Raftery, D.; Wishart, D.S. The future of NMR-based metabolomics. *Curr. Opin. Biotechnol.* **2017**, *43*, 34–40. [CrossRef]

Mamas, M.; Dunn, W.B.; Neyses, L.; Goodacre, R. The role of metabolites and metabolomics in clinically applicable biomarkers of disease. *Arch. Toxicol.* **2011**, *85*, 5–17. [CrossRef]

Heyman, H.M.; Meyer, J.J.M. NMR-Based metabolomics as a quality control tool for herbal products. *S. Afr. J. Bot.* **2012**, *82*, 21–32. [CrossRef]

Mncwangi, N.P.; Viljoen, A.M.; Zhao, J.; Vermaak, I.; Chen, W.; Khan, I. What the devil is in your phytomedicine? Exploring species substitution in *Harpagophytum* through chemometric modeling of 1H-NMR and UHPLC-MS datasets. *Phytochemistry* **2014**, *106*, 104–115. [CrossRef]

Nguyen, H.T.; Lee, D.-K.; Choi, Y.-G.; Min, J.-E.; Yoon, S.J.; Yu, Y.-H.; Lim, J.; Lee, J.; Kwon, S.W.; Park, J.H. A ^{1}H NMR-based metabolomics approach to evaluate the geographical authenticity of herbal medicine and its application in building a model effectively assessing the mixing proportion of intentional admixtures: A case study of *Panax ginseng*: Metabolomics for the authenticity of herbal medicine. *J. Pharm. Biomed. Anal.* **2016**, *124*, 120–128.

Windarsih, A.; Rohman, A.; Swasono, R.T. Application of ^{1}H-NMR based metabolite fingerprinting and chemometrics for authentication of *Curcuma longa* adulterated with *C. heyneana*. *J. Appl. Res. Med. Aromat. Plants* **2019**, *13*, 100203. [CrossRef]

Suil, S.J.M.; Zulaikha Zolkeflee, N.K.; Ramli, N.S.; Mohd Faudzi, S.M.; Tham, C.L.; Maulidiani, M.; Jamaludin, F.; Mediani, A.; Abas, F. Untargeted NMR metabolomics analysis of *Ficus racemosa* fruit extracts and UHPLC-MS/MS-molecular networking analysis of the active extract. *LWT* **2025**, *225*, 117931. [CrossRef]

Graziani, V.; Scognamiglio, M.; Belli, V.; Esposito, A.; D'Abrosca, B.; Chambery, A.; Russo, R.; Panella, M.; Russo, A.; Ciardiello, F.; et al. Metabolomic approach for a rapid identification of natural products with cytotoxic activity against human colorectal cancer cells. *Sci. Rep.* **2018**, *8*, 5309. [CrossRef]

Gargiulo, E.; Roscetto, E.; Galdiero, U.; Sirignano, C.; Yerbanga, S.R.; Chianese, G.; Catania, M.R.; Taglialatela-Scafati, O. A combined metabolomics and biochemometrics approach for the rapid identification of antibacterial naphthyl derivatives from *Guiera senegalensis* leaves. *J. Pharm. Biomed. Anal.* **2025**, *265*, 117069. [CrossRef] [PubMed]

Nothias-Scaglia, L.-F.; Esposito, M.; Costa, J.; Paolini, J.; Touboul, D.; Litaudon, M. Les réseaux moléculaires, une approche bio-informatique globale pour interpréter les données de spectrométrie de masse tandem. *Spectra Anal.* **2015**, *307*, 73–78.

Wang, M.; Carver, J.J.; Phelan, V.V.; Sanchez, L.M.; Garg, N.; Peng, Y.; Nguyen, D.D.; Watrous, J.; Kapono, C.A.; Luzzatto-Knaan, T.; et al. Sharing and community curation of mass spectrometry data with Global Natural Products Social Molecular Networking. *Nat. Biotechnol.* **2016**, *34*, 828–837. [CrossRef]

Aron, A.T.; Gentry, E.C.; McPhail, K.L.; Nothias, L.-F.; Nothias-Esposito, M.; Bouslimani, A.; Petras, D.; Gauglitz, J.M.; Sikora, N.; Vargas, F.; et al. Reproducible molecular networking of untargeted mass spectrometry data using GNPS. *Nat. Protoc.* **2020**, *15*, 1954–1991. [CrossRef] [PubMed]

Bosser, J.; Guého, J.; Jeffrey, C. *Flore des Mascareignes: La Réunion, Maurice, Rodrigues. 109. Composées;* The Sugar Industry Research Institute: Port Louis, Mauritius; L'Institut Français de Recherche Scientifique pour le Développement en Coopération (ORSTOM): Paris, France; The Royal Botanic Gardens, Kew: London, UK, 1993; p. 261.

Mahadeo, K.; Grondin, I.; Kodja, H.; Soulange Govinden, J.; Jhaumeer Laulloo, S.; Frederich, M.; Gauvin-Bialecki, A. The genus *Psiadia*: Review of traditional uses, phytochemistry and pharmacology. *J. Ethnopharmacol.* **2018**, *210*, 48–68. [CrossRef]

Robin, V.; Boustie, J.; Amoros, M.; Girre, L. *In vitro* antiviral activity of seven *Psiadia* species, Asteraceae: Isolation of two antipoliovirus flavonoids from *Psiadia dentata*. *J. Pharm. Pharmacol.* **1998**, *4*, 61–64.

https://doi.org/10.3390/molecules31060973

17. Fortin, H.; Tomasi, S.; Jaccard, P.; Robin, V.; Boustie, J. A Prenyloxycoumarin from *Psiadia dentata*. *Chem. Pharm. Bull.* **2001**, *4* 619–621. [CrossRef]
18. Fortin, H.; Vigor, C.; Devehat, F.L.-L.; Robin, V.; Bosse, B.L.; Boustie, J.; Amoros, M. *In vitro* antiviral activity of thirty-six plan from La Réunion Island. *Fitoterapia* **2002**, *73*, 346–350. [CrossRef] [PubMed]
19. Robin, V.; Irurzun, A.; Amoros, M.; Boustie, J.; Carrasco, L. Antipoliovirus flavonoids from *Psiadia dentata*. *Antivir. Che Chemother.* **2001**, *12*, 283–291. [CrossRef]
20. Robin, V. Criblage Antiviral de Sept Plantes Endémiques de L'île de La Réunion. Étude de L'activite Antipoliovirus de *Psiad dentata* (cass.) DC, Asteraceae: Isolement et Identification des Composés Actifs (flavonoïdes) Recherche, par Biologie Moléculair du Mécanisme de L'action Inhibitrice. Ph.D. Thesis, University of Rennes 1, Rennes, France, 1999.
21. Jonville, M.C.; Kodja, H.; Strasberg, D.; Pichette, A.; Ollivier, E.; Frederich, M.; Angenot, L.; Legault, J. Antiplasmodial, an inflammatory and cytotoxic activities of various plant extracts from the Mascarene Archipelago. *J. Ethnopharmacol.* **2011**, *13* 525–531. [CrossRef]
22. Jonville, M.C. Étude de la Composition Chimique et des Potentialités Antipaludiques de Plantes Utilisées en Médecine Traditio nelle au Cambodge et Dans L'archipel des Mascareignes. Ph.D. Thesis, University of Liège, Liège, Belgium, 2011.
23. Jakobsen, T.H.; Marcussen, H.V.; Adsersen, A.; Strasberg, D.; Smitt, U.W.; Jaroszewski, J.W. 3-Methoxyflavones and a nov coumarin from *Psiadia dentata*. *Biochem. Syst. Ecol.* **2001**, *29*, 963–965. [CrossRef]
24. Bero, J.; Frédérich, M.; Quetin-Leclercq, J. Antimalarial compounds isolated from plants used in traditional medicine. *J. Phar Pharmacol.* **2009**, *61*, 1401–1433. [CrossRef] [PubMed]
25. Bero, J.; Quetin-Leclercq, J. Natural products published in 2009 from plants traditionally used to treat malaria. *Planta Med.* **201** *77*, 631–640. [CrossRef]
26. Yang, J.Y.; Sanchez, L.M.; Rath, C.M.; Liu, X.; Boudreau, P.D.; Bruns, N.; Glukhov, E.; Wodtke, A.; de Felicio, R.; Fenner, A.; et a Molecular networking as a dereplication strategy. *J. Nat. Prod.* **2013**, *76*, 1686–1699. [CrossRef]
27. Ruttkies, C.; Schymanski, E.L.; Wolf, S.; Hollender, J.; Neumann, S. MetFrag relaunched: Incorporating strategies beyond in silic fragmentation. *J. Cheminform.* **2016**, *8*, 3. [CrossRef]
28. Allen, F.; Greiner, R.; Wishart, D. Competitive fragmentation modeling of ESI-MS/MS spectra for putative metabolite identificatio *Metabolomics* **2015**, *11*, 98–110. [CrossRef]
29. Wang, Y.; Hamburger, M.; Gueho, J.; Hostettmann, K. Antimicrobial flavonoids from *Psidia trinervia* and their methylated an acetylated derivatives. *Phytochemistry* **1989**, *28*, 2323–2327. [CrossRef]
30. Marie, D.; Gurib-Fakim, A.; Gray, A.; Waterman, P. Constituents of *Psiadia terebinthina* AJ Scott, an endemic Asteraceae fro Mauritius. *Nat. Prod. Res.* **2006**, *20*, 1169–1175. [CrossRef]
31. Canonica, L.; Rindone, B.; Scolastico, C.; Ferrari, G.; Casagrande, C. Structure and stereochemistry of psiadiol, a new diterpenoi *Tetrahedron Lett.* **1967**, *8*, 2639–2643. [CrossRef]
32. Ogweno Midiwo, J.; Owuor, F.A.O.; Juma, B.F.; Waterman, P.G. Diterpenes from the leaf exudate of *Psiadia punctulata*. *Phytochem istry* **1997**, *45*, 117–120. [CrossRef]
33. Sarı, A.; Keçeci, Z. Phytochemical investigations on chemical constituents of *Taraxacum bessarabicum* (Hornem.) Hand.-Maz subsp. *bessarabicum* (Hornem.) Hand.-Mazz. *Iran. J. Pharm. Res.* **2019**, *18*, 400–405.
34. Stein, A.C.; Álvarez, S.; Avancini, C.; Zacchino, S.; von Poser, G. Antifungal activity of some coumarins obtained from species c *Pterocaulon* (Asteraceae). *J. Ethnopharmacol.* **2006**, *107*, 95–98. [CrossRef] [PubMed]
35. Lu, A.-J.; Lu, Y.-L.; Tan, D.-P.; Qin, L.; Ling, H.; Wang, C.-H.; He, Y.-Q. Identification of pyrrolizidine alkaloids in *Senecio* plants b Liquid Chromatography-Mass Spectrometry. *J. Anal. Methods Chem.* **2021**, *2021*, 1957863. [CrossRef]
36. Cavalcanti, É.B.V.S.; de Paulo Emerenciano, V.; Scotti, L.; Scotti, M.T. Chapter 7—Flavonoids from Asteraceae as multitarge source of compounds against protozoal diseases. In *Multi-Scale Approaches in Drug Discovery*; Speck-Planche, A., Ed.; Elsevie Amsterdam, The Netherlands, 2017; pp. 149–190.
37. Moon, H.-I.; Lee, J.-H.; Lee, Y.-C.; Kim, K.-S. Antiplasmodial and cytotoxic activity of coumarin derivatives from dried roots c *Angelica gigas* Nakai *in vitro*. *Immunopharmacol. Immunotoxicol.* **2011**, *33*, 663–666. [CrossRef]
38. Oketch-Rabah, H.A.; Mwangi, J.W.; Lisgarten, J.; Mberu, E.K. A new antiplasmodial coumarin from *Toddalia asiatica* root *Fitoterapia* **2000**, *71*, 636–640. [CrossRef]
39. Zakaria, I.; Ahmat, N.; Jaafar, F.M.; Widyawaruyanti, A. Flavonoids with antiplasmodial and cytotoxic activities of *Macarang triloba*. *Fitoterapia* **2012**, *83*, 968–972. [CrossRef]
40. Bianchi, S.E.; Kaiser, S.; Pittol, V.; Doneda, E.; De Souza, K.C.B.; Bassani, V.L. Semi-preparative isolation and purification o phenolic compounds from *Achyrocline satureioides* (Lam) D.C. by high-performance counter-current chromatography. *Phytochen Anal.* **2019**, *30*, 182–192. [CrossRef]
41. Castillo, Q.; Triana, J.; Eiroa, J.; Padrón, J.; Plata, G.; Abel-Santos, E.; Báez, L.; Rodríguez, D.; Jiménez, M.; Pérez-Pujols, M Flavonoids from *Eupatorium illitum* and their antiproliferative activities. *Pharmacogn. J.* **2015**, *7*, 178–181. [CrossRef]

https://doi.org/10.3390/molecules3106097

Rossi, M.H.; Yoshida, M.; Soares Maia, J.G. Neolignans, styrylpyrones and flavonoids from an *Aniba* species. *Phytochemistry* **1997**, *45*, 1263–1269. [CrossRef]

Çitoğlu, G.S.; Sever, B.; Antus, S.; Baitz-Gács, E.; Altanlar, N. Antifungal flavonoids from *Ballota glandulosissima*. *Pharm. Biol.* **2003**, *41*, 483–486. [CrossRef]

Al-Oqail, M.; Hassan, W.H.B.; Ahmad, M.S.; Al-Rehaily, A.J. Phytochemical and biological studies of *Solanum schimperianum* Hochst. *Saudi Pharm. J.* **2012**, *20*, 371–379. [CrossRef] [PubMed]

Bai, N.; He, K.; Zhou, Z.; Lai, C.-S.; Zhang, L.; Quan, Z.; Shao, X.; Pan, M.-H.; Ho, C.-T. Flavonoids from *Rabdosia rubescens* exert anti-inflammatory and growth inhibitory effect against human leukemia HL-60 cells. *Food Chem.* **2010**, *122*, 831–835. [CrossRef]

Morales, G.; Sierra, P.; Paredes, A.; Loyola, L.; Borquez, J. Constituents of *Helenium atacamense* Cabr. *J. Chil. Chem. Soc.* **2006**, *51*, 769–772. [CrossRef]

Wang, J.; Gao, H.; Zhao, J.; Wang, Q.; Zhou, L.; Han, J.; Yu, Z.; Yang, F. Preparative separation of phenolic compounds from *Halimodendron halodendron* by high-speed counter-current chromatography. *Molecules* **2010**, *15*, 5998–6007. [CrossRef]

Zdero, C.; Ahmed, A.A.; Bohlmann, F.; Mungai, G.M. Diterpenes and sesquiterpene xylosides from east african *Conyza* species. *Phytochemistry* **1990**, *29*, 3167–3172. [CrossRef]

Yang, J.-Y.; Park, J.-H.; Lee, M.-J.; Lee, J.-H.; Lee, H.-S. Antimicrobial effects of 7,8-dihydroxy-6-methoxycoumarin and 7-hydroxy-6-methoxycoumarin analogues against foodborne pathogens and the antimicrobial mechanisms associated with membrane permeability. *J Food Prot.* **2017**, *80*, 1784–1790. [CrossRef]

Villaflores, O.B.; Macabeo, A.P.G.; Gehle, D.; Krohn, K.; Franzblau, S.G.; Aguinaldo, A.M. Phytoconstituents from *Alpinia purpurata* and their *in vitro* inhibitory activity against *Mycobacterium tuberculosis*. *Pharmacogn. Mag.* **2010**, *6*, 339–344. [CrossRef] [PubMed]

de Paula, V.; Cruz, M.P. Chemical constituents of *Bombacopsis glabra* (Bombacaceae). *Quim. Nova* **2006**, *29*, 213–215. [CrossRef]

Numonov, S.; Usmanova, S.; Aisa, H. A triterpenoid and flavonoids from *Dracocephalum heterophyllum*. *Chem. Nat. Compd.* **2013**, *48*, 1109–1110. [CrossRef]

Brown, G.D.; Liang, G.-Y.; Sy, L.-K. Terpenoids from the seeds of *Artemisia annua*. *Phytochemistry* **2003**, *64*, 303–323. [CrossRef]

Souza, F.; Silva, S.; Faria, L.; Alves, C.; Muller, A.; Guilhon, G.M.S.P.; Brasil, D. Spectroscopic data of labdane diterpenes: A theoretical analysis via NMR and DFT. *Quím. Nova* **2015**, *38*, 645–650. [CrossRef]

Zhang, Z.; Li, S. Cytotoxic triterpenoid saponins from the fruits of *Aesculus pavia* L. *Phytochemistry* **2007**, *68*, 2075–2086. [CrossRef] [PubMed]

Murakami, T.; Nakamura, J.; Kageura, T.; Matsuda, H.; Yoshikawa, M. Bioactive saponins and glycosides. XVII. Inhibitory effect on gastric emptying and accelerating effect on gastrointestinal transit of tea saponins: Structures of assamsaponins F, G, H, I, and J from the seeds and leaves of the tea plant. *Chem. Pharm. Bull.* **2000**, *48*, 1720–1725. [CrossRef]

Jakupovic, J.; Baruah, R.N.; Bohlmann, F.; King, R.M.; Robinson, H. New alicyclic diterpenes and *ent*-labdanes from *Gutierrezia solbrigii*. *Tetrahedron* **1985**, *41*, 4537–4544. [CrossRef]

Koo, K.A.; Sung, S.H.; Kim, Y.C. A new neuroprotective pinusolide derivative from the leaves of *Biota orientalis*. *Chem. Pharm. Bull.* **2002**, *50*, 834–836. [CrossRef]

Steck, W.; Mazurek, M. Identification of natural coumarins by NMR spectroscopy. *Lloydia* **1972**, *35*, 418–439. [PubMed]

Khan, A.; Kunesch, G.; Chuilon, S.; Ravisé, A. Structure and biological activity of Xanthyletin: A new phytoalexin of *Citrius*. *Fruits* **1985**, *40*, 807–811.

Carman, R.M.; Duffield, A.R. The biosynthesis of labdanoids. The optical purity of naturally occurring manool and abienol. *Aust. J. Chem.* **1993**, *46*, 1105–1114. [CrossRef]

Carman, R. Optical rotation and structure in the labdane series of diterpenoids. *Aust. J. Chem.* **1966**, *19*, 629–642. [CrossRef]

Wang, Y.-Z.; Tang, C.-P.; Ke, C.-Q.; Weiss, H.-C.; Gesing, E.-R.; Ye, Y. Diterpenoids from the pericarp of *Platycladus orientalis*. *Phytochemistry* **2008**, *69*, 518–526. [CrossRef]

Chen, L.; Zhu, H.; Wang, R.; Zhou, K.; Jing, Y.; Qiu, F. *Ent*-labdane diterpenoid lactone stereoisomers from *Andrographis paniculata*. *J. Nat. Prod.* **2008**, *71*, 852–855. [CrossRef]

Zhang, B.-Y.; Yin, H.-X.; Zhang, D.-J. Two new *ent*-labdane diterpenes from the roots of *Euphorbia yinshanica*. *Chem. Nat. Compd.* **2017**, *53*, 295–298. [CrossRef]

Li, D.-Z.; Tang, C.; Quinn, R.J.; Feng, Y.; Ke, C.-Q.; Yao, S.; Ye, Y. *Ent*-labdane diterpenes from the stems of *Mallotus japonicus*. *J. Nat. Prod.* **2013**, *76*, 1580–1585. [CrossRef]

Al-Musayeib, N.M.; Mohamed, G.A.; Ibrahim, S.R.M.; Ross, S.A. New thiophene and flavonoid from *Tagetes minuta* leaves growing in Saudi Arabia. *Molecules* **2014**, *19*, 2819–2828. [CrossRef] [PubMed]

da Saliva Filho, A.A.; Resende, D.O.; Fukui, M.J.; Santos, F.F.; Pauletti, P.M.; Cunha, W.R.; Silva, M.L.A.; Gregório, L.E.; Bastos, J.K.; Nanayakkara, N.P.D. *In vitro* antileishmanial, antiplasmodial and cytotoxic activities of phenolics and triterpenoids from *Baccharis Dracunculifolia* D. C. (Asteraceae). *Fitoterapia* **2009**, *80*, 478–482. [CrossRef]

https://doi.org/10.3390/molecules31060973

69. Mahadeo, K.; Herbette, G.; Grondin, I.; Jansen, O.; Kodja, H.; Soulange, J.; Jhaumeer-Laulloo, S.; Clerc, P.; Gauvin-Bialecki, A. Frederich, M. Antiplasmodial diterpenoids from *Psiadia arguta*. *J. Nat. Prod.* **2019**, *82*, 1361–1366. [CrossRef] [PubMed]
70. Kenmogne, M.; Prost, E.; Harakat, D.; Jacquier, M.-J.; Frédérich, M.; Sondengam, L.B.; Zèches, M.; Waffo-Téguo, P. Five labdan diterpenoids from the seeds of *Aframomum zambesiacum*. *Phytochemistry* **2006**, *67*, 433–438. [CrossRef]
71. Duker-Eshun, G.; Jaroszewski, J.W.; Asomaning, W.A.; Oppong-Boachie, F.; Olsen, C.E.; Christensen, S.B. Antiplasmodial activit of labdanes from *Aframomum latifolium* and *Aframomum sceptrum*. *Planta Med.* **2002**, *68*, 642–644. [CrossRef]
72. Su, Q.; Dalal, S.; Goetz, M.; Cassera, M.B.; Kingston, D.G.I. New antiplasmodial diterpenes from *Gutierrezia sarothrae*. *Nat. Pro Commun.* **2016**, *11*, 719–721. [CrossRef] [PubMed]
73. Liu, K.C.; Yang, S.L.; Roberts, M.F.; Elford, B.C.; Phillipson, J.D. Antimalarial activity of *Artemisia annua* flavonoids from whol plants and cell cultures. *Plant Cell Rep.* **1992**, *11*, 637–640. [CrossRef] [PubMed]
74. Garayev, E.; Herbette, G.; Di Giorgio, C.; Chiffolleau, P.; Roux, D.; Sallanon, H.; Ollivier, E.; Elias, R.; Baghdikian, B. Nev sesquiterpene acid and inositol derivatives from *Inula montana* L. *Fitoterapia* **2017**, *120*, 79–84. [CrossRef]
75. Leal, L.K.A.M.; Canuto, K.M.; da Silva Costa, K.C.; Nobre-Júnior, H.V.; Vasconcelos, S.M.; Silveira, E.R.; Ferreira, M.V.P.; Fontenel J.B.; Andrade, G.M.; de Barros Viana, G.S. Effects of amburoside A and isokaempferide, polyphenols from *Amburana cearensis*, o rodent inflammatory processes and myeloperoxidase activity in human neutrophils. *Basic Clin. Pharmacol. Toxicol.* **2009**, *10* 198–205. [CrossRef]
76. Abad, M.J.; Bermejo, P.; Alvarez, M.; Guerra, J.A.; Silván, A.M.; Villar, A.M. Flavonoids and a sesquiterpene lactone fron *Tanacetum microphyllum* inhibit anti-inflammatory mediators in LPS-stimulated mouse peritoneal macrophages. *Planta Med.* **200** *70*, 34–38. [PubMed]
77. Qi, L.; Jiang, H.; Luo, W.; Yin, H. Labdane and norlabdane diterpenoids from *Amomum sericeum* rhizomes and their potentia anti-inflammatory activities. *Bioorg. Chem.* **2025**, *164*, 108918. [CrossRef] [PubMed]
78. Girón, N.; Través, P.G.; Rodríguez, B.; López-Fontal, R.; Boscá, L.; Hortelano, S.; de las Heras, B. Supression of inflammator responses by labdane-type diterpenoids. *Toxicol. Appl. Pharmacol.* **2008**, *228*, 179–189. [CrossRef] [PubMed]
79. Trang, N.M.; Vinh, L.B.; Phong, N.V.; Yang, S.Y. Anti-inflammatory activity of labdane and norlabdane diterpenoids from *Leonuru sibiricus* related to modulation of MAPKs signaling pathway. *Planta Med.* **2024**, *91*, 29–39. [CrossRef]
80. Banskota, A.H.; Tezuka, Y.; Prasain, J.K.; Matsushige, K.; Saiki, I.; Kadota, S. Chemical constituents of Brazilian propolis and the cytotoxic activities. *J. Nat. Prod.* **1998**, *61*, 896–900. [CrossRef]
81. Salimi, M.; Ardestaniyan, M.H.; Mostafapour Kandelous, H.; Saeidnia, S.; Gohari, A.R.; Amanzadeh, A.; Sanati, H.; Sepahdar, Z Ghorbani, S.; Salimi, M. Anti-proliferative and apoptotic activities of constituents of chloroform extract of *Juglans regia* leaves. *Ce Prolif.* **2014**, *47*, 172–179. [CrossRef]
82. Chambers, M.C.; Maclean, B.; Burke, R.; Amodei, D.; Ruderman, D.L.; Neumann, S.; Gatto, L.; Fischer, B.; Pratt, B.; Egertson, J et al. A cross-platform toolkit for mass spectrometry and proteomics. *Nat. Biotechnol.* **2012**, *30*, 918–920. [CrossRef]
83. Shannon, P.; Markiel, A.; Ozier, O.; Baliga, N.S.; Wang, J.T.; Ramage, D.; Amin, N.; Schwikowski, B.; Ideker, T. Cytoscape: software environment for integrated models of biomolecular interaction networks. *Genome Res.* **2003**, *13*, 2498–2504. [CrossRef
84. Kim, H.; Choi, Y.; Verpoorte, R. NMR-Based metabolomic analysis of plants. *Nat. Protoc.* **2010**, *5*, 536–549. [CrossRef]
85. Jansen, O.; Tits, M.; Angenot, L.; Nicolas, J.-P.; Mol, P.; Jean-baptiste, N.; Frederich, M. Anti-plasmodial activity of *Dicom tomentosa* (Asteraceae) and identification of urospermal A-15-O-acetate as the main active compound. *Malar. J.* **2012**, *11*, 28 [CrossRef]
86. Etsè, K.S.; Etsè, K.D.; Zaragoza, G.; Mouithys-Mickalad, A. Structural description, IR, TGA, antiradical, HRP activity inhibitio and molecular docking exploration of *N*-cyclohexyl-*N*-tosylformamide. *J. Mol. Struct.* **2022**, *1269*, 133731. [CrossRef]
87. Iskandar, S.; Gnansounou, S.M.; Robin, M.; Lorquin, J.; Di Giorgio, C.; Piccerelle, P. Antioxidant, anti-inflammatory an neuroprotective activities of a plant extract derived from traditional chinese medicine: SuHeXiang Wan (AT000). *Chem. Ad Mater.* **2018**, *3*, 36–59.
88. Hosry, L.; Boyer, L.; Garayev, E.; Mabrouki, F.; Bun, S.; Debrauwer, L.; Auezova, L.; Cheble, E.; Elias, R. Chemical compositior antioxidant and cytotoxic activities of roots and fruits of *Berberis libanotica*. *Nat. Prod. Commun.* **2016**, *11*, 645–648. [CrossRef [PubMed]
89. Gerometta, E.; Herbette, G.; Garayev, E.; Marvilliers, A.; Naubron, J.-V.; Di Giorgio, C.; Campos, P.-E.; Clerc, P.; Ledoux, A Frederich, M.; et al. Cytotoxic metabolites from *Calophyllum tacamahaca* Willd.: Isolation and detection through Feature-Basec Molecular Networking. *Metabolites* **2023**, *13*, 582. [CrossRef]

https://doi.org/10.3390/molecules3106097

molecules

MDPI

ticle

White Teff Flour Ethanolic Extract: Phytochemical Profile, ntioxidant and Anti-Inflammatory Activity

ana Ferențiu [1], Tiberia Ioana Pop [2,*], Alina Elena Pârvu [1], Iulia Ioana Morar [1], Dinu Bolunduț [1,3], Marcel Pârvu [4], orica Ranga [5], Ciprian Ovidiu Dalai [6], Mădălina Țicolea [1] and Raluca Maria Pop [7]

1 Pathophysiology, Department of Morphofunctional Sciences, Faculty of Medicine, "Iuliu Hațieganu" University of Medicine and Pharmacy Cluj-Napoca, 400012 Cluj-Napoca, Romania; abrudan.ferentiu.ioana@elearn.umfcluj.ro (I.F.); parvualinaelena@umfcluj.ro (A.E.P.); iulia.morar@umfcluj.ro (I.I.M.); bolundut_dinu@elearn.umfcluj.ro (D.B.); madalinaticolea@umfcluj.ro (M.Ț.)

2 Department of Technical and Soil Sciences, Faculty of Agriculture, University of Agricultural Sciences and Veterinary Medicine Cluj-Napoca, 400372 Cluj-Napoca, Romania

3 Department of Medical Oncology, "Ion Chiricuță" Institute of Oncology, 400015 Cluj-Napoca, Romania

4 Department of Biology, Babeș-Bolyai University, 400015 Cluj-Napoca, Romania; marcel.parvu@ubbcluj.ro

5 Food Science and Technology, Department of Food Science, University of Agricultural Sciences and Veterinary Medicine Cluj-Napoca, 400372 Cluj-Napoca, Romania; florica.ranga@usamvcluj.ro

6 Department of Dental Medicine, Faculty of Medicine and Pharmacy, University of Oradea, 410073 Oradea, Romania; cipridalai@gmail.com

7 Pharmacology, Toxicology and Clinical Pharmacology, Department 2—Functional Sciences, Faculty of Medicine, "Iuliu Hațieganu" University of Medicine and Pharmacy Cluj-Napoca, 400012 Cluj-Napoca, Romania; raluca.pop@umfcluj.ro

* Correspondence: tiberia.pop@usamvcluj.ro

Abstract

Teff (*Eragrostis tef*) is a gluten-free cereal increasingly promoted as a functional food, yet the bioactive profile and mechanistic evidence of some varieties remain limited. This study characterized an ethanolic extract of white teff flour and evaluated its antioxidant and anti-inflammatory potential *in vitro* and in a rat model of acute inflammation. White teff flour was extracted by cold repercolation (1 g/mL; 70% ethanol). Total polyphenols and flavonoids were quantified spectrophotometrically, and phenolics were profiled by HPLC-DAD-ESI-MS. Antioxidant activity was assessed using DPPH, FRAP, H_2O_2 scavenging, and NO scavenging assays. *In vivo*, acute inflammation was induced with intramuscular turpentine in Wistar rats, testing teff extract therapeutically (post-induction) and prophylactically (10-day pretreatment), with diclofenac and Trolox as comparators. Serum oxidative stress biomarkers (TOS, TAC, OSI, AOPP, MDA, NOx, 3-NT, thiols) and inflammatory mediators (NFκB-p65, IL-1β, IL-18, caspase-1, IL-10) were measured. The extract showed low total polyphenols (0.044 ± 0.002 mg GAE/g d.w.) and higher flavonoids (11.83 ± 1.10 mg QE/100 g d.w.). Eighteen phenolics were identified (total 398.30 ± 1.48 μg/mL), dominated by flavone derivatives (notably apigenin- and luteolin-glycosides), while phenolic acids accounted for ~33.21%. *In vitro* antioxidant capacity was robust (DPPH 286.17 ± 11.52 μg TE/g d.w.; FRAP 263.17 ± 20.09 μg TE/g d.w.; H_2O_2 214.12 ± 18.22 mg TE/g d.w.; NO 300.77 ± 28.71 mg QE/g d.w.). *In vivo*, turpentine provoked marked oxidative stress and inflammatory activation; teff, particularly at the highest concentration and in prophylaxis, reduced nitro-oxidative damage markers (AOPP, MDA, NOx, 3-NT) and lowered NFκB-p65, IL-1β, IL-18, and caspase-1, while IL-10 was not significantly altered. White teff flour ethanolic extract contains a flavone-rich phenolic profile and exerts measurable antioxidant and anti-inflammatory effects in an acute inflammation model, supporting its potential development as a nutraceutical candidate.

Keywords: *Eragrostis tef*; phenolic compounds; flavone glycosides; antioxidant activity; anti-inflammatory effects

1. Introduction

Teff (*Eragrostis tef*) is a cereal originally domesticated in Ethiopia [1]. It belongs to the *Poaceae* family (the grass family), within the genus *Eragrostis*, which includes roughly 350 species; teff is the only species widely cultivated for food [2–4]. Beyond Ethiopia, cultivation has expanded to multiple regions, including Australia, parts of Africa, North America, and Europe [1]. Recent work also suggests that teff can be grown under Mediterranean climatic conditions [4,5]. In Ethiopian cuisine, teff flour is traditionally used to prepare injera, a fermented flatbread [6].

Teff is commonly marketed as white, brown, or mixed varieties [7]. Because teff flour is typically milled as a whole grain, it retains bran-derived components [8]. Multiple studies report that teff is naturally gluten-free, and provides slowly digestible carbohydrates (often with a low glycemic index), dietary fiber, essential amino acids, vitamins, minerals, and diverse polyphenols [7–10]. Teff's nutritional value is superior compared to that of other gluten-free grains, like quinoa and rice, with teff being richer in fiber, iron, and calcium than flours and other gluten-free products [11]. Consequently, teff has been proposed as a functional food/nutraceutical in a range of chronic conditions, and it is a suitable staple for individuals requiring lifelong gluten avoidance, including patients with celiac disease [12–15].

The World Health Organization describes functional foods as natural foods that contain bioactive components (including phytochemicals) that may contribute to the prevention and management of chronic non-communicable diseases [12,16]. The term nutraceutical (from "nutrition" and "pharmaceutical") refers to naturally derived products obtained from plant- or animal-based food that may exert pharmacologically relevant effects [17]. In the context of inflammation, functional foods and nutraceuticals are typically expected to provide antioxidant and/or anti-inflammatory activity [12,16], and current research focuses on their mechanisms of action, safety profiles, and clinical evidence [18].

Given the growing interest in teff as a nutrient-dense gluten-free grain [1,9] and the limited characterization of some varieties [3,19], this study evaluated the bioactive profile of a white teff flour ethanolic extract. We further aimed to generate evidence supporting the development of a nutraceutical based on white teff extract by assessing its antioxidant and anti-inflammatory mechanisms in an experimental model of inflammation.

2. Results

2.1. Phytochemical Analysis

The white teff ethanolic extract exhibited a relatively low total polyphenol content, while total flavonoids were comparatively higher (Table 1).

Table 1. Total polyphenol content and total flavonoid content of white teff ethanolic extract.

Plant Extract	Total Polyphenol Content (mg GAE/g d.w. Plant Material)	Total Flavonoid Content (μg QE/g d.w. Plant Material)
White teff 1 g/1 mL	0.044 ± 0.002	11.83 ± 1.10

GAE—gallic acid equivalents; QE—quercetin equivalents.

HPLC-DAD-ESI-MS profiling revealed appreciable levels of phenolic acids and flavones. Eighteen compounds were tentatively identified based on retention time, UV

https://doi.org/10.3390/molecules3104064

spectra, and MS signals. Phenolic acids included three hydroxybenzoic acid derivatives (2,3-dihydroxybenzoic acid, 2-hydroxybenzoic acid, protocatechuic acid) and one hydroxycinnamic acid (chlorogenic acid), together accounting for approximately 33.21% of the total identified phenolics. Most remaining peaks corresponded to flavones; apigenin derivatives represented 16.62% and luteolin derivatives 16.59% of the identified profile (Figure 1; Table 2).

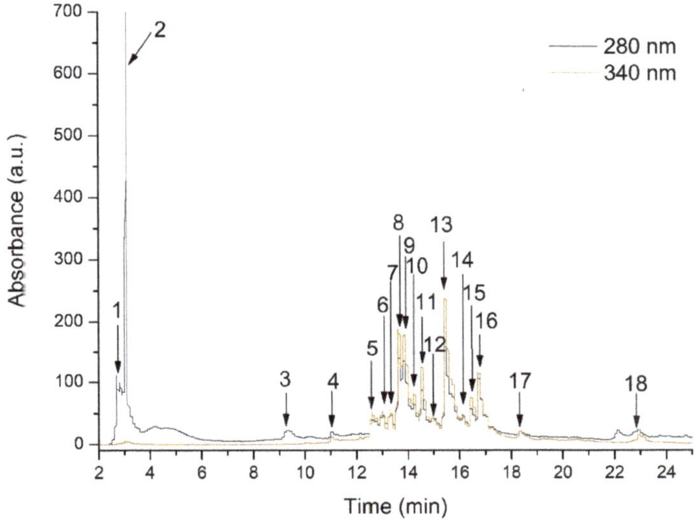

Figure 1. HPLC chromatogram of phenolic compounds from white teff ethanolic extract measured at 280 and 340 nm. The peak identification is provided in Table 2.

Table 2. Liquid Chromatography–Diode Array Detection–Electrospray Ionization Mass Spectrometry for phenolic-compound tentative identification from white teff ethanolic extract.

Peak No.	R_t (min)	UV λ_{max} (nm)	$[M + H]^+$ (m/z)	Compounds	Concentration (µg/mL)
1	2.71	280	139	2-Hydroxybenzoic acid [1]	20.16 ± 0.05
2	3.07	275	155	2,3-Dihydroxybenzoic acid [1]	101.06 ± 0.51
3	9.33	280	155	Protocatechuic acid [1]	11.09 ± 0.09
4	11.07	330	355	5-Caffeoylquinic acid (Chlorogenic acid) [2]	6.06 ± 0.27
5	12.59	350.270	611.287	Luteolin-diglucoside [3]	6.68 ± 0.08
6	12.99	350.270	595.287	Luteolin-rutinoside [3]	9.12 ± 0.29
7	13.29	350.270	433.287	Luteolin-rhamnoside [3]	9.74 ± 0.19
8	13.61	340.270	595.271	Apigenin-diglucoside [3]	25.21 ± 0.08
9	13.85	350.270	757.287	Luteolin-glucosyl-neohesperidoside [3]	40.57 ± 1.33
10	14.22	340.270	565.271	Apigenin-arabinosyl-galactoside [3]	13.73 ± 0.26
11	14.53	350.270	449.287	Luteolin-glucoside [3]	25.99 ± 0.10
12	14.93	340.270	565.271	Apigenin-arabinosyl-glucoside [3]	11.38 ± 0.09
13	15.44	340.270	433.271	Apigenin-glucoside [3]	64.71 ± 0.11
14	16.11	340.270	447.271	Apigenin-methylglucoside [3]	9.42 ± 0.15
15	16.45	340.270	775.271	Apigenin-syringyl-diglucoside [3]	13.27 ± 0.13
16	16.74	340.270	579.271	Apigenin-neohesperidoside [3]	19.79 ± 0.92
17	18.34	340.270	475.271	Apigenin-acetyl-glucoside [3]	5.86 ± 0.32
18	22.93	340.270	271	Apigenin [3]	4.23 ± 0.11
				Total phenolics	398.30 ± 1.48

[1]—compounds belonging to hydroxybenzoic acid subclass; [2]—compounds belonging to hydroxycinnamic acid subclass; [3]—compounds belonging to flavone subclass.

https://doi.org/10.3390/molecules31040644

In vitro antioxidant capacity, assessed by DPPH, FRAP, H_2O_2 scavenging, and NO scavenging assays, indicated robust activity for teff. Compared with Trolox (TX), TEFF showed stronger effects in the DPPH, FRAP, and H_2O_2 assays ($p < 0.001$). NO scavenging was also significant relative to quercetin ($p < 0.001$) (Table 3).

Table 3. *In vitro* antioxidant activity of the white teff ethanolic extract.

Sample	DPPH (µg TE/g d.w. Plant Material)	FRAP (µg TE/g d.w. Plant Material)	H_2O_2 Scavenging Activity (mg TE/g d.w. Plant Material)	NO Scavenging Activity (mg QE/g d.w. Plant Material)
White teff extract (1 gr/1 mL)	286.17 ± 11,52	263.17 ± 20.09	214.12 ± 18,22	300.77 ± 28.71
Trolox IC50	12.01 ± 1.6	13.21 ± 1.81	22.19 ± 2.06	-
Quercetin IC50			-	19.59 ± 2.32
p-value	0.001	0.001	0.001	0.001

DPPH—DPPH free-radical scavenging capacity; FRAP—ferric reducing antioxidant power; H_2O_2—hydrogen peroxide scavenging capacity; NO—nitric oxide radical scavenging assay; TE—Trolox equivalent; QE—quercetin equivalent.

2.2. *In Vivo Antioxidant Activity of the White Teff Ethanolic Extract*

Systemic redox status was evaluated by measuring oxidant markers (TOS, OSI, AOPP, MDA, NOx, 3-NT) and antioxidant markers (TAC, total thiols, SH) (Table 4).

Compared with SHAM, the INFL group exhibited a pronounced pro-oxidative shift, reflected by increased TOS and OSI (both $p < 0.001$) and reduced TAC ($p < 0.001$). This was accompanied by elevated AOPP, MDA, NOx, and 3-NT (all $p < 0.001$), and a significant decrease in SH ($p < 0.001$) (Table 4).

Diclofenac attenuated oxidative stress, with reductions in TOS ($p < 0.05$) and OSI ($p < 0.01$) and a marked increase in TAC ($p < 0.001$). AOPP ($p < 0.001$), MDA ($p < 0.01$), NOx ($p < 0.01$), and 3-NT ($p < 0.001$) were also lowered relative to INFL (Table 4).

Trolox (TX) decreased OSI ($p < 0.05$), driven by a modest reduction in TOS ($p < 0.05$) and a significant increase in TAC ($p < 0.001$). Oxidative damage markers (AOPP, MDA, NOx, and 3-NT) decreased (*p*-values as shown in Table 4), and TAC elevation was paralleled by increased SH ($p < 0.001$). Overall, no significant differences were detected between diclofenac and Trolox in their effects on oxidative stress (Table 4).

After inflammation induction, a clear reduction in TOS and OSI was observed primarily in INFL/TEFF100 ($p < 0.01$). TAC increased moderately across teff-treated groups ($p < 0.05$). Among oxidant markers, MDA was not significantly affected, whereas NOx, 3-NT, and AOPP decreased in a dose-dependent manner, with the highest teff concentration showing the strongest effect ($p < 0.001$) (Table 4). SH did not change significantly with TEFF treatment ($p > 0.05$).

In the prophylactic protocol, TEFF100/INFL and TEFF50/INFL prevented the turpentine-induced rise in TOS and OSI ($p < 0.001$), and all teff dilutions increased TAC ($p < 0.001$). TOS reduction coincided with substantial decreases in AOPP and 3-NT across all prophylactic groups ($p < 0.001$). A smaller decrease in MDA and NOx was noted mainly in TEFF100/INFL ($p < 0.01$). SH increased significantly in TEFF50/INFL and TEFF25/INFL ($p < 0.01$) (Table 4).

https://doi.org/10.3390/molecules31040644

Table 4. *In vivo* antioxidant activity of the white teff ethanolic extract.

Groups	TOS (μmol H_2O_2 Equiv./L)	TAC (mmol Trolox Equiv./L)	OSI	AOPP (μmol/L)	MDA (nmol/L)	NOx (μmol/L)	3-NT (ng/mL)	SH (μmol/L)
CONTROL	5.70 ± 1.10 ***	1.09 ± 0.00 ***	5.50 ± 1.04 ***	81.36 ± 15.17 ***	4.89 ± 0.41 ***	30.52 ± 3.96 ***	26.00 ± 2.84 ***	582.00 ± 10.97 ***
INFL	16.16 ± 1.84	1.08 ± 0.00	16.23 ± 1.85	140.51 ± 14.80	6.83 ± 0.44	79.79 ± 2.79	55.95 ± 3.48	317.40 ± 34.70
INFL/DICLO	9.62 ± 1.65 *	1.09 ± 0.00 ***	8.57 ± 1.59 **	95.22 ± 10.23 **	5.12 ± 0.81 *	51.78 ± 2.06 **	22.89 ± 2.92 ***	373.40 ± 25.08
INFL/TX	8.36 ± 1.09 *	1.09 ± 0.00 ***	8.55 ± 1.16 **	92.78 ± 9.87 **	5.16 ± 0.18 *	54.21 ± 8.70 **	23.68 ± 4.16 ***	641.33 ± 54.06 ***
INFL/TEFF100	8.21 ± 1.78 **	1.09 ± 0.00 *	7.47 ± 1.82 **	57.65 ± 6.11 ***	6.00 ± 0.23	25.11 ± 1.96 ***	22.34 ± 3.42 ***	374.33 ± 13.00
INFL/TEFF50	12.50 ± 2.57	1.09 ± 0.00 *	12.49 ± 1.88	89.41 ± 2.95 **	6.57 ± 0.38	29.45 ± 2.85 **	25.18 ± 3.64 **	358.33 ± 31.77
INFL/TEFF25	14.55 ± 2.66	1.09 ± 0.00 *	12.95 ± 1.63	107.51 ± 7.07 *	6.56 ± 0.12	56.83 ± 4.07 **	32.07 ± 4.25 *	346.33 ± 26.39
TEFF100/INFL	7.66 ± 1.53 ***	1.09 ± 0.00 ***	6.94 ± 1.62 **	66.08 ± 4.77 ***	5.80 ± 0.13 *	55.35 ± 2.57 **	21.06 ± 1.16 ***	398.60 ± 19.74
TEFF50/INFL	9.63 ± 2.86 ***	1.09 ± 0.00 ***	8.59 ± 1.84 **	68.55 ± 3.45 ***	6.15 ± 0.31	70.70 ± 5.51	18.27 ± 1.68 ***	496.67 ± 22.66 **
TEFF25/INFL	13.76 ± 2.55	1.09 ± 0.00 ***	13.55 ± 2.18	70.43 ± 5.95 ***	6.19 ± 0.30	72.98 ± 6.99	11.85 ± 1.22 ***	509.67 ± 39.45 ***

INFL—inflammation; DICLO—diclofenac; TX—Trolox; TEFF100—white teff 100%; TEFF50—white teff 50%; TEFF25—white teff 25%; TOS—total oxidative status; TAC—total antioxidant capacity; OSI—oxidative stress index; AOPP—advanced oxidation protein products; MDA—malondialdehyde; NOx—nitrites and nitrates; 3-NT—3-nitrotyrosine; SH—total thiols. Vs INFL: * $p < 0.05$; ** $p < 0.01$; *** $p < 0.001$.

Principal component analysis (PCA) was performed to investigate the relationships between oxidative stress and inflammatory biomarkers and to evaluate the global effect of therapeutic interventions. Because the correlation matrix was not positive definite, reflecting strong collinearity between OSI and its constituent parameters (TOS and TAC), PCA was interpreted using an oblique rotation (Direct Oblimin). Communalities after extraction were generally high (range 0.469–0.988), indicating that most variables were well represented by the retained components. Four principal components were retained, together explaining 80.0% of the total variance (PC1 = 43.3%, PC2 = 16.4%, PC3 = 11.4%, PC4 = 9.0%).

The rotated structure showed that PC1 captured a major oxidative–inflammatory axis, with high positive loadings for IL-1β, MDA, NF-κB, IL-18, 3-NT, and NO and negative loadings for SH and TAC, reflecting concomitant oxidative damage and antioxidant depletion. PC2 was mainly associated with AOPP, NO, and caspase activity, indicating protein oxidation and apoptosis-related processes. PC3 was defined primarily by IL-10, suggesting regulatory or counter-inflammatory signaling, while PC4 was dominated by TOS and OSI, representing global oxidant status. Overall, PCA demonstrates that therapeutic administration modulates several interrelated but partially independent oxidative and inflammatory pathways rather than a single biological dimension (Figure 2).

https://doi.org/10.3390/molecules31040644

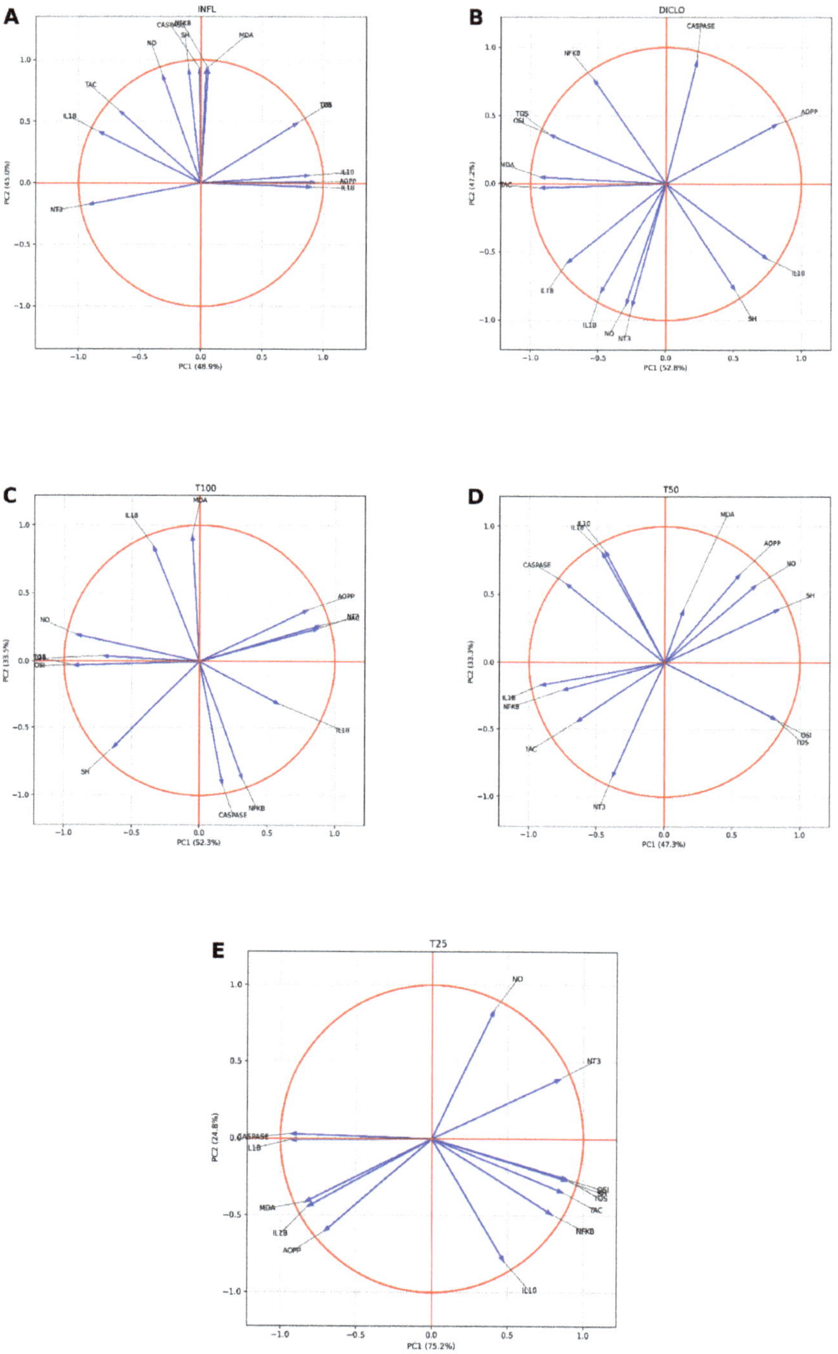

Figure 2. PCA correlation circles of oxidative stress and inflammatory markers in experimental groups. Vectors represent the contribution and correlation of each biomarker with the first two principal components (PC1 and PC2). (**A**) INFL group; (**B**) DICLO group; (**C**) T100% group; (**D**) T50% group; (**E**) T25% group. The direction and length of vectors indicate the weight of each variable in discriminating between groups.

https://doi.org/10.3390/molecules31040644

2.3 In Vivo Inflammatory Markers

Inflammatory response was assessed by measuring NFkB-p65, IL-1β, IL-18, and caspase-1 (as indirect markers of NLRP3 inflammasome activity), together with IL-10 as an anti-inflammatory cytokine. Relative to SHAM, INFL animals showed significant increases in NFkB-p65, IL-1β, IL-18, and caspase-1 ($p < 0.01$), while IL-10 did not change significantly ($p > 0.05$) (Table 5).

Compared with INFL, diclofenac reduced NFkB-p65, IL-1β, IL-18, and caspase-1 (all $p < 0.001$). Trolox also reduced NFkB-p65 ($p < 0.05$), IL-1β ($p < 0.001$), IL-18 ($p < 0.001$), and caspase-1 ($p < 0.05$) (Table 5).

Post-induction TEFF treatment significantly lowered NFkB-p65, IL-1β, IL-18, and caspase-1 ($p < 0.001$). Relative to diclofenac, TEFF showed a weaker reduction in NFkB-p65, IL-1β, and caspase-1, but a stronger reduction in IL-18. Compared with Trolox, TEFF had smaller effects on NFkB-p65, IL-1β, and IL-18, while exhibiting a stronger reduction in caspase-1 (Table 5).

Pre-treatment with teff for 10 days mitigated the turpentine-associated rise in NFkB-p65 ($p < 0.001$), IL-1β ($p < 0.01$), IL-18 ($p < 0.001$), and caspase-1 ($p < 0.001$). Overall, prophylactic administration tended to produce stronger inhibition of inflammatory markers than therapeutic administration. Aside from caspase-1, TEFF100 and TEFF50 showed a more pronounced preventive effect than TEFF25; for the other markers, the three dilutions showed comparable efficacy (Table 5).

IL-10 levels were not significantly influenced by any of the tested interventions ($p > 0.05$) (Table 5).

Table 5. *In vivo* anti-inflammatory activity of the white teff ethanolic extract.

Groups	NfkB-p65 (pg/mL)	IL-1b (pg/mL)	IL-18 (pg/mL)	Caspase-1 (pg/mL)	IL-10 (pg/mL)
CONTROL	139.72 ± 25.92 ***	26.37 ± 1.85 ***	8.03 ± 1.14 ***	51.40 ± 3.63 ***	61.24 ± 4.44
INFL	589.83 ± 63.19	64.52 ± 2.70	22.61 ± 2.84	252.07 ± 15.69	65.21 ± 5.55
INFL/DICLO	157.73 ± 35.19 ***	33.88 ± 2.29 ***	12.95 ± 2.46 ***	12.85 ± 1.90 ***	65.65 ± 2.99
INFL/TX	155.06 ± 34.42 ***	34.97 ± 4.29 ***	8.52 ± 2.17 ***	65.09 ± 8.83 ***	63.74 ± 4.46
INFL/TEFF100	203.70 ± 66.20 ***	52.59 ± 5.29 **	8.98 ± 2.41 ***	32.64 ± 4.19 ***	66.82 ± 3.99
INFL/TEFF50	255.09 ± 35.52 ***	45.26 ± 3.83 **	9.59 ± 1.27 ***	31.72 ± 2.75 ***	62.71 ± 2.99
INFL/TEFF25	178.55 ± 27.31 ***	45.97 ± 4.82 **	8.86 ± 2.47 ***	38.60 ± 3.50 ***	65.94 ± 2.17
TEFF100/INFL	137.68 ± 24.52 ***	44.41 ± 8.32 **	7.24 ± 1.75 ***	25.25 ± 4.05 ***	64.47 ± 1.56
TEFF50/INFL	126.53 ± 8.42 ***	46.15 ± 5.37 **	7.50 ± 0.70 ***	58.49 ± 3.18 ***	64.91 ± 5.34
TEFF25/INFL	170.46 ± 53.10 ***	41.61 ± 5.31 **	8.30 ± 1.69 ***	125.19 ± 16.72 **	65.35 ± 9.72

INFL—inflammation; DICLO—diclofenac; TX—Trolox; TEFF100—white teff 100%; TEFF50—white teff 50%; TEFF25—white teff 25%; NfkB-p65—nuclear factor-κB; IL-1b—interleukine 1-b; IL-18—interleukine 18; IL-18—interleukine 10; ** $p < 0.01$; *** $p < 0.001$.

Principal component analysis (PCA) was performed to explore the multivariate relationships among oxidative stress and inflammatory biomarkers following prophylactic administration. Based on eigenvalues > 1 and inspection of the scree plot, three principal components were retained, together explaining 76.7% of the total variance (PC1: 52.8%, PC2: 13.5%, PC3: 10.5%). Communalities after extraction were generally high, indicating adequate representation of most variables within the retained component space. The rotated solution revealed that PC1 was characterized by high positive loadings of NF-κB, caspase activity, OSI/TOS, MDA, IL-1β, IL-18, NO, and 3-NT, together with negative loadings for SH and TAC, reflecting a dominant axis integrating oxidative stress burden, inflammatory signaling, and reduced antioxidant capacity. PC2 was primarily associated with AOPP and NO, indicating variability related to protein oxidation and nitric oxide-dependent processes, while PC3 was mainly defined by IL-10, representing an independent regulatory inflammatory dimension. Correlations between components were low, confirming that

https://doi.org/10.3390/molecules31040644

these axes describe largely distinct biological patterns. Specific correlation circles illustrated a progressive redistribution of biomarkers across the PCA space. The INFL group showed tight clustering of oxidative and inflammatory markers along PC1. In contrast, the T100 group displayed a marked displacement of several pro-oxidant and inflammatory variables away from PC1, with relative preservation of antioxidant-related parameters. The T50 group exhibited an intermediate configuration, whereas the T25 group remained closer to the inflamed profile. Overall, PCA demonstrates a dose-dependent modulation of the oxidative–inflammatory biomarker network following therapeutic intervention (Figure 3).

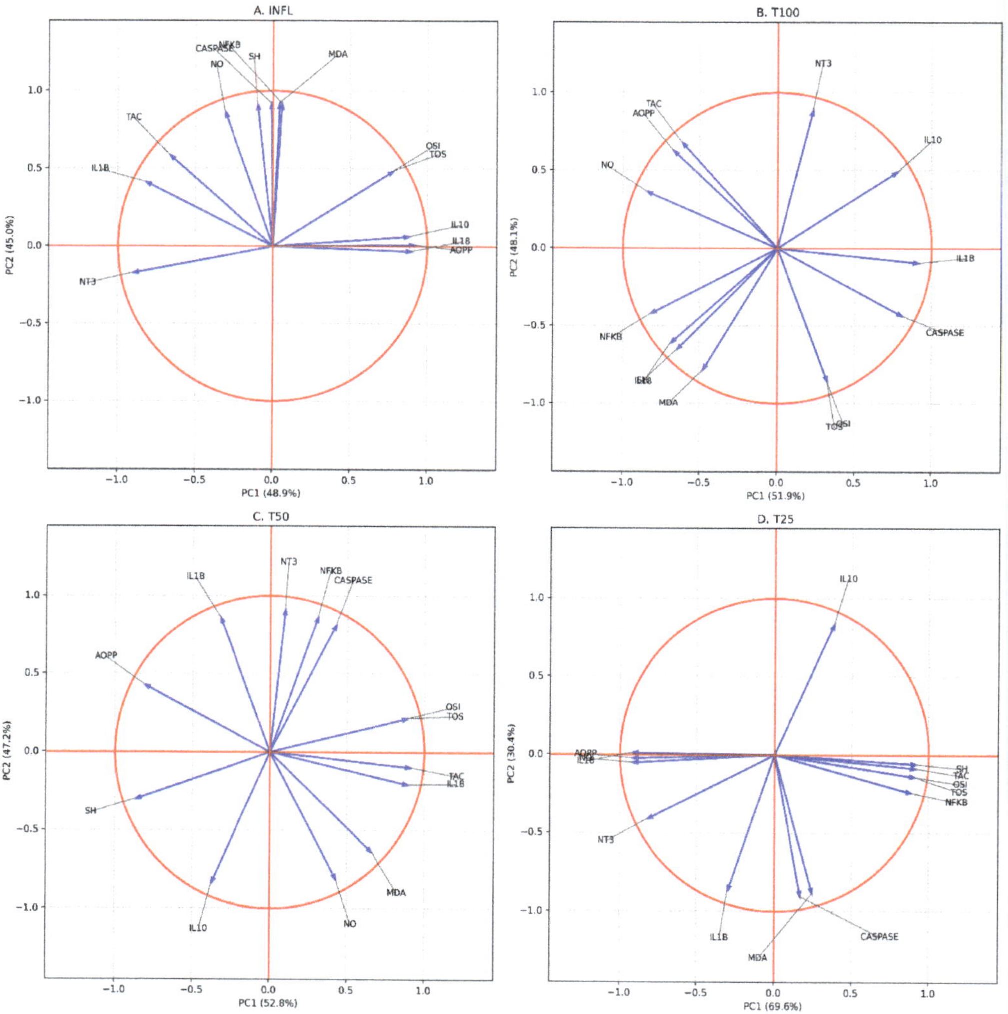

Figure 3. PCA correlation circles of oxidative stress and inflammatory markers in the prophylactic experimental setting. Vectors represent the contribution and correlation of each biomarker with the first two principal components (PC1 and PC2). (**A**) Inflammation group (INFL); (**B**) TEFF extract 100% group (T100); (**C**) TEFF extract 50% group (T50); and (**D**) TEFF extract 25% group (T25). The orientation and length of the vectors indicate the degree of association of each variable with the principal components and their relative contribution to group discrimination.

https://doi.org/10.3390/molecules31040644

3. Discussion

Teff (*Eragrostis tef*) is a long-cultivated cereal from the Horn of Africa that has recently gained international visibility because of its nutritional quality and potential health benefits. Teff is naturally gluten-free [3] and provides protein, dietary fiber, and micronutrients such as iron, calcium, zinc, and magnesium [4,7,17]. Beyond basic nutrition, teff contains bioactive constituents (e.g., polyphenols, flavonoids, phytosterols, and saponins) that have been associated with antioxidant and anti-inflammatory effects [4,20]. Although teff has a long history of cultivation, systematic characterization of its phytochemical profile has expanded mainly in recent years [4].

Polyphenols comprise a diverse group of plant secondary metabolites, and many are recognized for their capacity to counter oxidative and inflammatory processes. Flavonoids represent an important polyphenol subclass and are present in teff. Both preclinical and clinical evidence suggests that sustained intake of polyphenol-rich diets may be associated with lower risk or improved management of several chronic conditions, including cardiovascular and neurodegenerative diseases, diabetes, cancer, and inflammatory disorders [18].

The phenolic profile of teff is influenced by cultivar and genetics, agronomic practices, environmental conditions, and extraction methodology. Reports for methanolic teff extracts describe TPC values of 46–133 mg GAE/100 g and TFC values of 15–113 mg QE/100 g, with brown varieties often exhibiting higher phenolics than white varieties, due to observable variations in the agroecological zones and the genetic make-up of the samples [1,21]. In the present work, ethanol-extracted TPC and TFC were lower than some values reported for methanol or ultrasound-assisted extraction [8,22]. When compared with other grains, like white fonio seeds, teff had a higher TPC [23].

In our HPLC-DAD-ESI+ analysis, white teff flour extract displayed a broad range of phenolic constituents. This is consistent with earlier reports showing that teff flours can contain higher phenolic levels than other refined cereal flours, and that soluble versus bound phenolics may differ between white and brown teff varieties [1,24,25].

Phenolic acids (e.g., ferulic, caffeic, and p-coumaric acids) and flavonoids (e.g., luteolin, apigenin, quercetin, rutin) have been described in teff, with substantial amounts located in the bran fraction [4,12,22]. These compounds can contribute to antioxidant capacity through radical scavenging, metal chelation, and modulation of enzyme activity [12].

In the present ethanolic extract, we detected hydroxybenzoic acid derivatives (2,3-dihydroxybenzoic acid, 2-hydroxybenzoic acid, and protocatechuic acid) together with the hydroxycinnamic acid chlorogenic acid. Such phenolic acids are frequently linked to antioxidant activity and have also been discussed in the context of anti-inflammatory, antimicrobial, and anticancer effects [18]. The dominant phenolics in our profile were flavones, mainly glycosylated derivatives of luteolin and apigenin. These flavonoids have been reported to interfere with NF-κB signaling and other pro-inflammatory pathways, thereby limiting cytokine release [4]. Luteolin, widely distributed in plants, has been associated with Nrf2-related antioxidant responses and NF-κB-related inflammatory responses [26], as well as metabolic effects in models of insulin resistance and diabetic complications [27,28]. In our extract, a luteolin-glucosyl-arabinoside derivative was the most abundant luteolin-related compound. Among apigenin derivatives, apigenin-glucoside and apigenin-diglucoside predominated. Apigenin has been discussed as a potential antidiabetic agent through mechanisms including stimulation of insulin secretion, inhibition of α-glucosidase, and ROS scavenging [8].

Oxidative stress reflects an imbalance between pro-oxidant species and antioxidant defenses, which can promote cellular injury and sustain inflammatory signaling. Persistent oxidative stress has been implicated in the pathogenesis of cardiovascular disease,

https://doi.org/10.3390/molecules31040644

neurodegeneration, diabetes, and malignancy. Antioxidants mitigate oxidative damage by neutralizing reactive species. They include endogenous systems (e.g., enzyme-based defenses) and exogenous antioxidants obtained from one's diet. Whole grains, fruits, and vegetables are important dietary contributors to exogenous antioxidant intake.

To characterize bioactivity, nutraceutical candidates are commonly evaluated *in vitro* and *in vivo* [18]. Although polyphenols are widely studied as antioxidants, under certain *in vivo* conditions they may also exhibit pro-oxidant behavior. This shift can depend on factors such as solubility, metal chelation/reduction properties, and local pH [29]. Accordingly, we first assessed the antioxidant potential of white teff extract using DPPH, FRAP, H_2O_2, and NO scavenging assays.

The DPPH and FRAP results supported notable antioxidant capacity, consistent with the presence of multiple phenolic compounds. However, total phenolics (TPC/TFC) did not show a strong linear relationship with radical scavenging activity, which is plausible because antioxidant performance depends on both concentration and the chemical structures of individual phenolics. Previous work comparing white and brown teff has reported measurable levels of phenolics and flavonoids as well as antioxidant activity in both free and bound fractions. For example, there were reported flavonoids in the range of 0.52–1.02 mg RE/g and total phenolics of 0.90–1.42 mg GAE/g, with higher antioxidant activity in the free phenolic fraction (ABTS: 1.70–4.37 mmol TEAC/g) [21]. Other studies also report higher FRAP values in red/brown teff compared with white teff, likely reflecting differences in phenolic profiles [22,30].

H_2O_2 and NO scavenging assays further indicated that white teff extract can neutralize specific oxidants/reactive species. These observations suggest a potential role for teff-derived compounds in limiting nitro-oxidative stress, a process associated with aging and chronic disease development.

Oxidative stress arises when reactive oxygen species (ROS) generation exceeds the capacity of antioxidant defenses. ROS include species such as superoxide, hydrogen peroxide, hydroxyl radicals, and peroxynitrite, which can damage lipids, proteins, and nucleic acids when present at high levels [31].

Organisms counter ROS and reactive nitrogen species through coordinated enzymatic defenses (e.g., superoxide dismutase, catalase, peroxidases) and non-enzymatic antioxidants (e.g., glutathione, ascorbate, thiol groups, and plasma proteins) [32]. Because oxidative injury can manifest across multiple biomolecule classes, combined assessment of oxidant and antioxidant markers can provide a more robust view of redox status [33].

Although polyphenols are widely studied as antioxidants, under certain *in vivo* conditions they may also exhibit pro-oxidant behavior. This shift can depend on factors such as solubility, metal chelation/reduction properties, and local pH [29].

Given the complexity of oxidative stress, using a panel of oxidant and antioxidant biomarkers is considered a sound approach.

TOS provides an integrated estimate of circulating oxidants, whereas TAC captures the cumulative antioxidant capacity of a sample. Interpreting these together is informative: lower TOS and/or higher TAC suggests a more favorable redox balance. OSI, calculated from TOS and TAC, summarizes this balance as a single index [29,34]. In our model, turpentine-induced inflammation produced an oxidant shift (increased TOS/OSI and reduced TAC). Both diclofenac and Trolox improved these parameters, supporting the link between inflammatory signaling and redox imbalance. White teff extract also reduced oxidative stress in both therapeutic and prophylactic protocols, with a dose-dependent pattern and the undiluted extract (TEFF100) performing comparably to diclofenac and Trolox.

To explore mechanisms beyond global indices, we examined protein and lipid oxidation markers. Because many ROS are short-lived, downstream oxidation products in

https://doi.org/10.3390/molecules31040644

proteins and lipids can serve as practical indicators of oxidative injury [35]. AOPP are oxidatively modified proteins (often albumin) generated by chlorinated oxidants such as hypochlorous acid produced during neutrophil activation [35,36]. AOPP may also propagate inflammation by promoting cytokine release and endothelial dysfunction, creating a reinforcing loop between oxidative stress and inflammation [37]. In our study, Trolox lowered AOPP, and white teff extract reduced AOPP even more markedly than diclofenac and Trolox in both experimental schemes.

MDA, a widely used lipid peroxidation marker, reflects oxidative degradation of polyunsaturated fatty acids and is often used to assess dietary or supplement effects on membrane lipid damage [38]. Teff polyphenols have been discussed as inhibitors of lipid peroxidation [12]. In our experiments, white teff extract reduced MDA after and before inflammation induction in a dose-dependent manner, although the magnitude of effect was smaller than with diclofenac or Trolox.

Nitric oxide is produced from L-arginine by nitric oxide synthases: NOS3 in endothelium, NOS2 in activated immune cells, and NOS1 in neurons. In activated macrophages, concurrent production of superoxide and NO can lead to peroxynitrite and other reactive nitrogen species (RNS) [39]. RNS can modify protein tyrosine residues, forming 3-nitrotyrosine (3NT), a stable indicator of nitrosative stress [40]. In our study, teff administration after inflammation induction significantly lowered NOx, with TEFF100 and TEFF50 outperforming diclofenac and Trolox in this parameter. Prophylactic benefit on NOx was evident mainly with TEFF100. Teff also reduced 3NT in both protocols, with effects comparable to diclofenac and Trolox. To our knowledge, these findings provide early evidence that teff extract may attenuate RNS-related injury during acute inflammation.

Total thiols represent the combined pool of free and protein-bound sulfhydryl groups, including glutathione and cysteine-containing proteins. Because thiols can directly quench reactive species and help maintain redox balance, they are often treated as an antioxidant capacity marker. In our study, SH increased significantly in TEFF50/INFL and TEFF25/INFL groups, which may relate to dietary antioxidant components and micronutrients present in teff [12,41].

Beyond direct scavenging, teff constituents may also enhance endogenous antioxidant responses. For example, teff grain extracts have been reported to activate the Nrf2 pathway in THP-1 cells, increasing expression of genes involved in glutathione synthesis and recycling [42].

Taken together, the antioxidant activity of teff likely reflects both direct chemical scavenging and indirect upregulation of endogenous defense pathways, which may translate into reduced oxidative injury in conditions linked to redox imbalance [4].

Inflammation is essential for host defense and tissue repair, yet persistent inflammation contributes to multiple chronic diseases. Diet is one modifiable factor influencing inflammatory tone, and whole grains with high fiber and bioactive phytochemicals are increasingly studied for their anti-inflammatory potential. Teff may contribute through fiber, resistant starch, minerals, and antioxidant compounds [4].

Dietary fiber can support beneficial gut microbiota and increase short-chain fatty acid production, which may help regulate immune responses. Teff also contains resistant starch, which has been associated with improved glycemic control and lower inflammatory markers. Its mineral content (especially magnesium) may additionally modulate inflammatory pathways; magnesium deficiency has been linked to higher C-reactive protein levels [4].

Because teff generally has a low glycemic index, it may help avoid rapid postprandial glucose excursions that can activate inflammatory signaling [13].

NF-κB is a central transcription factor in inflammatory regulation. In resting cells, it is retained in the cytoplasm by IκB proteins; upon stimulation (e.g., cytokines, oxidative

https://doi.org/10.3390/molecules31040644

stress, PAMPs/DAMPs), IκB is degraded and NF-κB translocates to the nucleus to induce pro-inflammatory genes. NF-κB activation can also enhance iNOS expression and NO production, linking inflammation and nitrosative stress [43]. Direct studies on teff and NF-κB signaling are limited, but teff contains flavonoids (luteolin, apigenin, quercetin) and other polyphenols that are known in other systems to suppress NF-κB-regulated mediators such as TNF-α and IL-1β. By lowering oxidative stress, these compounds may reduce upstream triggers that activate inflammatory cascades [4,24].

Additional anti-inflammatory mechanisms may involve microbiota-derived metabolites (e.g., butyrate from fiber fermentation) and mineral status. Magnesium can inhibit NF-κB signaling, whereas zinc deficiency may enhance it [44].

In our experiments, teff reduced NF-κB-p65 in both therapeutic and prophylactic protocols, with stronger effects in the prophylactic setting. TEFF100 and TEFF50 showed greater inhibition than diclofenac or Trolox in this preventive model. Nonetheless, human data linking teff intake with NF-κB activity are limited, and clinical studies are needed to validate these observations.

The NLRP3 inflammasome is an intracellular protein complex that senses diverse danger signals and promotes caspase-1 activation, leading to processing of IL-1β and IL-18 and, in some contexts, pyroptotic cell death. Dysregulated or persistent NLRP3 activation contributes to inflammatory pathology across multiple diseases [45,46].

IL-18 differs from IL-1β in that its precursor may be present in several healthy tissues, including blood monocytes and gastrointestinal epithelium. After caspase-1 activation, mature IL-18 is released, while a substantial proportion of the precursor can remain intracellular [47].

Excessive NLRP3 signaling has been implicated in disorders such as gout, atherosclerosis, type 2 diabetes, neurodegenerative disease, and autoimmune conditions. Dietary antioxidants and anti-inflammatory phytochemicals have been proposed as modulators of inflammasome activity [48].

Although direct evidence linking teff intake to NLRP3 modulation is limited, teff's antioxidant profile could reduce oxidative stress, a known trigger for inflammasome activation. In our model, we assessed indirect NLRP3-related markers by measuring IL-1β, IL-18, and caspase-1. White teff extract reduced IL-1β in both protocols (with smaller effects than diclofenac and Trolox), inhibited IL-18 (with stronger preventive than therapeutic effects), and lowered caspase-1 (less than diclofenac). Overall, the biomarker pattern is consistent with partial attenuation of NLRP3-related inflammatory activation by white teff extract.

IL-10 is a pleiotropic immunoregulatory cytokine that helps constrain inflammation, in part by suppressing NF-κB target gene expression and downregulating pro-inflammatory mediators [49,50]. Evidence directly assessing teff's influence on IL-10 is limited. In our acute turpentine model, IL-10 changes were not significant, and teff did not modify IL-10 levels in either protocol.

Together, our results indicate that phenolic acids and flavonoids in white teff extract are associated with measurable anti-inflammatory activity, reflected by reductions in NF-κB-p65, IL-1β, IL-18, and caspase-1 without a detectable increase in IL-10. This aligns with prior discussions suggesting teff polyphenols may support immune and inflammatory balance [12].

In summary, white teff extract showed concurrent antioxidant and anti-inflammatory effects in an acute inflammation model. By reducing oxidative stress and dampening pro-inflammatory signaling, teff may offer supportive value for conditions in which chronic inflammation and redox imbalance coexist. These findings reinforce teff's potential role as

https://doi.org/10.3390/molecules31040644

a functional food component [4,16] and support further investigation of its contribution to cardiometabolic risk reduction when consumed as part of a balanced diet [44].

4. Materials and Methods

4.1. Chemicals

Folin–Ciocâlteu reagent, quercetin, gallic acid, standard chlorogenic acid (>98% HPLC), luteolin, and apigenin (>99% HPLC) were purchased from Sigma-Aldrich (St. Louis, MO, USA); Trolox (6-hydroxy-2.5.7.8-tetramethylchroman-2-carboxylic acid) was obtained from Alfa-Aesar (Karlsruhe, Germany); acetonitrile of HPLC purity was obtained from Merck (Darmstadt, Germany); ultrapure water for the HPLC analysis was purified using the Direct-Q UV system from Millipore (Billerica, MA, USA); methanol, ethanol, ferrous ammonium sulfate, vanadium chloride (III) (VCl3), sulfanylamide (SULF), N-(1-Naphthyl)ethylenediaminedihydrochloric acid (NEDD), sodium nitroprusside (SNP), hydrochloric acid, hydrogen peroxide (H_2O_2), sulphanilic acid, ortho dianisidine dihydrochloric acid (3-3′-dimethoxybenzidine), o-phthalaldehyde, thiobarbituric acid, and xylenol orange [ocresosulfonphthalein-3.3-bis (sodium methyliminodiacetate)] were obtained from Merck (Darmstadt, Germany); the rat ELISA kits were obtained from Elabscience Bionovation Inc. (Houston, Texas) and MyBiosource (San Diego, CA, USA).

4.2. Plant Material Collection and Extract Preparation

Eragrostis tef flour (AGRO DELIVERY SRL, Chitila, Romania) was extracted with 70% (*v*/*v*) ethanol using a modified Squibb cold-repercolation procedure performed at room temperature. The resulting extract was prepared at a 1:1 ratio (1 g flour per 1 mL solvent; *w*:*v*) [51].

4.3. Phytochemical Analysis

4.3.1. Total Polyphenol Content

Total polyphenol content (TPC) was determined by a modified Folin–Ciocâlteu assay. Briefly, 2 mL of teff ethanolic extract was diluted 25 times. Then, 1 mL Folin–Ciocâlteu reagent and 10 mL distilled water were added, and the mixture was brought to 25 mL with 290 g/L sodium carbonate solution. After 30 min incubation in the dark, absorbance was recorded at 760 nm. The total phenolic content (TPC) was determined using a gallic acid calibration curve (10-50-100-150-200-250-300-350-400-450-500 μg/mL) (R^2 = 0.999). Results were expressed as mg gallic acid equivalents per g of dry-weight plant flour (mg GAE/g d.w.) [51].

4.3.2. Total Flavonoid Content

Total flavonoid content (TFC) was assessed as previously reported. To 1 mL of teff ethanol extract, 0.3 mL of 5% $NaNO_2$ was added, followed by 0.3 mL of 10% $AlCl_3$. Next, 2 mL of 1 M NaOH was added and the volume was adjusted to 10 mL with distilled water. After 15 min incubation, absorbance was measured at 510 nm. TFC was quantified via a quercetin standard curve (1-5-10-15-20-25-30-35-40-45-50 μg/mL) (R^2 = 0.999). Values were expressed as mg quercetin equivalents per 100 g of dry-weight plant material (mg QE/100 g d.w.) [51].

4.3.3. High-Performance Liquid Chromatography Coupled with Electrospray Ionization Mass Spectrometry (HPLC-ESI MS) Analysis

The teff ethanolic extract was profiled by HPLC-DAD coupled to single-quadrupole MS (Agilent 1200 HPLC with DAD; Agilent 6110 MS, Santa Clara, CA, USA). Separation was performed on an Eclipse XDB-C18 column (4.6 × 150 mm, 5 μm) at room temperature. Mobile phase A consisted of 0.1% acetic acid in water with acetonitrile (99:1, *v*/*v*), and

https://doi.org/10.3390/molecules31040644

mobile phase B consisted of 0.1% acetic acid in acetonitrile (*v*/*v*). The gradient was as follows: 95% A (0–2 min), 95–60% A (2–18 min), 60–10% A (18–20 min), 10% A (20–24 min), return to 95% A in 1 min and hold for 5 min [15]. The flow rate was 0.5 mL/min. DAD chromatograms were recorded at 280 nm (phenolic acids) and 340 nm (flavonoids). MS detection used ESI in positive mode (350 °C, 3000 V capillary voltage, nitrogen 8 L/min) with scanning from 100 to 1000 *m*/*z*. Compounds were tentatively identified using UV–Vis spectra, retention times, co-chromatography with available standards (Sigma-Aldrich, St. Louis, MO, USA), mass spectra, and literature data. Prior to injection, the lyophilized extract was dissolved in MeOH. Quantification relied on external calibration curves constructed from five concentrations of standards in methanol: chlorogenic acid for phenolic acids (R^2 = 0.9937; LOD 0.41 µg/mL; LOQ 1.64 µg/mL), luteolin for flavones (R^2 = 0.9972; LOD 0.26 µg/mL; LOQ 0.95 µg/mL), and rutin for flavonols (R^2 = 0.9981; LOD 0.21 µg/mL; LOQ 0.84 µg/mL) [51].

4.4. In Vitro Antioxidant Activity Analysis

4.4.1. 2,2-Diphenyl-1-Picrylhydrazyl (DPPH) Radical Scavenging Capacity

DPPH radical scavenging activity was assessed using a standard protocol. In brief, 3 mL of teff extract were mixed with 1 mL of 0.1 mM DPPH solution in methanol and incubated for 30 min in the dark at room temperature. Absorbance was measured at 517 nm. Antioxidant activity was calculated as AA% = [(Acontrol − Asample)/Acontrol] × 100. A Trolox standard curve (5-10-20-30-40-50-60-70-80-90-100 µg/mL) (R^2 = 0.999) was used, and IC50 values were expressed as µg Trolox equivalents per g of dry-weight flour extract (µg TE/g d.w.) [51].

4.4.2. Ferric Reducing Antioxidant Power (FRAP) Assay

Ferric reducing antioxidant power (FRAP) was measured as described previously [52]. A 100 µL aliquot of teff ethanolic extract was mixed with 3.4 mL FRAP reagent. After 30 min, absorbance was read at 593 nm, and results were reported as µg Trolox equivalents per g of dry-weight plant flour extract (µg TE/g d.w.) using a Trolox standard curve (25-50-75-100-125-150-175-200 mg/mL) (R^2 = 0.999).

4.4.3. Hydrogen Peroxide (H_2O_2) Scavenging Activity

Hydrogen peroxide (H_2O_2) scavenging activity was evaluated following a published method. Teff extract was added to an H_2O_2 solution and, after 10 min, absorbance was measured at 230 nm against phosphate buffer. Scavenging percentage was calculated as (Acontrol − Asample)/Acontrol × 100. Results were reported as IC50 in mg Trolox equivalents per g of dry-weight plant flour extract (mg TE/g d.w.) using a Trolox standard curve (10-20-30-40-50-60-70-80-90-100 µg/mL) (R^2 = 0.999) [51].

4.4.4. Nitric Oxide (NO) Radical Scavenging Assay

Nitric oxide (NO) scavenging was determined using sodium nitroprusside (SNP) as the NO donor. Teff extract (0.5 mL) was combined with SNP solution (2 mL SNP + 0.5 mL PBS, pH 7.4) and incubated for 2.5 h at 25 °C. Then, 0.5 mL of the reaction mixture was mixed with 1 mL sulphanilic acid; after 5 min, 1 mL N-(1-naphthyl)ethylenediamine dihydrochloride was added. Samples were vortexed and incubated in the dark for 30 min, and absorbance was measured at 546 nm. Inhibition was calculated as (Ablank − Asample)/Ablank × 100. Results were expressed as IC50 in µg quercetin equivalents per g of dry-weight plant flour (µg QE/mL d.w.) [51].

All *in vitro* assays were performed in triplicate. Spectrophotometric measurements were obtained using a UV–Vis spectrophotometer (Jasco V-350, Jasco International Co., Ltd., Tokyo, Japan).

https://doi.org/10.3390/molecules3104064

4.5 *In Vivo Experimental Design*

4.5.1. Experimental Protocol

Adult male Wistar rats (200–250 g) were obtained from the Establishment for Breeding and Use of Laboratory Animals, "Iuliu Hațieganu" University of Medicine and Pharmacy (Cluj-Napoca, Romania). Animals were housed under standard conditions (25 ± 1 °C; 55 ± 5% relative humidity; 12 h light/dark cycle) with ad libitum access to standard chow and water. Rats were randomized into 10 groups (n = 9). The extract was tested at three doses of dry teff flour extract per mL: 100 mg/mL (ab-breviated as TEFF100), 50 mg/mL (abbreviated as TEFF50), and 25 mg/mL (abbreviated as TEFF25). All treatments were administered by oral gavage in a volume of 1 mL/rat/day and prepared fresh each day. The SHAM group served as a healthy control without treatment. For the treatment protocol in six groups, acute inflammation was induced on day 1 by intramuscular turpentine oil (6 mL/kg b.w.). For the subsequent 10 days, treatments were administered orally (gavage) as follows: INFL group received tap water (1 mL/rat/day); DICLO group received diclofenac (10 mg/kg b.w./day); TX group received Trolox (50 mg/kg b.w./day); INFL/TEFF100 (500 mg/kg/day), INFL/TEFF50 (250 mg/kg/day), and INFL/TEFF25 (125 mg/kg/day) [7]. For the prophylactic protocol (TEFF100/INFL, TEFF50/INFL, TEFF25/INFL), the corresponding teff flour extract doses were given for 10 days, and inflammation was induced on day 11. On day 12, rats were anesthetized with ketamine (60 mg/kg b.w.) and xylazine (15 mg/kg b.w.) [52]; blood was collected by retro-orbital puncture, serum was separated, and samples were stored at −80 °C until analysis.

The project was approved by the Veterinary Sanitary Direction and Food Safety Cluj-Napoca (No. 372/04.07.2023). All the procedures comply with Directive 2010/63/EU and Romanian national law 43/2014 for the protection of animals used for scientific purposes.

4.5.2. Oxidative Stress Biomarkers Assessment

Total Oxidative Status (TOS)

Total oxidative status (TOS) was quantified based on the oxidation of ferrous ions to ferric ions in an acidic environment in the presence of reactive oxygen species. Absorbance was measured at 560 nm and results were expressed as µmol H_2O_2 equivalents/L (µM H_2O_2 equiv/L) [52].

Total Antioxidant Capacity (TAC)

Total antioxidant capacity (TAC) was assessed by measuring suppression of hydroxyl radical generation in a Fenton-type reaction by antioxidants present in serum. Absorbance was recorded at 444 nm and results were reported as mmol Trolox equivalents/L (mM TE/L) [53].

Oxidative Stress Index (OSI)

Oxidative stress index (OSI) was calculated as the ratio between oxidants and antioxidants: OSI (arbitrary units) = TOS (mM H_2O_2 equiv/L)/TAC (mM Trolox equiv/L). This index provides an integrated estimate of oxidative stress burden [54].

Advanced Oxidation Protein Products (AOPP)

Advanced oxidation protein products (AOPP) were measured as a protein oxidation marker using a spectrophotometric assay [55]. Samples and chloramine-T (blank) were diluted to 10% in PBS, and then potassium iodide and glacial acetic acid were added. Absorbance was read at 340 nm and concentrations were expressed as µM chloramine-T equivalents/L.

https://doi.org/10.3390/molecules31040644

Malondialdehyde (MDA)

Malondialdehyde (MDA), as an index of lipid peroxidation, was determined using a thiobarbituric acid-based method [56]. In brief, 0.1 mL serum was mixed with 0.1 mL 40% trichloroacetic acid and 0.2 mL 0.67% thiobarbituric acid, heated for 30 min in a boiling water bath, cooled in ice, and centrifuged for 5 min at 3.461 g. Absorbance was measured at 532 nm and MDA concentration was expressed as nM/mL serum.

Nitric Oxide Synthesis (NO)

Nitric oxide production was estimated by determining total nitrites and nitrates (NOx) via the Griess reaction. Serum proteins were removed by extraction with methanol/diethyl ether (3:1, *v*/*v*), nitrates were reduced to nitrites with vanadium (III) chloride, and absorbance was measured at 540 nm. Results were expressed as nitrite µM/L [56].

3-Nitrotyrosine (3NT)

3-Nitrotyrosine (3NT), a marker of peroxynitrite-mediated oxidative damage [56], was quantified using an ELISA kit (E-EL-0040) according to the manufacturer's protocol. Results were expressed as ng/mL.

Total Thiols (SH)

Total thiols (SH) were determined with Ellman's reagent [57]. Absorbance was read at 412 nm, and concentrations were expressed as mM glutathione equivalents per mL (mM GSH/mL).

4.5.3. Inflammatory Biomarkers Assessment

Systemic inflammatory response was evaluated by ELISA quantification of NFkB-p65 (E-EL-RO674), IL-1β (E-EL-0012), IL-18 (E-EL-R0567), caspase-1 (E-EL-R0371), and IL-10 (E-EL-R0016). Assays were performed according to manufacturers' instructions and results were reported as pg/mL.

Spectrophotometric measurements were performed on a UV–Vis spectrophotometer (Jasco V-350, Jasco International Co., Ltd., Tokyo, Japan). ELISA procedures used a Biotek microplate 50 TS washer and an 800 TS microplate reader (Agilent Technologies Inc., Santa Clara, CA, USA).

4.6. Statistical Analysis

Data are presented as mean ± standard deviation (three independent determinations) when normally distributed. Group comparisons were performed using one-way ANOVA followed by Tukey's post hoc test. Pearson correlation and principal component analysis (PCA) were applied to explore relationships between oxidative stress and inflammatory markers. Statistical significance was set at $p < 0.05$. Analyses were conducted in SPSS Statistics v26.0 (SPSS, Chicago, IL, USA).

5. Conclusions

In conclusion, the white teff flour ethanolic extract shows promise as a nutraceutical candidate, with the potential to support inflammatory conditions through simultaneous reductions in inflammatory biomarkers and nitro-oxidative stress.

As research on teff expands, additional health-relevant mechanisms may be clarified, particularly in relation to oxidative stress pathways and chronic disease risk. At present, teff remains an attractive gluten-free ancient grain that combines nutritional density with bioactive constituents, supporting its emerging role in functional food development.

https://doi.org/10.3390/molecules31040644

Further well-designed human studies are needed to confirm these effects, clarify the bioavailability and metabolic fate of teff antioxidants, and define how teff-based dietary strategies can be integrated into the management of inflammatory diseases.

Author Contributions: Conceptualization: I.F., A.E.P. and R.M.P.; methodology: I.F., R.M.P., M.P., F.R. and T.I.P.; software: D.B., T.I.P. and M.Ţ.; validation: A.E.P., M.P. and R.M.P.; formal analysis: I.F., A.E.P. and R.M.P.; investigation: I.F., I.I.M. and C.O.D.; resources: I.F., A.E.P., F.R. and R.M.P.; data curation: A.E.P. and R.M.P.; writing—original draft preparation: I.F., T.I.P. and D.B.; writing—review and editing, supervision: A.E.P. All authors have read and agreed to the published version of the manuscript.

Funding: This research received no external funding.

Institutional Review Board Statement: The animal study protocol was approved by the Ethics Committees from both "Iuliu Hatieganu" University of Medicine and Pharmacy, Cluj-Napoca, and the Romanian National Sanitary Veterinary and Food Safety Authority (Approval Code No. 372/04 July 2023).

Informed Consent Statement: Not applicable.

Data Availability Statement: Data will be available in the PhD thesis of the first author after the thesis debate at the "Iuliu Haţieganu" University of Medicine and Pharmacy, Cluj-Napoca.

Acknowledgments: The authors would like to offer special thanks to the chemist Ana Uifălean for helping with the biochemical tests.

Conflicts of Interest: The authors declare no conflicts of interest.

eferences

Reta, C.; Atlabachew, M.; Asmellash, T.; Hilluf, W.; Yayinie, M.; Wubieneh, T.A. Polyphenol and Flavonoid Content in Major Teff [*Eragrostis tef* (Zuccagni) Trotter] Varieties in Amhara Region, Ethiopia. *PLoS ONE* **2022**, *17*, e0272010. [CrossRef]

Cheng, A.; Mayes, S.; Dalle, G.; Demissew, S.; Massawe, F. Diversifying Crops for Food and Nutrition Security—A Case of Teff: A Case of Teff. *Biol. Rev.* **2017**, *92*, 188–198. [CrossRef]

Gebremariam, M.M.; Zarnkow, M.; Becker, T. Teff (*Eragrostis tef*) as a Raw Material for Malting, Brewing and Manufacturing of Gluten-Free Foods and Beverages: A Review. *J. Food Sci. Technol.* **2014**, *51*, 2881–2895. [CrossRef] [PubMed]

Abebe, B.S.; Aprodu, I.; Istrati, D.I.; Vizireanu, C. Nutritional Characteristics, Health-Related Properties, and Food Application of Teff (*Eragrostis tef*): An Overview. *Int. J. Mol. Sci.* **2025**, *26*, 9293. [CrossRef]

Adepoju, M.; Verheecke-Vaessen, C.; Pillai, L.R.; Phillips, H.; Cervini, C. Unlocking the Potential of Teff for Sustainable, Gluten-Free Diets and Unravelling Its Production Challenges to Address Global Food and Nutrition Security: A Review. *Foods* **2024**, *13*, 3394. [CrossRef]

Tessema, B.; Gonfa, G.; Hailegiorgis, S.M.; Sundramurthy, V.P. Characterization of Teff Straw from Selected Teff Varieties from Ethiopia. *Heliyon* **2023**, *9*, e17422. [CrossRef] [PubMed]

El-Alfy, T.S.; Ezzat, S.M.; Sleem, A.A. Chemical and Biological Study of the Seeds of *Eragrostis tef (Zucc.) Trotter*. *Nat. Prod. Res.* **2012**, *26*, 619–629. [CrossRef] [PubMed]

Zhu, F. Chemical Composition and Food Uses of Teff (*Eragrostis tef*). *Food Chem.* **2018**, *239*, 402–415. [CrossRef]

Tiguh, E.E.; Delele, M.A.; Ali, A.N.; Kidanemariam, G.; Fanta, S.W. Assessment of Harvest and Postharvest Losses of Teff (*Eragrostis tef* (Zucc.)) and Methods of Loss Reduction: A Review. *Heliyon* **2024**, *10*, e30398. [CrossRef]

. Martín-Diana, A.B.; Blanco Espeso, B.; Jimenez Pulido, I.J.; Acebes Martínez, P.J.; Rico, D. Twin-Screw Extrusion as Hydrothermal Technology for the Development of Gluten-Free Teff Flours: Effect on Antioxidant, Glycaemic Index and Techno-Functional Properties. *Foods* **2022**, *11*, 3610. [CrossRef]

. Amtataw, M.; Kassahun, E.; Tibebu, S.; Andargie, T.; Ayanaw, T.; Agazie, A.; Wogayehu, M.; Teshome, A.; Jemal, S.; Teferi, D. Physicochemical and Sensory Analysis of Sorghum, Rice, and Teff Flours Blending with Flaxseed Flour for Better Quality *Injera*. *Food Sci. Nutr.* **2025**, *13*, e4680. [CrossRef]

. Fekete, M.; Varga, P.; Ungvari, Z.; Fekete, J.T.; Buda, A.; Szappanos, Á.; Lehoczki, A.; Mózes, N.; Grosso, G.; Godos, J.; et al. The Role of the Mediterranean Diet in Reducing the Risk of Cognitive Impairment, Dementia, and Alzheimer's Disease: A Meta-Analysis. *Geroscience* **2025**, *47*, 3111–3130. [CrossRef]

https://doi.org/10.3390/molecules31040644

13. Habte, M.L.; Beyene, E.A.; Feyisa, T.O.; Admasu, F.T.; Tilahun, A.; Diribsa, G.C. Nutritional Values of Teff (*Eragrostis tef*) i Diabetic Patients: Narrative Review. *Diabetes Metab. Syndr. Obes.* **2022**, *15*, 2599–2606. [CrossRef]
14. Dingeo, C.; Difonzo, G.; Paradiso, V.M.; Rizzello, C.G.; Pontonio, E. Teff Type-I Sourdough to Produce Gluten-Free Muffi *Microorganisms* **2020**, *8*, 1149. [CrossRef]
15. Zhang, Z.; Deng, Y.; Zhang, W.; Wu, Y.; Messing, J. Towards Coeliac-safe Bread. *Plant Biotechnol. J.* **2020**, *18*, 1056–1065. [CrossRe
16. Herdiana, Y. Functional Food in Relation to Gastroesophageal Reflux Disease (GERD). *Nutrients* **2023**, *15*, 3583. [CrossRef]
17. Ooi, S.-L.; Pak, S.-C. Nutraceuticals in Immune Function. *Molecules* **2021**, *26*, 5310. [CrossRef]
18. Chandra, S.; Saklani, S.; Kumar, P.; Kim, B.; Coutinho, H.D.M. Nutraceuticals: Pharmacologically Active Potent Dietar Supplements. *BioMed Res. Int.* **2022**, *2022*, 2051017. [CrossRef] [PubMed]
19. Woldeyohannes, A.B.; Iohannes, S.D.; Miculan, M.; Caproni, L.; Ahmed, J.S.; De Sousa, K.; Desta, E.A.; Fadda, C.; Pè, M.E Dell'Acqua, M. Data-Driven, Participatory Characterization of Farmer Varieties Discloses Teff Breeding Potential under Curre and Future Climates. *eLife* **2022**, *11*, e80009. [CrossRef] [PubMed]
20. Ronis, M.J.J.; Pedersen, K.B.; Watt, J. Adverse Effects of Nutraceuticals and Dietary Supplements. *Annu. Rev. Pharmacol. Toxic* **2018**, *58*, 583–601. [CrossRef] [PubMed]
21. Sumczynski, D.; Kotásková, E.; Družbíková, H.; Mlček, J. Determination of Contents and Antioxidant Activity of Free and Boun Phenolics Compounds and *In Vitro* Digestibility of Commercial Black and Red Rice (*Oryza sativa* L.) Varieties. *Food Chem.* **201** *211*, 339–346. [CrossRef] [PubMed]
22. Yisak, H.; Tukue, H.; Redi-Abshiro, M.; Chandravanshi, B.S.; Yaya, E.E. Development of Green UV-Vis Method for Dire Determination of Total Sugars in the Aqueous Extract of Teff (*Eragrostis tef* (Zuccagni) Trotter) Grains and Other Cereals. *Int. Anal. Chem.* **2022**, *2022*, 5129510. [CrossRef]
23. Ananth, D.A.; Mahalakshmi, V.; Yermiyahu, U.; Van Oss Pinhasi, R.; Klipcan, L.; Tietel, Z. Tef and Fonio-morphology, Determin tion of Amino Acids, Phenolic Compounds and Antioxidant Capacity of Two Gluten-free Grains. *Int. J. Food Sci. Technol.* **2023**, *5* 5947–5961. [CrossRef]
24. Jaroszewska, A.; Jedrejek, D.; Sobolewska, M.; Kowalska, I.; Dzięcioł, M. Mineral, Nutritional, and Phytochemical Compositio and Baking Properties of Teff and Watermelon Seed Flours. *Molecules* **2023**, *28*, 3255. [CrossRef]
25. Kotásková, E.; Sumczynski, D.; Mlček, J.; Valášek, P. Determination of Free and Bound Phenolics Using HPLC-DAD, Antioxida Activity and *In Vitro* Digestibility of *Eragrostis tef*. *J. Food Compos. Anal.* **2016**, *46*, 15–21. [CrossRef]
26. Li, B.; Du, P.; Du, Y.; Zhao, D.; Cai, Y.; Yang, Q.; Guo, Z. Luteolin Alleviates Inflammation and Modulates Gut Microbiota i Ulcerative Colitis Rats. *Life Sci.* **2021**, *269*, 119008. [CrossRef] [PubMed]
27. Xiong, C.; Wu, Q.; Fang, M.; Li, H.; Chen, B.; Chi, T. Protective Effects of Luteolin on Nephrotoxicity Induced by Long-Ter Hyperglycaemia in Rats. *J. Int. Med. Res.* **2020**, *48*, 0300060520903642. [CrossRef]
28. Jayawickreme, D.K.; Ekwosi, C.; Anand, A.; Andres-Mach, M.; Wlaź, P.; Socała, K. Luteolin for Neurodegenerative Diseases: Review. *Pharmacol. Rep.* **2024**, *76*, 644–664. [CrossRef]
29. Sánchez-Rodríguez, M.A.; Mendoza-Núñez, V.M. Oxidative Stress Indexes for Diagnosis of Health or Disease in Human *Oxidative Med. Cell. Longev.* **2019**, *2019*, 4128152. [CrossRef]
30. Shumoy, H.; Lauwens, S.; Gabaza, M.; Vandevelde, J.; Vanhaecke, F.; Raes, K. Traditional Fermentation of Tef Injera: Impact on *I Vitro* Iron and Zinc Dialysability. *Food Res. Int.* **2017**, *102*, 93–100. [CrossRef]
31. Rudrapal, M.; Khairnar, S.J.; Khan, J.; Dukhyil, A.B.; Ansari, M.A.; Alomary, M.N.; Alshabrmi, F.M.; Palai, S.; Deb, P.K.; Devi, Dietary Polyphenols and Their Role in Oxidative Stress-Induced Human Diseases: Insights Into Protective Effects, Antioxida Potentials and Mechanism(s) of Action. *Front. Pharmacol.* **2022**, *13*, 806470. [CrossRef]
32. Sachdev, S.; Ansari, S.A.; Ansari, M.I.; Fujita, M.; Hasanuzzaman, M. Abiotic Stress and Reactive Oxygen Species: Generatio Signaling, and Defense Mechanisms. *Antioxidants* **2021**, *10*, 277. [CrossRef]
33. Romanov, A.V.; Graf, T.N.; Vidar, W.; Jima, D.D.; Oberlies, N.H.; Jia, Z.; Ligaba-Osena, A. Studying the Molecular Mechanisms o Tef (*Eragrostis tef*) Antioxidant Properties in Human THP-1 Monocytes. *J. Funct. Foods* **2025**, *135*, 107092. [CrossRef] [PubMed]
34. Paulis, G.; Paulis, A.; De Giorgio, G.; Quattrocchi, S. Measurement of Oxidative Stress Index (OSI) in Penile Corpora Cavernos and Peripheral Blood of Peyronie's Disease Patients: A Report of 49 Cases. *Metabolites* **2024**, *14*, 55. [CrossRef]
35. Bagyura, Z.; Takács, A.; Kiss, L.; Dósa, E.; Vadas, R.; Nguyen, T.D.; Dinya, E.; Soós, P.; Szelid, Z.; Láng, O.; et al. Level of Advance Oxidation Protein Products Is Associated with Subclinical Atherosclerosis. *BMC Cardiovasc. Disord.* **2022**, *22*, 5. [CrossRef]
36. Arimura, N.; Watanabe, H.; Kato, H.; Imafuku, T.; Nakano, T.; Sueyoshi, M.; Chikamatsu, M.; Tokumaru, K.; Nagasaki, T.; Maed H.; et al. Advanced Oxidation Protein Products Contribute to Chronic-Kidney-Disease-Induced Adipose Inflammation throug Macrophage Activation. *Toxins* **2023**, *15*, 179. [CrossRef]
37. Azouaou Toualbi, L.; Mounir, A.; Wafa, B.; Medina, A.; Abderrezak, K.; Chahine, T.; Henni, C.; Abdelghani, B.; Atmane, Implications of Advanced Oxidation Protein Products and Vitamin E in Atherosclerosis Progression. *Arch. Med. Sci. Atheroscle Dis.* **2021**, *6*, 135–144. [CrossRef] [PubMed]

https://doi.org/10.3390/molecules3104064

Milei, J.; Forcada, P.; Fraga, C.G.; Grana, D.R.; Iannelli, G.; Chiariello, M.; Tritto, I.; Ambrosio, G. Relationship between Oxidative Stress, Lipid Peroxidation, and Ultrastructural Damage in Patients with Coronary Artery Disease Undergoing Cardioplegic Arrest/Reperfusion. *Cardiovasc. Res.* **2007**, *73*, 710–719. [CrossRef]

Thomson, L. 3-Nitrotyrosine Modified Proteins in Atherosclerosis. *Dis. Markers* **2015**, *2015*, 708282. [CrossRef]

Teixeira, D.; Fernandes, R.; Prudêncio, C.; Vieira, M. 3-Nitrotyrosine Quantification Methods: Current Concepts and Future Challenges. *Biochimie* **2016**, *125*, 1–11. [CrossRef]

Slavin, J.L. Dietary Fiber and Body Weight. *Nutrition* **2005**, *21*, 411–418. [CrossRef] [PubMed]

Cotter, T.G.; Bledsoe, A.C.; Sweetser, S. Colon Ischemia. *Mayo Clin. Proc.* **2016**, *91*, 671–677. [CrossRef]

Kim, J.H.; Kim, D.H.; Jo, S.; Cho, M.J.; Cho, Y.R.; Lee, Y.J.; Byun, S. Immunomodulatory Functional Foods and Their Molecular Mechanisms. *Exp. Mol. Med.* **2022**, *54*, 1–11. [CrossRef]

Lemecha, M.; Morino, K.; Seifu, D.; Imamura, T.; Nakagawa, F.; Nagata, A.; Okamato, T.; Sekine, O.; Ugi, S.; Maegawa, H. Improved Glucose Metabolism by *Eragrostis tef* Potentially through Beige Adipocyte Formation and Attenuating Adipose Tissue Inflammation. *PLoS ONE* **2018**, *13*, e0201661. [CrossRef]

Seoane, P.I.; Lee, B.; Hoyle, C.; Yu, S.; Lopez-Castejon, G.; Lowe, M.; Brough, D. The NLRP3–Inflammasome as a Sensor of Organelle Dysfunction. *J. Cell Biol.* **2020**, *219*, e202006194. [CrossRef]

O'Keefe, M.E.; Dubyak, G.R.; Abbott, D.W. Post-Translational Control of NLRP3 Inflammasome Signaling. *J. Biol. Chem.* **2024**, *300*, 107386. [CrossRef]

Dinarello, C.A. Overview of the IL-1 Family in Innate Inflammation and Acquired Immunity. *Immunol. Rev.* **2018**, *281*, 8–27. [CrossRef]

Chen, B.; Lu, Y.; Chen, Y.; Cheng, J. The Role of Nrf2 in Oxidative Stress-Induced Endothelial Injuries. *J. Endocrinol.* **2015**, *225*, R83–R99. [CrossRef] [PubMed]

Saxton, R.A.; Tsutsumi, N.; Su, L.L.; Abhiraman, G.C.; Mohan, K.; Henneberg, L.T.; Aduri, N.G.; Gati, C.; Garcia, K.C. Structure-Based Decoupling of the pro- and Anti-Inflammatory Functions of Interleukin-10. *Science* **2021**, *371*, eabc8433. [CrossRef]

Mishra, B.; Bachu, M.; Yuan, R.; Wingert, C.; Chaudhary, V.; Brauner, C.; Bell, R.; Ivashkiv, L.B. IL-10 Targets IRF Transcription Factors to Suppress IFN and Inflammatory Response Genes by Epigenetic Mechanisms. *Nat. Immunol.* **2025**, *26*, 748–759. [CrossRef] [PubMed]

Ţicolea, M.; Pop, R.M.; Pârvu, M.; Usatiuc, L.-O.; Uifălean, A.; Brito, V.A.; Fischer-Fodor, E.; Ranga, F.; Rusu, C.C.; Crisan, M.; et al. The Phytochemical and Functional Characterization of the Aerial Parts of *Artemisia alba Turra (Asteraceae)* Grown in Romania. *Foods* **2025**, *14*, 1389. [CrossRef] [PubMed]

Erel, O. A New Automated Colorimetric Method for Measuring Total Oxidant Status. *Clin. Biochem.* **2005**, *38*, 1103–1111. [CrossRef] [PubMed]

Erel, O. A Novel Automated Method to Measure Total Antioxidant Response Against Potent Free Radical Reactions. *Clin. Biochem.* **2004**, *37*, 112–119. [CrossRef]

Harma, M.; Harma, M.; Erel, O. Increased Oxidative Stress in Patients with Hydatidiform Mole. *Swiss Med. Wkly.* **2003**, *133*, 563–566. [CrossRef]

Morar, I.I.; Pop, R.M.; Peitzner, E.; Ranga, F.; Orăsan, M.S.; Cecan, A.D.; Chera, E.I.; Bonci, T.I.; Usatiuc, L.O.; Ţicolea, M.; et al. Phytochemical Composition and Antioxidant Activity of Manuka Honey and Ohia Lehua Honey. *Nutrients* **2025**, *17*, 276. [CrossRef]

Usatiuc, L.-O.; Pop, R.M.; Adrian, S.; Pârvu, M.; Ţicolea, M.; Uifălean, A.; Vălean, D.; Gavrilaş, L.-I.; Csilla-Enikő, S.; Leopold, L.F.; et al. Multitargeted Effects of *Plantago ovata* Ethanol Extract in Experimental Rat Streptozotocin-Induced Diabetes Mellitus and Letrozole-Induced Polycystic Ovary Syndrome. *Int. J. Mol. Sci.* **2025**, *26*, 4712. [CrossRef]

Erel, O.; Neselioglu, S. A Novel and Automated Assay for Thiol/Disulphide Homeostasis. *Clin. Biochem.* **2014**, *47*, 326–332. [CrossRef]

https://doi.org/10.3390/molecules31040644

molecules

MDPI

Article

Serum Starvation Enhances the Antitumor Activity of Natural Matrices: Insights into Bioactive Molecules from Dromedary Urine Extracts

Maria Noemi Sgobba [1,*,†], Biagia Musio [2,†], Carlos Iglesias Pastrana [3], Stefano Todisco [2], Nikola Schlosserovà [1], Federica Mastropirro [1], Maria Favia [4], Antonio Radesco [5], Iola F. Duarte [6], Anna De Grassi [1], Mariateresa Volpicella [1], Vito Gallo [2,7], Ciro Leonardo Pierri [8,*], Elena Ciani [1] and Lorenzo Guerra [1]

[1] Department of Biosciences, Biotechnologies and Environment, University of Bari "Aldo Moro", Via Orabona 4, 70125 Bari, Italy; f.mastropirro@phd.uniba.it (F.M.); anna.degrassi@uniba.it (A.D.G.); mariateresa.volpicella@uniba.it (M.V.); elena.ciani@uniba.it (E.C.); lorenzo.guerra1@uniba.it (L.G.)

[2] Department of Civil, Environmental, Land, Building Engineering and Chemistry (DICATECh), Polytechnic University of Bari, Via Orabona 4, 70125 Bari, Italy; biagia.musio@poliba.it (B.M.); stefano.todisco@poliba.it (S.T.); vito.gallo@poliba.it (V.G.)

[3] Faculty of Veterinary Sciences, Department of Genetics, University of Córdoba, 14071 Córdoba, Spain; ciglesiaspastrana@gmail.com

[4] Department of Translational Biomedicine and Neurosciences (DiBraiN), University of Bari "Aldo Moro", Piazza Giulio Cesare, 70124 Bari, Italy; maria.favia@uniba.it

[5] Istituto Tumori "Giovanni Paolo II" I.R.C.C.S., Viale Orazio Flacco 65, 70124 Bari, Italy

[6] Department of Chemistry, CICECO—Aveiro Institute of Materials and LAQV-REQUIMTE, University of Aveiro, 3810-193 Aveiro, Portugal; ioladuarte@ua.pt

[7] Innovative Solutions S.r.l.—Spin-Off Company of the Polytechnic University of Bari, Zona H 150/B, 70015 Noci, Italy

[8] Department of Pharmacy—Pharmaceutical Sciences, University of Bari Aldo Moro, Via Orabona 4, 70125 Bari, Italy

* Correspondence: maria.sgobba@uniba.it (M.N.S.); ciro.pierri@uniba.it (C.L.P.)

† These authors contributed equally to this work.

Abstract: Natural matrices have historically been a cornerstone in drug discovery, offering a rich source of structurally diverse and biologically active compounds. However, research on natural products often faces significant challenges due to the complexity of natural matrices, such as urine, and the limitations of bioactivity assessment assays. To ensure reliable insights, it is crucial to optimize experimental conditions to reveal the bioactive potential of samples, thereby improving the validity of statistical analyses. Approaches in metabolomics further strengthen this process by identifying and focusing on the most promising compounds within natural matrices, enhancing the precision of bioactive metabolite prioritization. In this study, we assessed the bioactivity of 17 dromedary urine samples on human renal cells under serum-reduced conditions (1%FBS) in order to minimize possible FBS-derived interfering factors. Using viability assays and Annexin V/PI staining, we found that the tumor renal cell lines Caki-1 and RCC-Shaw were more sensitive to the cytotoxic effects of the small molecules present in dromedary urine compared to non-tumor HK-2 cells. Employing NMR metabolomics analysis combined with detected in vitro activity, our statistical model highlights the presence of bioactive compounds in dromedary urine, such as azelaic acid and phenylacetyl glycine, underscoring its potential as a sustainable source of bioactive molecules within the framework of green chemistry and circular economy initiatives.

https://doi.org/10.3390/molecules30040821

Keywords: natural matrices; dromedary urine extracts; serum-reduced conditions; human renal cells; cell viability; apoptosis; NMR metabolomics; azelaic acid

1. Introduction

Natural matrices have long been esteemed as a source of metabolites, providing a wide range of bioactive compounds with potential therapeutic benefits [1,2]. In this context, the urine of various animals, such as goats, sheep, buffalo, elephants, horses, camels, and donkeys, have historically been utilized as a natural remedy for numerous health issues, including abdominal tumors, tuberculosis, anemia, and infectious diseases [3–5]. In particular, dromedary urine has been traditionally used in various Arabian countries for its medicinal properties and is now gaining scientific attention for its unique metabolite profile [3,5]. Dromedary urine is considered the best candidate for urotherapy, as it appears to possess more beneficial effects than other animal urine [6,7]. Indeed, researchers have demonstrated that dromedary urine can exhibit antimicrobial, antibacterial, antifungal, anticancer, and antiviral properties, making it a subject of interest in natural product research [3–5,8–10]. However, the efficacy and reproducibility of these studies can vary significantly depending on the animal species and the environment in which they live; thus, the metabolome of a biological system is defined by a set of end products deriving from physiological and environmental stimuli [11]. This variability underscores the importance of considering both biological and environmental factors affecting the therapeutic potential of natural products [7,12,13]. In this regard, metabolomics offers the opportunity to gain insights into metabolites that can vary based on a variety of factors, such as diet, intestinal microbiota, physical activity, and environmental exposures. Metabolomics analysis can specifically contribute to the decomposition of natural matrices' complexity and can facilitate the identification of bioactive compounds within dromedary urine by matching the information of dromedary urine biological activities with their chemical composition [14].

Additionally, in vitro studies are affected by a further source of variability [15], such as fetal bovine serum (FBS), a widely used cell culture medium supplement that supports cell growth. The serum is a critical factor that significantly influences bioactivity assay outcomes, especially when dealing with complex natural matrices like dromedary urine. Indeed, FBS is known to contain a wide range of growth factors, hormones, and other macromolecules that support cell growth and maintenance. However, its variability and potential to influence experimental outcomes necessitate the careful optimization of its concentration in culture media [16–18]. Studies have shown that the presence of FBS can impact the pharmacodynamics of tested compounds, potentially masking or enhancing their bioactivity and affecting their availability and interaction with target cells [19,20]. Moreover, FBS can influence cellular responses, such as proliferation, differentiation, and metabolic activity, further complicating the interpretation of results [17]. Furthermore, the natural matrix under investigation necessitates careful consideration of its unique physico-chemical properties, such as osmolarity [3,10], thus highlighting the importance of optimizing experimental conditions to accurately assess the bioactive potential of these animal-derived metabolites.

In this work, we tested the bioactive potential of a set of 17 urine samples from dromedaries of different sex, age, and physiological status on three human renal cell lines, the non-tumor HK2 and the tumor Caki-1 and RCC-Shaw cell lines, by carefully considering the possible influence of FBS concentration in the in vitro bioassays, thus performing all the assays under conditions of serum starvation (1% FBS). Moreover, in order to explore the presence of potential bioactive compounds, the chemical composition of dromedary urine was deeply investigated using a non-targeted Nuclear Magnetic Resonance (NMR)-based approach.

2. Results

2.1. Impact of FBS Concentration on Renal Cell Viability

The use of serum, especially fetal bovine serum (FBS), is essential to in vitro animal cell cultures, though no standard method exists to assess the potential adverse effects of unknown factors present in the complex composition of FBS in the experiments [16]. Bearing in mind the issues related to the employment of FBS, we first assessed whether reducing the FBS concentration from 10% to 1% could affect HK-2, Caki-1, and RCC-Shaw cell viability. The Alamar Blue assay was performed after 24 and 48 h of exposure of cells to 1% FBS-supplemented cell culture medium, and the results showed, in all the three considered cell lines, no significant alterations in cells' metabolic activity or viability, compared to the control condition, which is represented by the standard FBS concentration of 10% *v/v* (Figure 1). Considering the minimal impact on cell viability observed with 1% FBS, all experiments were conducted under serum-reduced conditions to explore the potential bioactivity of dromedary urine on human renal cells in a low-interference experimental condition. The Alamar Blue assay highlights no significant alteration in cell viability in either non-tumor HK-2 cells or the tumor Caki-1 and tumor RCC-Shaw cell lines.

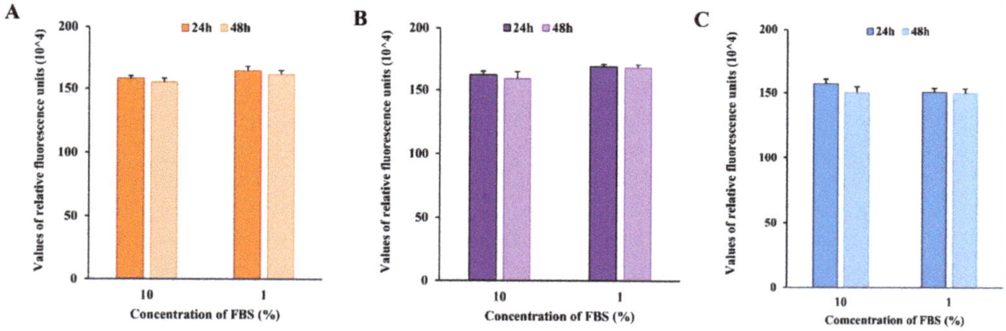

Figure 1. Alamar Blue assay after exposure to serum-reduced cell culture medium. Human renal cells (**A**) HK-2, (**B**) Caki-1, (**C**) RCC-Shaw were exposed to a 10% FBS-supplemented cell culture medium (control) and to a 1% FBS-supplemented medium (serum-reduced condition) for 24 and 48 h. No significant impact on renal cell viability was observed in serum-reduced conditions compared to the control. Cell viability was measured using the values of relative fluorescence units (RFU). The statistical significance of the RFU of the treated cells compared to the control ones was assessed using the Mann–Whitney test (*n* for each condition is included within 25 and 33 observations from at least 3 independent experiments).

2.2. Impact of Hyperosmotic Stress on Renal Cell Viability in Low-Serum Conditions

Considering the strong water reabsorption capacity of dromedary kidneys [21] and the highly concentrated urine that this species can excrete [13], the osmolarity of the tested dromedary urine samples was measured before cell exposure. We found great variability in the urine concentration, with values between 424 and 1976 mOsm/L, as reported in Table 1.

Table 1. Measurement of dromedary urine samples' osmolarity (mOsm/L) through a vapor pressure osmometer.

Dromedary ID	Sex, Age (years)	Urine Osmolarity (mOsm/L)
01MA	Male, 15	1112
02MC	Male, 13	1648
03FD	Female, 36	1000
04MF	Male, 2.5	1725
05FG	Female, 12.5	541
06FG	Female, 6	784
07FG	Female, 5	691
08FK	Female, 1.5	1095
09FL	Female, 12	1653
10MM	Male, 30	1727
11MM	Male, 14	424
12MP	Male, 34	1417
13FP	Female, 5	1917
14FS	Female, 20	1465
15MS	Male, 31	1513
16MS	Male, 5	1976
17MT	Male, 21	1915

The in vitro exposure of cells to hyperosmotic stress induces a cascade of cellular responses. These include cell shrinkage, oxidative stress, protein carbonylation, mitochondrial depolarization, DNA damage, and cell cycle arrest, consequently impeding cell proliferation and enhancing cell susceptibility to apoptosis [22–25]. In order to exclude the possible impact of urine hyperosmolarity on the viability of the investigated cell lines, a series of cell viability experiments were carried out to identify the tolerance threshold of the non-tumor renal cell line HK-2 and the tumor renal cells Caki-1 and RCC-Shaw when exposed to solutions with osmolarities exceeding the physiological value of about 290 mOsm/L in serum-reduced conditions. To simulate hyperosmolarity, D-(-)-mannitol was used as an inert, and cell-impermeable osmolyte [26] was dissolved at various concentrations in the 1% FBS-supplemented cell culture medium. In this way, the complete cell culture medium osmolarity was increased from the physiological value of 290 mOsm/L up to 400, 450, 500, 600, and 700 mOsm/L.

HK-2, Caki-1, and RCC-Shaw cells were seeded in 96 multi-well plates to undergo the Alamar Blue cell viability assay after 24 and 48 h of exposure to D-(-)-mannitol-composed hyperosmotic solutions and to the control condition, represented by the 1% FBS cell culture medium. A statistically significant reduction in cell viability was observed, compared to the control condition, in all three renal cell lines for the 500, 600, and 700 mOsm/L solutions after both 24 and 48 h of treatment (Figure 2). Notably, after 48 h of incubation at the highest concentration (700 mOsm/L), cell viability was approximately 40% in HK2 and Caki-1 cells and 23% in RCC cells (Figure 2). The observed results suggest a value of 450 mOsm/L as the hyperosmolarity tolerance threshold for the investigated cellular models. Therefore, hyperosmolar dromedary urine samples (as reported in Table 1) were opportunely diluted to make them remain under this threshold.

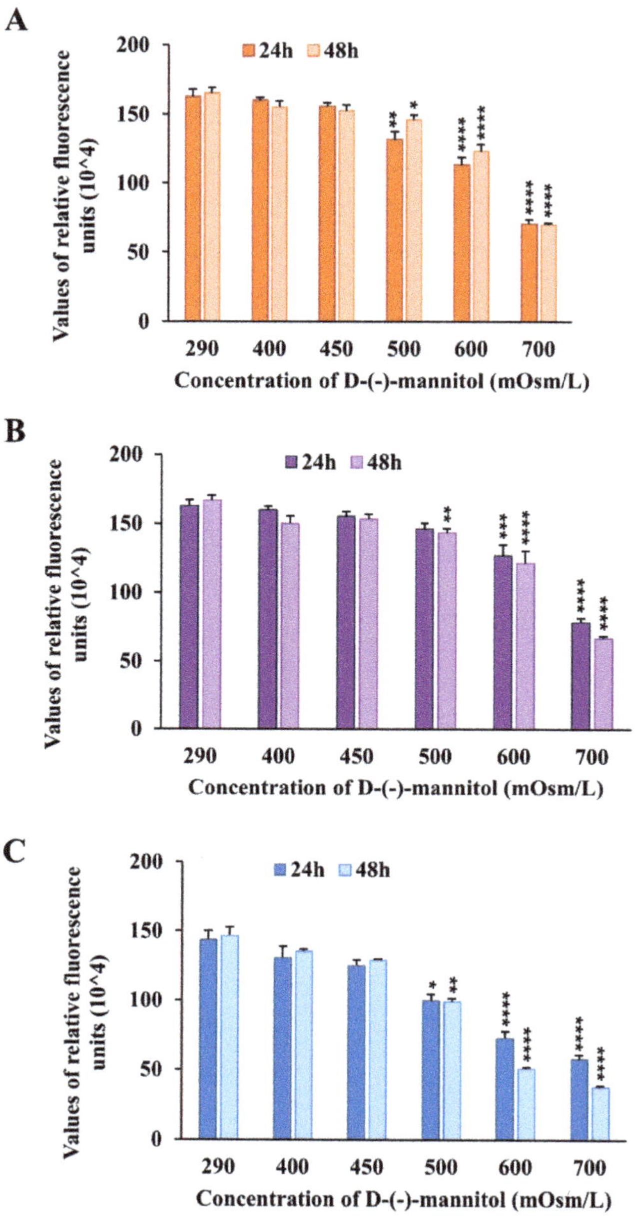

Figure 2. Alamar Blue assay after treatment with a D-(-)-mannitol-composed medium. Human renal cells (**A**) HK-2, (**B**) Caki-1, and (**C**) RCC-Shaw were exposed to 1% FBS-supplemented hyperosmolar solutions for 24 and 48 h. All three considered cell lines exhibited no significant decline in cell viability decline, compared to the control condition, up to 450 mOsm/L, thus identifying this value as the hyperosmolarity tolerance threshold. Cell viability was measured using the values of relative fluorescence units (RFU). The statistical significance of the RFU of the treated cells compared to the control ones was assessed through the Mann–Whitney test (n for each condition is included within 9 and 15 observations, from at least 3 independent experiments; * $p < 0.05$, ** $p < 0.01$, *** $p < 0.001$, **** $p < 0.0001$).

2.3. Impact of Dromedary Urine Solutions on Renal Cell Viability

Based on the aforementioned results, all dromedary urine samples were diluted in a 1% FBS-supplemented cell culture medium to reduce the osmolarity values below the identified tolerance threshold of 450 mOsm/L. Hence, the cell lines HK-2, Caki-1, and RCC-Shaw were seeded in 96 multi-well plates for exposure to dromedary urine solutions for 24 and 48 h, and the Alamar Blue assay was performed at each time point. The results from the viability assay performed on the non-tumor HK-2 cells, as summarized in Figure 3 (and detailed in Table S1), indicated that 8 of the 17 tested urine samples exhibited a statistically significant reduction in cell viability at 24 and 48 h post-treatment. Some of the bioactive urine samples, i.e., 03FD, 14FS, and 17MT, induced a viable decline of roughly 10%, while the strongest impact on HK-2 was observed for the sample 12MP, which reduced the non-tumor cell viability by up to 25% at 48 h (Figure 3, Table S1).

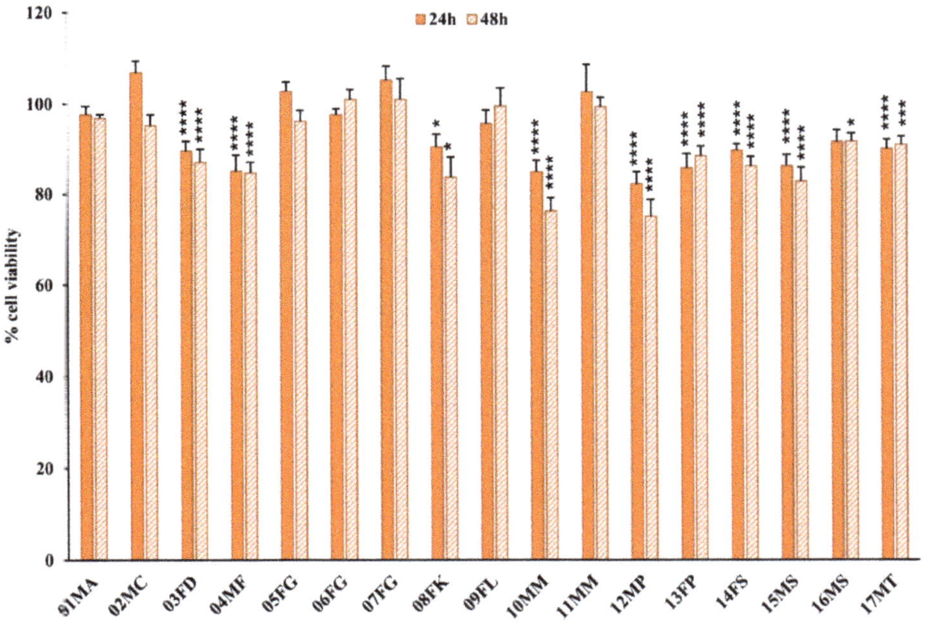

Figure 3. Cell viability assay on non-tumor HK-2 cells after treatment with dromedary urine samples. The non-tumor cell line after 24 and 48 h of exposure to dromedary urine exhibited a statistically significant viable decline in 8 out of the 17 tested samples. Data are reported as the percentage of the mean and the standard error of the mean (SEM) compared to the respective control conditions (non-exposed cells), and samples were labeled as (i) inactive when $p > 0.01$, (ii) mildly bioactive when $0.01 < p < 0.0001$, and (iii) bioactive when $p < 0.0001$ (Mann–Whitney test, XLSTAT, $n = 9$–15; * $p < 0.05$, *** $p < 0.001$, **** $p < 0.0001$).

In the case of Caki-1 tumor renal cells, the Alamar Blue assay revealed a statistically significant reduction in cell viability following 24 h of exposure to all the urine samples, except for 07FG and 11MM, as reported in Figure 4 (and detailed in Table S2). However, extending the incubation of Caki-1 cells up to 48 h with the same urine solutions, a significant decrease in cell viability persisted only for 8 of the 16 bioactive urine solutions, i.e., from 04MF, 05FG, 08FK, 09FL, 10MM, 12MP, 14FS, and 15MS dromedaries, while 01MA, 02MC, 03FD, 06FG, 13FP, 16MS, and 17MT lost their effect (Figure 4, Table S2).

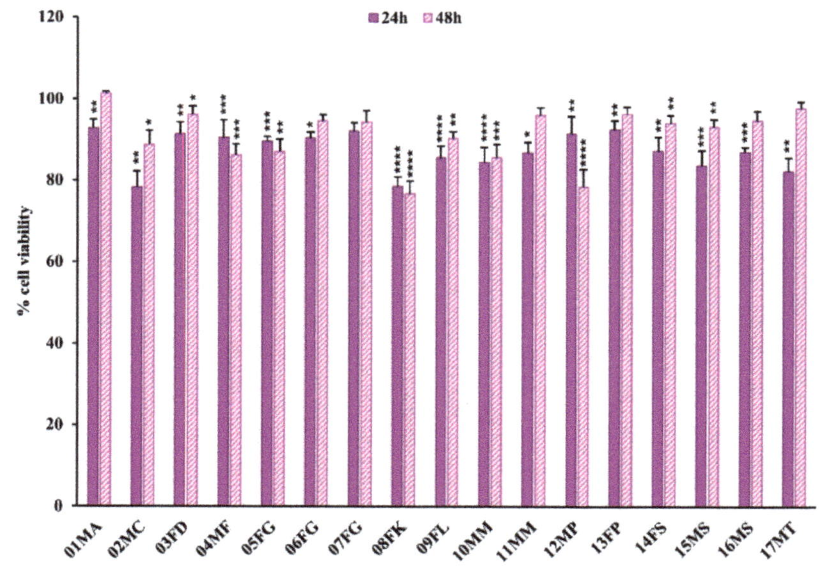

Figure 4. Cell viability assay on tumor Caki-1 cells after treatment with dromedary urine samples. The tumor renal cells Caki-1 showed a weak but statistically significant decline in viability in almost all of the tested urine samples after 24 h of incubation. The observed cytotoxic effect remained only for 8 of the bioactive urine samples after 48 h of exposure. Data are reported as the percentage of the mean and the standard error of the mean (SEM) compared to the respective control conditions (non-exposed cells), and samples were labeled as (i) inactive when $p > 0.01$, (ii) mildly bioactive when $0.01 < p < 0.0001$, and (iii) bioactive when $p < 0.0001$ (Mann–Whitney test, XLSTAT, n = 9–18; * $p < 0.05$, ** $p < 0.01$, *** $p < 0.001$, **** $p < 0.0001$).

When the same experiments were performed on RCC-Shaw cells, a primary tumor-derived renal cell line, a remarkable reduction in viability was observed. In particular, after 24 h of incubation with 13 out of the 17 tested dromedary urine solutions, a significant cytotoxic effect was detected, with a decrease in cell viability of up to 79%, as observed for the urine solution from the 10MM dromedary compared to the control condition (Figure 5, Table S3). At 48 h of incubation with the same urine solutions, an even greater decline in viability was observed compared to the first time point. All the tested urine solutions exhibited a strong and statistically significant cytotoxic effect, with noticeably low residual viability, i.e., 14.5%, 12.6%, 12.7%, and 15.5% for 04MF, 10MM, 12MP, and 17MT, respectively, as shown in Figure 5 (and detailed in Table S3).

The observed cytotoxic effects of urine solutions on RCC-Shaw cells were corroborated by optical microscopy observations. Figure 6 depicts the brightfield microscopy analysis of the treated and control tumor renal cells after 48 h of incubation with dromedary urine solutions. The samples exhibiting cytotoxic effects through the Alamar Blue assay demonstrated a significant reduction in cellular density when microscopically examined compared to the control condition. For all the tested urine samples, a decrease in cell number was readily apparent with 100× magnification compared to the control condition (CTR, in the upper left panel). Moreover, alterations in the RCC-Shaw morphology after exposure to dromedary urine were observed: the polygonal RCC-Shaw cells of the control condition (CTR) became round-shaped (as shown for 04MF, 10MM samples), with the appearance of spikes (as can be seen in 03FD, 09FL, and 13FP), extroflections (14FS, 15MS), and debris (12MP, 17MT), suggesting the occurrence of cell death processes after exposure to dromedary urine.

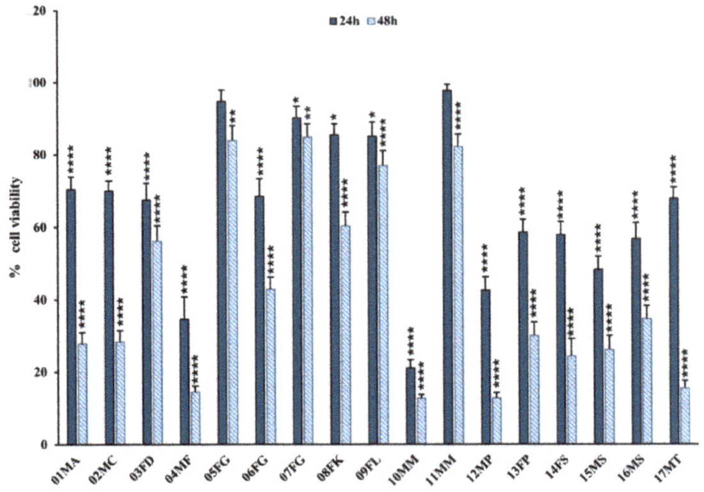

Figure 5. Cell viability assay on tumor RCC-Shaw cells after treatment with dromedary urine samples. Dromedary urine drastically impacted tumor renal cell viability; thus, all the tested samples induced a strong and statistically significant decrease in cell viability after 48 h of incubation. In particular, urine samples from 04MF, 10MM, 12MP, and 17MT demonstrated a residual cell viability of less than 20%. Data are reported as the percentage of the mean and the standard error of the mean (SEM) compared to the respective control conditions (non-exposed cells), and samples were labeled as (i) inactive when $p > 0.01$, (ii) mildly bioactive when $0.01 < p < 0.0001$, and (iii) bioactive when $p < 0.0001$ (Mann–Whitney test, XLSTAT, $n = 9$–12; * $p < 0.05$, ** $p < 0.01$, **** $p < 0.0001$).

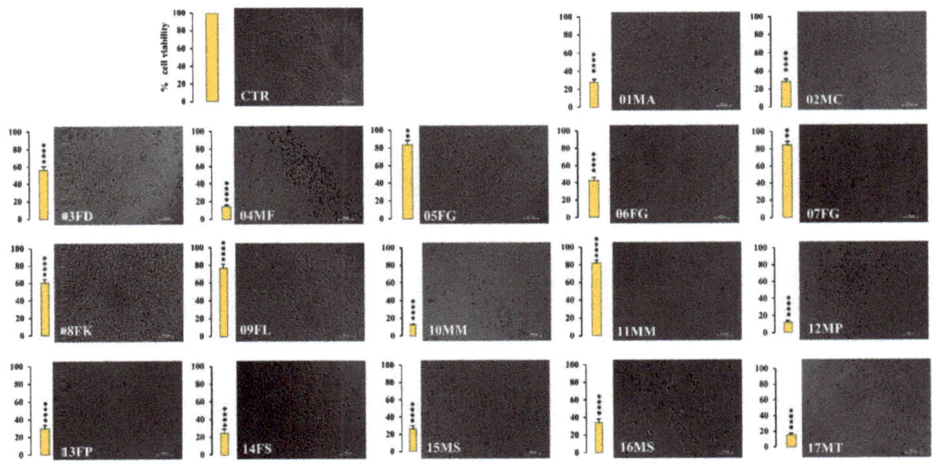

Figure 6. Optical microscopy analysis of RCC-Shaw cells after cell viability assessment. After the Alamar Blue viability assay, RCC-Shaw tumor renal cells, seeded in 96-multi-well plates, were observed under an inverted microscope with 100× magnification, and brightfield images of the replicates were acquired. Here, a representative field of each condition is reported. For all the tested urine samples, a decrease in cell number was readily apparent compared to the control condition (CTR). Moreover, the polygonal tumor renal cells of the CTR became round-shaped, with the appearance of spikes (03FD, 09FL, and 13FP), extroflections (14FS, 15MS), and debris (12MP, 17MT), suggesting the occurrence of cell death processes. In each graph, data are reported as the percentage of the mean and the standard error of the mean (SEM) compared to the respective control conditions (non-exposed cells) assessed through the Mann–Whitney test, XLSTAT ($n = 9$–12; ** $p < 0.01$, **** $p < 0.0001$). Scale bar: 100 μm.

2.4. Dromedary Urine Extracts Induce Apoptosis in Tumor Renal Cells

In order to further investigate the molecular pathway underlying the observed death of RCC-Shaw tumor renal cells after exposure to dromedary urine, Annexin V-FITC/Propidium iodide (PI) double staining was performed. Briefly, RCC-Shaw cells were seeded in black 96 multi-well plates and exposed to dromedary urine solutions. After 24 h of treatment, cells underwent the Annexin V-FITC/PI assay and were analyzed through fluorescence microscopy. Figure 7 reports representative brightfield images and their respective green (Annexin V-FITC) and red (PI) fluorescence fields from the control condition, as well as from cells exposed to urine from 04MF and 12MP dromedaries, which exhibited a strong cytotoxic effect on RCC-Shaw cells according to the Alamar Blue assay, and from 05FG and 11MM, the latter of which only slightly affected cell viability. As shown in Figure 7, cells in the control (CTR) panels displayed negative results for both Annexin V-FITC and PI staining, indicating almost no apoptosis or necrosis, and similar results could be observed for cells exposed to urine from 05FG and 11MM dromedaries, which showed a weak Annexin V stain. However, after exposure to urine solutions from 04MF and 12MP animals for 24 h, cells exhibited strong green and red fluorescence, suggesting that apoptosis and necrosis are induced in RCC-Shaw cells. Therefore, the results from the Annexin V-FITC and PI double staining are in line with the cell viability outcomes. Cells in the control (CTR) panels display negative results for both Annexin V-FITC and PI staining, indicating almost no apoptosis or necrosis, and similar results can be observed for cells exposed to urine from 05FG and 1MM dromedaries (Figure 7). After exposure to urine solutions from 04MF and 12MP dromedaries for 24 h, cells exhibited strong green and red fluorescence, suggesting that apoptosis and necrosis were induced in RCC-Shaw cells (Figure 7).

2.5. NMR Metabolic Characterization of Dromedary Urine Samples

The chemical composition of the urine samples under investigation was defined through the study of the corresponding 1D ^{1}H NOESY spectra (see Table 2). Such identification was made possible by performing a comparison with reference compounds and homo- and heteronuclear 2D NMR experiments. Many organic acids were detected, including acetic, lactic, pyruvic, azelaic, valeric, malonic, citric, and formic acids. Furthermore, typical derivatives of benzoic acid were found, such as hippuric acid and 4-hydroxy hippuric acid (HHA). Allantoin, most likely derived from the metabolism of uric acid, was well represented in all the samples under investigation. Hypoxanthine derivatives derived from the metabolism of purine were identified. Creatinine was detected in all samples, most likely derived from the metabolism of proteins. A few amino acid derivatives were identified, including carnitine, glycine, phenylacetylglycine, and valine. Furthermore, trimethylamine-N-oxide (TMAO) was contained in the samples under investigation as a result of potential oxidized products of quaternary ammonium salts, such as choline, betaine, and carnitine.

Once the pool of the main metabolites was defined in the urine samples, a multivariate statistical analysis was attempted to determine the main variations in the metabolic compositions correlated with the observed bioactivity in the in vitro tests. As mentioned above, the urine samples were assigned to classes of active, mildly active, and inactive samples based on the calculated p value following the Mann–Whitney test. Specifically, the samples were considered active when $p < 0.0001$, mildly active when $0.0001 < p < 0.01$, and inactive when $p > 0.01$. The spectroscopic data collected from the NMR measurements of the urine samples were subjected to a statistical investigation to unveil specific differences in the metabolic composition within the group of bioactive samples (red scores in Figure 8) vs. the mildly active (blue scores in Figure 8), and inactive samples (green scores in Figure 8). Three separate supervised analyses (PLS-DA) were conducted on the same spectroscopic data, and the bioactivity of different urine samples towards the three cell lines was considered to indicate classes of interest, both at 24 h and 48 h, i.e., for HK-2

(Figure 8A,B), Caki-1 (Figure 8C,D), and RCC-Shawn cells (Figure 8E,F). The models were validated through cross-validation to exclude any overfitting, and the coefficients were computed considering the contribution of five components. In all cases under investigation, the samples were grouped according to the class belonging to Component 1. In the specific case of the treatment of Caki-1 cells with the urine solutions, those samples that were defined as mildly active based on the in vitro experiments clustered closer to the inactive samples than the active ones, regardless of the duration of the treatment (Figure 8C,D).

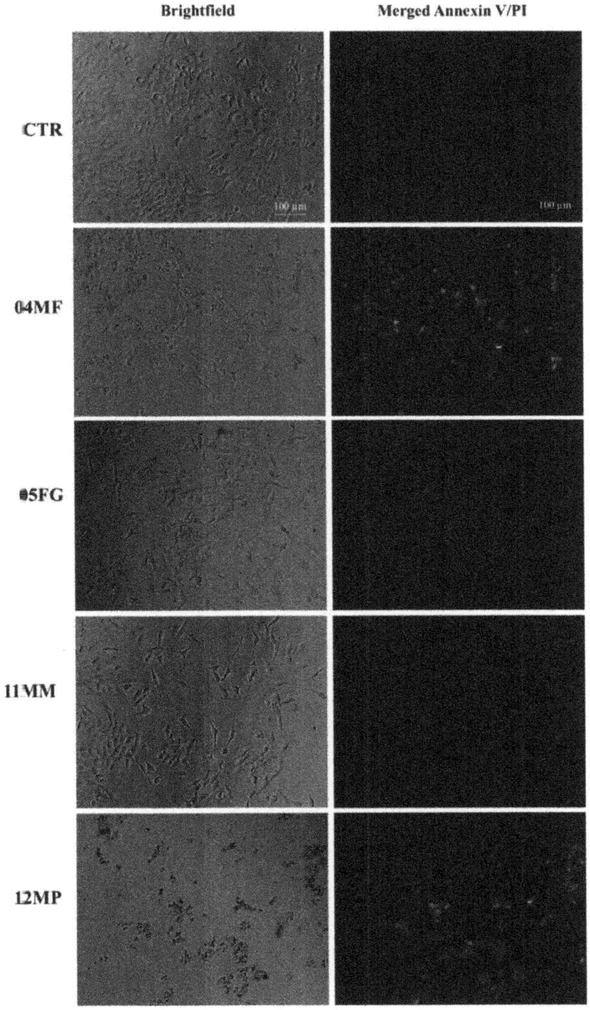

Figure 7. Annexin V-FITC/Propidium iodide double staining on RCC-Shaw cells after 24 h of treatment with dromedary urine solutions. Representative pictures of the brightfield images (on the left) and their respective green (Annexin V-FITC) and red (PI) fluorescence fields from the control condition (CTR), for cells exposed to urine from 04MF and 12MP dromedaries, and for cells exposed to 05FG and 11MM (on the right). Cells in the control (CTR) panels display negative results for both Annexin V-FITC and PI staining, and similar results can be observed for cells exposed to urine from 05FG and 1MM dromedaries. After exposure to urine solutions from 04MF and 12MP dromedaries for 24 h, cells exhibited strong green and red fluorescence, suggesting that apoptosis and necrosis are induced in RCC-Shaw cells. Scale bar: 100 μm.

Table 2. List of the main metabolites detected in the aqueous extracts of the analyzed urine samples.

Metabolite	δ (ppm) and Multiplicity Details [a]
Valine	0.96 (d, 7.0 Hz); 1.07 (d, 7.0 Hz)
Lactic ac.	1.34 (d, 6.9 Hz); 4.14 (q)
Acetic ac.	1.93 (s)
Pyruvic ac.	2.34 (s)
Azelaic ac.	1.29 (m); 1.54 (m); 2.16 (t, 3.2 Hz)
Valeric ac.	0.9 (t, overlap); 1.31 (m); 1.53 (m); 2.19 (t, overlap))
Creatinine	3.02 (s); 4.01 (s)
Malonic ac.	3.15 (s)
Carnitine	3.19 (s)
Trimethylamine *N*-oxide (TMAO)	3.26 (s)
Glycine	3.56 (s)
Allantoin	5.37 (s)
Citric ac.	2.55 (d, 15.8 Hz); 2.70 (d, 15.7 Hz)
Phenylacetyl-glycine	3.67 (s); 3.74 (s); 7.36 (m); 7.42 (m)
Hippuric acid	7.55 (m); 7.63 (m); 7.83 (m); 3.96 (s)
4-Hydroxyhippuric ac.	6.86 (d, 8.5 Hz); 7.17 (d, 8.5 Hz)
Benzoic acid	7.88 (m); 7.56 (m); 7.49 (m)
Hypoxanthine	8.14 (s); 8.16(s)
Formic ac.	8.47 (s)

[a] The resonance frequencies of the signals are indicated as delta (δ), and the units are given in part per million (ppm); multiplicity patterns are described as follows: (s) singlet, (d) doublet, (t) triplet, (q) quartet, (m) multiplet, and (dd) doublet of doublets.

The analysis of the Variable Importance in Projection (VIP) was carried out to identify the pool of metabolites that more heavily changed from one class to the other following the PLS-DA performed and considering the treatment with urine solutions of the three cell lines at the two time points, i.e., 24 h and 48 h (Figure 9A–F).

The urine solutions that were active for HK-2 showed a relatively higher content of both phenylacetylglycine and creatinine, both at 24 h and 48 h of treatment. In addition, at this latter time point, a relatively higher content of hypoxanthine and pyruvic acid was also observed (Figure 9A,B). On the other side, azelaic acid and TMAO were contained in relatively higher amounts in the urine solutions that were active on Caki-1 cells, regardless of the treatment duration. In the case of RCC Shawn cells, the pool of metabolites contained in the urine solutions that were active at 24 h was more similar to that found for the urine solutions active towards HK-2 cells, with higher contents of phenylacetylglycine and creatinine. Conversely, after a longer treatment time (48 h), the following metabolites were found to be more abundant: organic acids (azelaic, valeric, hippuric, citric, and malonic acid), phenylacetylglycine, and TMAO.

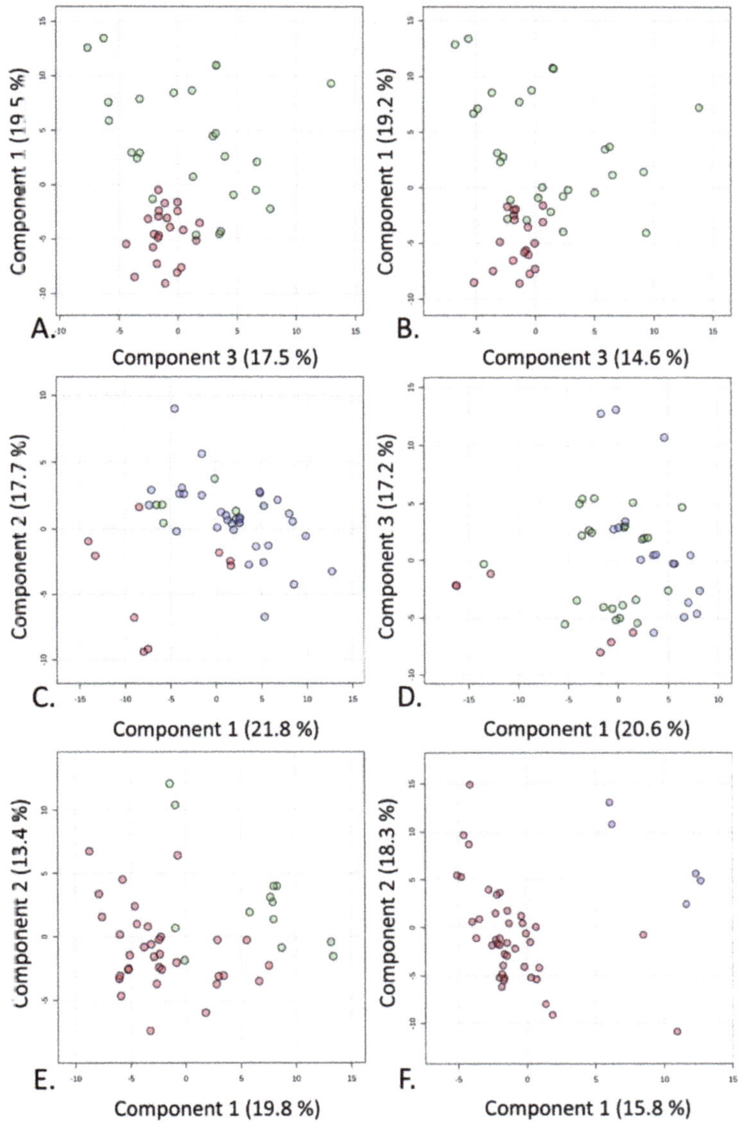

Figure 8. PLS-DA applied to the 46 spectra using UV-scaled 0.04 ppm-sized bucketing. The urine samples were assigned to the class of active ($p < 0.0001$), mildly active ($0.0001 < p < 0.01$), and inactive ($p > 0.01$) samples based on the calculated p value following the Mann–Whitney test. Score plots are shown between the selected components where the observations are indicated according to in vitro activity: "•" active, "•" mildly active, and "•" inactive. (**A**) HK-2 at 24 h. (**B**) HK-2 at 48 h. (**C**) Caki-1 at 24 h. (**D**) Caki-1 at 48 h. (**E**) RCC-Shawn at 24 h. (**F**) RCC-Shawn at 48 h.

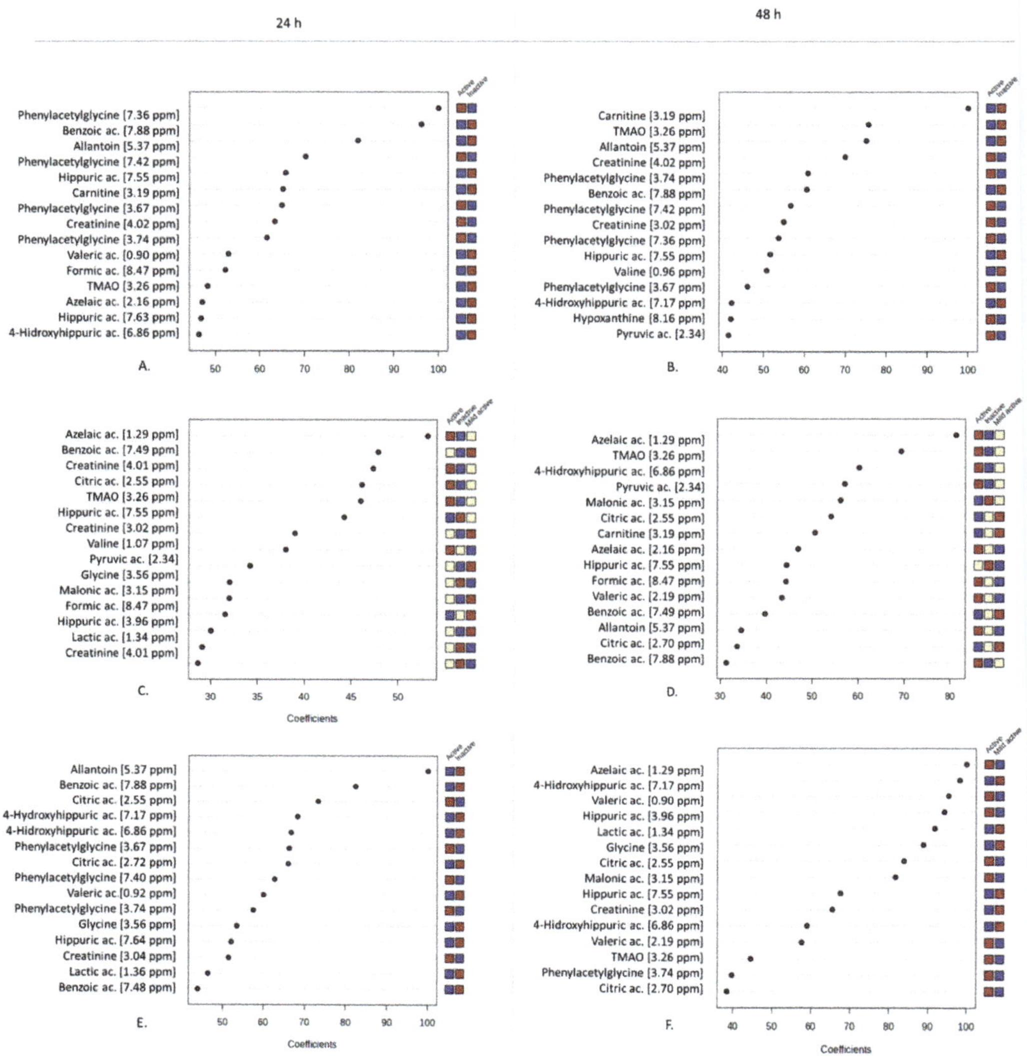

Figure 9. Analysis of the Variable Importance in Projection (VIP) identified by PLS-DA applied to the 46 spectra using UV-scaled 0.04 ppm-sized bucketing. The urine samples were assigned to classes of active ($p < 0.0001$), mildly active ($0.0001 < p < 0.01$), and inactive ($p > 0.01$) samples based on the calculated p value following the Mann–Whitney test. The colored boxes on the right indicate the relative concentrations of the corresponding metabolite in each group under study (■ high, intermediate, and ■ low). (**A**). HK-2 at 24 h. (**B**). HK-2 at 48 h. (**C**). Caki-1 at 24 h. (**D**). Caki-1 at 48 h. (**E**). RCC-Shawn at 24 h. (**F**). RCC-Shawn at 48 h.

3. Discussion

Natural matrices have historically been a valuable driver for drug discovery, even though natural product-based research is associated with some intrinsic issues related to the matrices' properties and to the assays employed for the bioactivity assessment [14,27]. The present study aimed to evaluate the antiproliferative effects of dromedary urine, optimizing the experimental conditions to highlight their real impact on human renal cell lines and disentangling the contribution of hyperosmolarity to cytotoxicity [10,13]. Indeed, the tested dromedary urine samples showed a wide range of osmolarity values, all of which exceeded the physiological threshold of 290 mOsm/L. These observations suggest differences in urine concentration, possibly related to the interindividual variability of the sampled animals (sex, age, physiological status, and different levels of water consumption). In this context, the hyperosmolarity of urine samples should be considered a critical factor itself, which can alter bioactivity assay results due to its potential cytotoxicity, thus affecting the viability of cells not used to the hyperosmolar environment, such as proximal tubular epithelial cells [10].

Special focus was given to the role of fetal bovine serum (FBS), which is commonly employed as a supplement in animal cell cultures due to its essential nutrient content, supporting cell growth and proliferation. However, it is a complex mixture (and a natural matrix itself), and the elements within FBS remain partially unidentified. Their effects on cultured cells are poorly understood, even though they can interfere with in vitro experiments and affect the quality, reproducibility, and reliability of the studies. Indeed, FBS content may affect various cellular processes, including cell proliferation, differentiation, the modulation of molecular and cellular mechanisms, cell viability, and cytokine production [16,17,28,29]. Moreover, during in vitro experiments, the proteins and lipids contained in FBS can interact through reversible chemical bonds with the tested small molecules or drugs, possibly altering their outcomes and concealing the impacts of the tested chemicals on cellular processes [19,20]. Considering these observations, the experiments were carried out under serum-reduced conditions in order to minimize possible interfering factors derived from FBS. First, it was verified that the growth of the investigated cellular models did not suffer from 1% FBS supplementation. Subsequently, the hyperosmolarity tolerance threshold of HK-2, Caki-1, and RCC-Shaw cell lines under this culture condition was established. In a 1% FBS-supplemented cell culture medium, the hyperosmolarity tolerance threshold of 450 mOsm/L was identified for all the considered cell lines. A higher value of the hyperosmolarity tolerance threshold, i.e., 500 mOsm/L, was reported for HK-2 and Caki-1 cell lines in the presence of 10% FBS [10]. The lower amount of serum adopted in the present study could be responsible for the different outcomes, as a higher FBS content (10%) is known to affect in vitro stress responses [16,17,19], including responses in complex mixtures such as dromedary urine. Indeed, when diluted in 1% FBS-supplemented cell culture medium and administered to human renal cells, the urine showed a higher overall toxicity compared to the antiproliferative effect already reported on human renal cells [10], possibly due to the higher availability of small molecules in the serum-reduced condition [17,19,20]. In particular, the number of urine samples significantly affecting the viability of non-tumor HK-2 cells was lower than the samples that resulted in bioactivity towards the two tumor renal cell lines. Such evidence is in line with previous studies showing the different behavior of non-tumor vs. tumor cell lines [10,30–32]. The differential effects of urine samples on the tested cell lines may be attributed to the metabolite composition of the investigated urine. Each compound, whether alone or in combination with other molecules in the sample, could selectively affect cancer cells, which depend more strongly on specific metabolic pathways, such as those involved in energy production. As a result, they are more vulnerable to metabolic disruptions. In contrast, non-cancerous cells have greater metabolic flexibility and can adapt to certain metabolic changes, making them more resistant to specific alterations [8,33–36]. Interestingly, Caki-1 and RCC-Shaw exhibited opposite responses to dromedary urine exposure. Indeed, the Caki-1 cell line, derived from skin metastasis, initially showed a weak but statistically significant decline in viability for almost all the tested samples, but after 48 h, 6 of the 16 bioactive samples lost their effect on tumor renal cells, which appeared to adapt to the urine chemical composition and proliferate, in line with the ability of tumor cells to survive in challenging conditions [35,37]. On the other hand, all the tested dromedary urine samples were able to induce a strong and time-dependent decline in viability in the primary tumor-derived RCC-Shaw cells by roughly 85% of the cell toxicity caused by 04MF, 10MM, and 12MP samples, all derived from male animals, after 48 h of exposure. As demonstrated by the Annexin/PI staining profile of the tumor renal cells, the observed toxicity was mediated by apoptosis activation.

The 1D ^{1}H NOESY analysis of urine samples revealed the presence of various organic acids, including pyruvic, acetic, lactic, malonic, formic, citric, hippuric, benzoic acids, valeric, azelaic, and 4-hydroxyhippuric acids. Also, valine, glycine, phenylacetylglycine, trimethylamine N-oxide (TMAO), carnitine, creatinine, allantoin, glycine, and hypoxanthine were found in the urine solutions. Following the chemometric analysis, relatively higher amounts of azelaic acid were found in urine samples that exhibited bioactivity against Caki-1 and RCC-Shaw cells but not against HK-2 cells. This observation can be explained by the relationship between azelaic acid and the NF-kB pathway. Azelaic acid (AA) is a naturally occurring nine-carbon dicarboxylic acid known for its antimicrobial and anticancer properties, particularly against various melanoma [38,39] and leukemia [40,41] cell lines. Additionally, it has anti-inflammatory [38] and radical oxygen species (ROS) scavenging activities [42]. Recent studies suggest that azelaic acid exerts its anti-inflammatory effect through binding to cyclooxygenase enzymes and its anticancer activity by targeting pirin, an iron-containing nuclear protein and transcription cofactor highly expressed in tumor cells [38,43], which is involved in the regulation of the NF-kB pathway [44]. A common gene mutation in clear cell renal carcinoma is the loss of function of the von Hippel-Lindau (VHL) gene, which is associated with NF-kB overexpression [45]. The inhibitory effect of azelaic acid on the NF-kB pathway might hinder the survival of RCC-Shaw and Caki-1 cells, thus explaining the observed results. In this context, the lower impact of dromedary urine on Caki-1 cells could be attributed to the fact that this cell line expresses a wild-type VHL gene [46–48], making it less affected by the activity of azelaic acid. Future research could explore AA efficacy as an adjuvant therapy in oncology, particularly in combination with existing treatments to enhance cancer cell sensitivity or reduce inflammatory side effects.

Among the other mentioned molecules highlighted by chemometric analysis, it was observed that relatively higher amounts of phenylacetyl glycine (PAGly) in urine samples might be responsible for the cytotoxicity against RCC-Shaw cells, but not against Caki-1 cells, despite affecting HK-2 renal cells. Phenylacetyl glycine (PAGly) is a phenylalanine derivative primarily synthesized by the gut microbiota [49,50]. Recently, PAGly has been proposed to target β2-adrenergic receptors (β2-AR), offering potential benefits for cerebral ischemia/reperfusion injury through the inhibition of receptor signaling [50]. Although the physiological role of β2-adrenergic receptors in the kidney remains underexplored, their protein and RNA expressions have been reported in various kidney subunits, including proximal tubules, glomeruli, and podocytes [51]. Moreover, blocking β2-adrenergic receptors has shown therapeutic benefits in clear cell renal cell carcinoma (ccRCC) with the VHL gene's loss of function by reducing inflammation and oxidative stress [52,53], thus possibly explaining the observed effects on RCC-Shaw and Caki-1 cell lines. However, β2-AR antagonism has also been reported to increase renal production of proinflammatory cytokines, possibly hampering HK-2 proximal tubular cells [54,55]. Future studies should investigate the precise mechanism of action of PAGly in renal cancer models and assess its potential as a metabolic biomarker for therapeutic response prediction.

Nonetheless, it should be noted that the studied cellular models were exposed to a complex mixture of compounds, each contributing to the overall effect through either synergistic or antagonistic interactions, thereby complicating the interpretation of the results. In this regard, in the broader context of natural product research, the inherent variability in the composition of the tested natural matrices, such as urine in our assays, presents a significant challenge in ensuring the reproducibility of results. Factors such as individual metabolic differences, dietary habits, and the health status of animals can lead to fluctuations in metabolite profiles, potentially affecting the consistency of observed biological effects.

4. Materials and Methods

4.1. Urine Collection and Processing for In Vitro Assays

All the animals included in this study were reared on the same farm located in the Doñana National Park (southwestern Spain) and fed on the same diet, consisting of alfalfa, beet pulp, calcium carbonate, salt, and selenium. Urine sampling was conducted as previously described [10,56]. Briefly, for each animal, a sterile, single-use plastic bag was placed in a collection cone holder to recover mid-stream, first-morning-void urine during natural urination. Samples were collected from 17 dromedary camels (as detailed in Table 1, Section 2.2.) on a single day in November 2022. Immediately after collection, urine was transferred to 15 mL tubes and centrifuged at 2500× *g* for 5 min at 4 °C. The supernatant was filtered using a 0.22 µm PES syringe filter (Primo®, Euroclone S.p.A., Milan, Italy) and aliquoted. All aliquots were frozen at −20 °C and subsequently freeze-dried overnight at −50 °C in a vacuum below 0.5 mbar using a 5-pascal freeze drier (Speedvac SPD 140, Thermo Fisher Scientific, Milan, Italy). Lyophilized urine samples were reformed in ultrapure water at their initial volume immediately before the in vitro bioactivity experiments were conducted.

4.2 Assessment of Osmolarity in Dromedary Urine Samples

The osmolarity of dromedary urine samples was measured by a VAPRO® vapor pressure osmometer 5600 (Wescor, Inc., Logan, UT, USA) with fresh 290 mmol/kg, 1000 mmol/kg, and 100 mmol/kg standard solutions, following the manufacturer's instructions. Briefly, a single Whatman filter paper disc (Wescor, ss-033) was placed in the central depression of the holder using metal forceps, and 10 µL of the samples, diluted 1:2 with ultrapure water, was applied onto the disc. Saturated discs were promptly transferred to the vapor pressure osmometer sample holder, and osmolarity was determined. All samples were assessed three times, and osmolarity was calculated as the mean of the three determinations.

4.3. Cell Cultures

The human proximal tubule epithelial cell line (HK2) [57] and the human renal carcinoma cell line (Caki-1) [58], which serve as a model system of proximal tubule epithelium, were purchased from the American Type Culture Collection (ATCC, Manassas, VA, USA). RCC-Shaw cells are primary renal carcinoma cell lines derived from primary kidney tissue explants [59]. All three cell lines were cultured in Dulbecco's modified Eagle's medium with high glucose levels (Elabscience Bionovation Inc., Houston, TX, USA, EP-CM-L0032), supplemented with 10% fetal bovine serum (Euroclone S.p.A., Milan, Italy), in 25 cm^2 culture flasks, and maintained at 37 °C and 5% CO_2. They were regularly passaged upon reaching 80% confluency using a 0.25% trypsin-EDTA solution (Elabscience Bionovation Inc., Houston, TX, USA, EP-CM-L0446).

4.4. Preparation of Hyperosmolar Solutions

Different osmolarity conditions of the cell culture medium increased and adjusted with mannitol, were assessed to determine the osmotic range at which no significant decrease in cell viability occurred. D-(-)-mannitol (Honeywell-Riedel-de Haën, Charlotte, North Carolina, USA, cat. no. 33440) was used as a reference osmolyte to investigate the tolerance limits of cultured cells to osmotic stress in serum-reduced conditions. Hyperosmolar solutions were prepared by adding the appropriate amount of D-(-)-mannitol powder to the culture medium supplemented with 1% FBS (whose osmolarity was 290 mOsm/L) to achieve osmolarities of 400, 450, 500, 600, and 700 mOsm/L, respectively. Once the D-(-)-mannitol was dissolved, the solutions were sterilized using 0.22 μm PES syringe filters (Primo®, Euroclone S.p.A., Milan, Italy).

4.5. Preparation of Urine Samples for In Vitro Bioactivity Test

All the freeze-dried urine samples were initially reconstructed in ultrapure water to restore their original volume. Subsequently, they were properly diluted using the 1% FBS-supplemented cell culture medium, which is referred to as a serum-reduced condition as it remained below the 450 mOsm/L threshold (as outlined in Results, Section 3). The obtained urine solutions were sterilized using 0.22 μm PES syringe filters (Primo®, Euroclone S.p.A., Milan, Italy).

4.6. Cell Viability Assay

HK-2, Caki-1, and RCC-Shaw cells were seeded in 96-well plates at a density of 15,000 cells per well in the growth medium, supplemented with 10% FBS and allowed to adhere overnight. Subsequently, the cells were treated with hyperosmolar media adjusted with D-(-)-mannitol or with urine solutions, both prepared in a 1% FBS-supplemented medium for 24 and 48 h at 37° C and 5% CO_2. Cell viability was assessed under control conditions (cells exposed to the 1% FBS-supplemented medium) and experimental conditions using the Alamar Blue assay (Immunological Sciences, Rome, Italy), according to the manufacturer's instructions. This assay relies on the ability of living cells to convert the blue non-fluorescent dye (resazurin) to a red fluorescent dye (resorufin) via mitochondrial reductase. Following a 45 min incubation with resazurin, the fluorescence of resorufin was measured at $\lambda_{Ex/Em}$ 535/590 nm using an FLUOstar® Omega microplate reader (BMG LABTECH GmbH, Ortenberg, Germany). All the experiments were performed in triplicate at least three times.

4.7. Annexin V-FITC/Propidium Iodide Double Staining

The Annexin V-FITC/PI assay can differentiate viable cells (Annexin V− and PI−) from early apoptosis-staged cells (Annexin V+ and PI−) or from late apoptosis (Annexin V+ and PI+) and necrosis (Annexin V− and PI+) stages. The Annexin V-FITC/PI Apoptosis Kit (Elabscience Inc, Houston, TX, USA, E-CK-A211) was performed following the manufacturer's instructions. Briefly, RCC-Shaw cells were seeded in black 96 multi-well plates (Greiner Bio-One S.r.l, Cassina de Pecchi, Italy, cod. 655090) at a density of 10,000 cells/well in the growth medium and allowed to become attached overnight. Afterward, cells were exposed to dromedary urine diluted in a serum-reduced cell culture medium for 24 h and were stained for Annexin V-FITC and PI for 15 min at room temperature. Subsequently, the plates were placed in a fluorescence microscope (TE2000 inverted microscope by Nikon, Turin, Italy) equipped with a Nikon Digital SIGHT camera and the NIS Elements software (version 3.00 SP7; Nikon, Turin, Italy) to capture fluorescence images, with excitation/emission wavelengths of 485/520 nm for Annexin V-FITC and 535/615 nm for PI with 100× magnification.

4.8. Nuclear Magnetic Resonance (NMR) Analysis

Freeze-dried urine samples (150 mg) were reconstituted in 2.0 mL phosphate buffer (100 mM, pH 7.4) containing 0.1 mM TSP-d4. Next, the samples were vortexed at

2500 rpm for 2 min (Advanced Vortex Mixer ZX3, VELP Scientifica Srl, Usmate, Italy). Next, three aliquots of 600 μL each were obtained and transferred into three different NMR tubes to achieve three analytical replicates. The 1D 1H NOESY spectra were recorded using a Bruker Avance I 400 MHz spectrometer equipped with an autosampler and a 5 mm inverse probe (Bruker BioSpin GmbH, Ettlingen, Germany). The 1D 1H NOESY spectra were acquired using the pulse program (noesygppr1d). The following parameters were applied: number of scans: 128; data points: 64 K; spectral width: 8013 Hz; 90° pulse angle: 8.16 μs; acquisition time: 4.09 s; mixing time: 10 ms; and recycle delay: 6 s. Each spectrum was acquired using the Topspin 2.1 software (Bruker BioSpin GmbH) under an automatic process that lasted around 22 min and included sample loading, temperature stabilization for 5 min at 298.2 K, tuning, matching, and shimming. The NMR raw data (Free Induction Decays, FIDs) were processed using the software MestReNova 11.0 (Mestrelab Research SL, Santiago de Compostela, Spain). The FIDs were zero-filled to a 128 K number of points and then underwent Fourier transformation through the application of an exponential multiplication function with a line broadening of 0.1 Hz. The phase and baseline were automatically corrected, and the TSP-d4 singlet signal set at $\delta = 0.00$ ppm was used as a chemical shift reference.

4.9. Chemometric Analysis

The raw data (FIDs) derived from the 1D 1H NOESY measurements were processed using Mestrelab and segmented into regular-sized (0.04 ppm) intervals (buckets) in the range of [10.50, 0.50] ppm. The underlying area of each bucket was calculated and normalized to the total intensity. The areas of the buckets in the region [5.13, 4.69] ppm, corresponding to the residual water signal, were set to 0. The data were stored in a table with one sample per row and one variable (bucket) per column. The data matrices were imported into MetaboAnalyst 5.0, and buckets were subjected to mean-centering and divided by the standard deviation of each variable (Unit Variance scaling). PLS-DA was used as a supervised method that uses multivariate regression techniques to extract information via a linear combination of original variables (X) that could predict the class membership (Y). The PLS regression was performed using the plsr function provided by the R package version 2.1-0 [60]. The classification and cross-validation were performed using the corresponding wrapper function offered by the caret package [61]. The caret package is available online: https://topepo.github.io/caret/ (accessed on 24 November 2024). To assess the significance of class discrimination, a permutation test was performed. In each permutation, a PLS-DA model was built between the data (X) and the permuted class labels (Y) using the optimal number of components determined by cross-validation for the model based on the original class assignment [62]. To estimate the predictive ability of the model, the performance measure Q2 was calculated via cross-validation (CV). In each CV, the predicted data were compared with the original data, and the sum of squared errors was calculated. The prediction error was then summed over all the samples (the Predicted Residual Sum of Squares or PRESS). The variable importance was measured in PLS-DA using the Variable Importance in Projection (VIP), which is a weighted sum of squares of the PLS loadings, taking into account the amount of explained Y-variation in each dimension.

4.10. Statistical Analysis

Data are presented as the means ± standard errors (SEM). The effect of the treatments, i.e., on (i) hyperosmotic media and (ii) urine solutions, was analyzed with XLSTAT software (version 2023.3.0.1415) using the Mann–Whitney test. Differences were considered statistically significant at $p < 0.05$. In particular, statistical significance was assessed compared to the control condition and indicated as follows: * $p < 0.05$, ** $p < 0.01$, *** $p < 0.001$, **** $p < 0.0001$. The number of observations (n) for each condition is indicated in the figure legends.

5. Conclusions

This study underscores the critical importance of determining optimal experimental conditions when evaluating the potential bioactivity of natural matrices in vitro, particularly dromedary urine, which presents specific challenges, such as osmolarity issues. The amount of fetal bovine serum, which can alter the sensitivity of the tested cell lines to the complex mixture of molecules present in urine, plays a pivotal role in the outcomes of cell-based assays, as demonstrated in our cytotoxicity evaluations. Furthermore, it is crucial to correlate in vitro evaluations with a thorough characterization of the chemical composition of the tested matrices. Indeed, the intrinsic variability of urine, which limits the applications of this natural matrix within the field of natural product-related drug discovery, can be reduced by ensuring appropriate experimental conditions, which is essential for enabling more reliable statistical analyses when comparing metabolic profiles. In this context, metabolomics approaches play a pivotal role in highlighting the most promising compounds present in this natural matrix, such as azelaic acid and phenylacetyl glycine, allowing us to propose dromedary urine as a sustainable source of bioactive molecules with potential biomedical applications. Overall, our findings on AA and PAGly, obtained by metabolomics analysis, emphasize the importance of identifying bioactive molecules from natural matrices in order to shape new therapeutic strategies, particularly in the context of precision medicine, where individual metabolic profiles may guide treatment decisions and enhance drug efficacy.

Moreover, repurposing camel urine—a by-product often considered a waste material—as a source of bioactive molecules aligns with the principles of green chemistry and circular economy, offering an innovative avenue for the value-added utilization of natural resources.

Supplementary Materials: The following supporting information can be downloaded at: https://www.mdpi.com/article/10.3390/molecules30040821/s1, Table S1: Cell viability assay on non-tumor HK-2 cells after treatment with dromedary urine samples. Data are reported as the percentage of the mean and the standard error of the mean (SEM) com-pared to the respective control conditions (non-exposed cells) and samples were labeled as (i) inactive when $p > 0.01$, (ii) mild bioactive when $0.01 < p < 0.0001$, (iii) bioactive when $p < 0.0001$ (Mann-Whitney Test, XLSTAT, n = 9–15). Table S2: Cell viability assay on tumor Caki-1 cells after treatment with dromedary urine samples. Data are reported as the percentage of the mean and the standard error of the mean (SEM) compared to the respective control conditions (non-exposed cells) and samples were labeled as (i) inactive when $p > 0.01$, (ii) mild bioactive when $0.01 < p < 0.0001$, (iii) bioactive when $p < 0.0001$ (Mann-Whitney Test, XLSTAT, n = 9–18). Table S3: Cell viability assay on tumor RCC-Shaw cells after treatment with dromedary urine samples. Data are reported as the percentage of the mean and the standard error of the mean (SEM) compared to the respective control conditions (non-exposed cells) and samples were labeled as (i) inactive when $p > 0.01$, (ii) mild bioactive when $0.01 < p < 0.0001$, (iii) bioactive when $p < 0.0001$ (Mann-Whitney Test, XLSTAT, n = 9–12).

Author Contributions: Conceptualization: L.G. and B.M. Data curation: M.N.S., B.M., S.T., F.M., N.S. and A.R. Formal analysis: M.N.S., B.M., S.T., F.M., N.S. and A.R. Funding acquisition: L.G., E.C. and C.L.P.; Investigation: M.N.S., B.M, S.T., M.F., F.M. and N.S.; Methodology: M.N.S., B.M., S.T., V.G., C.I.P., I.F.D. and M.F. Project administration: L.G., E.C., V.G., C.L.P., A.D.G. and M.V.; Resources: C.I.P., L.G., E.C., C.L.P., I.F.D. and M.F. Software: M.N.S., B.M., A.R. and C.L.P. Supervision: L.G., E.C., A.D.G., M.V., V.G. and C.L.P. Validation: C.I.P., M.N.S., B.M. and N.S.; Visualization: M.N.S., B.M., F.M., N.S. and A.R. Writing—Original Draft: M.N.S., B.M., N.S., C.I.P., L.G., E.C. and C.L.P.

Writing—Review and Editing: M.N.S., B.M., N.S., A.R., L.G., I.F.D., E.C., C.L.P., V.G., M.F., A.D.G. and M.V. All authors have read and agreed to the published version of the manuscript.

Funding: This research did not receive any dedicated grant. However, the study was carried out in the framework of the EU-funded project CAMEL-SHIELD (Camel breeding systems: actors in the sustainable economic development of the northern Sahara territories through innovative strategies for natural resource management and marketing), which assured the availability of the instrumental and biological resources necessary for this work. This work was also supported by CICECO—Aveiro Institute of Materials (UIDB/50011/2020, DOI 10.54499/UIDB/50011/2020; UIDP/50011/2020, DOI 10.54499/UIDP/50011/2020 and LA/P/0006/2020, DOI 10.54499/LA/P/0006/2020), and LAQV-REQUIMTE (LA/P/0008/202, DOI 10.54499/LA/P/0008/2020; UIDP/50006/2020, DOI 10.54499/UIDP/50006/2020; UIDB/50006/2020, DOI10.54499/UIDB/50006/2020), financed by national funds through FCT/MCTES.

Institutional Review Board Statement: Not applicable.

Informed Consent Statement: Not applicable.

Data Availability Statement: The data presented in this study are available upon request from the corresponding author.

Acknowledgments: The authors would like to thank the personnel of the Aires Africanos farm (https://airesafricanos.com/ (accessed on 30 January 2025)) located in the Doñana National Park (southwestern Spain) for their kind collaboration during urine collection.

Conflicts of Interest: Author Vito Gallo of the Polytechnic University of Bari contributed to the foundation of the company Innovative Solutions S.r.l. The remaining authors declare that the research was conducted in the absence of any commercial or financial relationships that could be construed as a potential conflict of interest. The authors declare that they have no known competing financial interests or personal relationships that could have appeared to influence the work reported in this paper.

References

Mathur, S.; Hoskins, C. Drug development: Lessons from nature (Review). *Biomed. Rep.* **2017**, *6*, 612–614. [CrossRef] [PubMed]

Harvey, A.L.; Edrada-Ebel, R.; Quinn, R.J. The re-emergence of natural products for drug discovery in the genomics era. *Nat. Rev. Drug Discov.* **2015**, *14*, 111–129. [CrossRef]

Iglesias Pastrana, C.; Delgado Bermejo, J.V.; Sgobba, M.N.; Navas Gonzalez, F.J.; Guerra, L.; Pinto, D.C.; Gil, A.M.; Duarte, I.F.; Lentini, G.; Ciani, E. Camel (*Camelus* spp.) Urine Bioactivity and Metabolome: A Systematic Review of Knowledge Gaps, Advances, and Directions for Future Research. *Int. J. Mol. Sci.* **2022**, *23*, 15024. [CrossRef] [PubMed]

Al-Abdalall, A. The inhibitory effect of camels urine on mycotoxins and fungal growth. *Afr. J. Agric. Res.* **2010**, *5*, 1331–1337.

Amina, R.; Habiba, R.; Abouddihaj, B. Camel urine as a potential source of bioactive molecules showing their efficacy against pathogens: A systematic review. *Saudi J. Biol. Sci.* **2024**, *31*, 103966. [CrossRef]

Salamt, N.; Idrus, R.B.H.; Kashim, M.I.A.M.; Mokhtar, M.H. Anticancer, antiplatelet, gastroprotective and hepatoprotective effects of camel urine: A scoping review. *Saudi Pharm. J.* **2021**, *29*, 740–750. [CrossRef]

Abdel Gader, A.G.M.; Alhaider, A.A. The unique medicinal properties of camel products: A review of the scientific evidence. *J. Taibah Univ. Med. Sci.* **2016**, *11*, 98–103. [CrossRef]

Al-Yousef, N.; Gaafar, A.; Al-Otaibi, B.; Al-Jammaz, I.; Al-Hussein, K.; Aboussekhra, A. Camel urine components display anti-cancer properties in vitro. *J. Ethnopharmacol.* **2012**, *143*, 819–825. [CrossRef]

Attas, A.; Khorshid, F.; Kao, M.; Bahieldin, A. Antiviral activity of extracted fraction from camel urine against corona and influenza a (H1N1) viruses. *Appl. Ecol. Environ. Res.* **2020**, *17*. [CrossRef]

Pastrana, C.I.; Sgobba, M.N.; González, F.J.N.; Bermejo, J.V.D.; Pierri, C.L.; Lentini, G.; Musio, B.; Osman, T.K.S.; Gallo, V.; Duarte, I.F.; et al. Factors influencing the bioactivity of natural matrices: The case of osmolarity-dependent modulation of cell viability by different dilutions of camel urines. *Res. Vet. Sci.* **2024**, *180*, 105419. [CrossRef]

Todisco, S.; Musio, B.; Pesce, V.; Cavalluzzi, M.M.; Petrosillo, G.; La Piana, G.; Sgobba, M.N.; Schlosserová, N.; Beltrame, L.C.; Di Lorenzo, E.; et al. Targeting mitochondrial impairment for the treatment of cardiovascular diseases: From hypertension to ischemia-reperfusion injury, searching for new pharmacological targets. *Biochem. Pharmacol.* **2023**, *208*, 115405. [CrossRef] [PubMed]

12. Alkhamees, O.A.; Alsanad, S.M. A Review of the Therapeutic Characteristics of Camel Urine. *Afr. J. Tradit. Complement. Altern. Med.* **2017**, *14*, 120–126. [CrossRef]
13. Gaughan, J. Which physiological adaptation allows camels to tolerate high heat load—And what more can we learn? *J. Camelid Sci.* **2011**, *4*, 85–88.
14. Atanasov, A.G.; Zotchev, S.B.; Dirsch, V.M.; Supuran, C.T. Natural products in drug discovery: Advances and opportunities. *Nat. Rev. Drug Discov.* **2021**, *20*, 200–216. [CrossRef]
15. Hirsch, C.; Schildknecht, S. In Vitro Research Reproducibility: Keeping Up High Standards. *Front. Pharmacol.* **2019**, *10*, 0148 Available online: https://www.frontiersin.org/journals/pharmacology/articles/10.3389/fphar.2019.01484/full (accessed on 29 November 2024). [CrossRef]
16. Liu, S.; Yang, W.; Li, Y.; Sun, C. Fetal bovine serum, an important factor affecting the reproducibility of cell experiments. *Sci. Rep.* **2023**, *13*, 1942. [CrossRef]
17. Santos, L.M.; Cardoso, P.e.S.; Diniz, E.A.; Rahhal, J.G.; Sipert, C.R. Different concentrations of fetal bovine serum affect cytokine modulation in Lipopolysaccharide-activated apical papilla cells in vitro. *J. Appl. Oral. Sci.* **2023**, *31*, e20230020. [CrossRef]
18. Chelladurai, K.S.; Christyraj, J.D.S.; Rajagopalan, K.; Yesudhason, B.V.; Venkatachalam, S.; Mohan, M.; Vasantha, N.C.; Christyraj, J.R.S.S. Alternative to FBS in Animal Cell Culture—An Overview and Future Perspective. *Heliyon* **2021**, *7*, e07686. Available online: https://www.cell.com/heliyon/abstract/S2405-8440(21)01789-8 (accessed on 1 December 2024).
19. Hestermann, E.V.; Stegeman, J.J.; Hahn, M.E. Serum Alters the Uptake and Relative Potencies of Halogenated Aromatic Hydrocarbons in Cell Culture Bioassays. *Toxicol. Sci.* 2000, 53, pp. 316–325. Available online: https://academic.oup.com/toxsci/article-abstract/53/2/316/1650421?redirectedFrom=fulltext (accessed on 1 December 2024).
20. Fischer, F.C.; Abele, C.; Droge, S.T.; Henneberger, L.; König, M.; Schlichting, R.; Scholz, S.; Escher, B.I. Cellular Uptake Kinetics of Neutral and Charged Chemicals in in Vitro Assays Measured by Fluorescence Microscopy. *Chem. Res. Toxicol.* **2018**, *31*, 646–65 [CrossRef]
21. Abdalla, M.A. Anatomical Features in the Kidney Involved in Water Conservation Through Urine Concentration in Dromedaries (*Camelus dromedarius*). *Heliyon* **2020**, *6*, e03139. Available online: https://www.cell.com/heliyon/abstract/S2405-8440(19)36798- (accessed on 23 August 2022). [CrossRef]
22. Kültz, D. Hyperosmolality triggers oxidative damage in kidney cells. *Proc. Natl. Acad. Sci. USA* **2004**, *101*, 9177–9178. [CrossRef]
23. Hosseiniyan Khatibi, S.M.; Zununi Vahed, F.; Sharifi, S.; Ardalan, M.; Mohajel Shoja, M.; Zununi Vahed, S. Osmolytes resist against harsh osmolarity: Something old something new. *Biochimie* **2019**, *158*, 156–164. [CrossRef] [PubMed]
24. Chen, X.; Zhou, C.; Yan, C.; Ma, J.; Zheng, W. Hyperosmotic stress induces cisplatin sensitivity in ovarian cancer cells by stimulating aquaporin-5 expression. *Exp. Ther. Med.* **2015**, *10*, 2055–2062. [CrossRef] [PubMed]
25. Grauso, M.; Lan, A.; Andriamihaja, M.; Bouillaud, F.; Blachier, F. Hyperosmolar environment and intestinal epithelial cells: Impact on mitochondrial oxygen consumption, proliferation, and barrier function in vitro. *Sci. Rep.* **2019**, *9*, 11360. [CrossRef]
26. Bálint, Z.; Krizbai, I.A.; Wilhelm, I.; Farkas, A.E.; Párducz, A.; Szegletes, Z.; Váró, G. Changes induced by hyperosmotic mannitol in cerebral endothelial cells: An atomic force microscopic study. *Eur. Biophys. J.* **2007**, *36*, 113–120. [CrossRef]
27. Atanasov, A.G.; Waltenberger, B.; Pferschy-Wenzig, E.M.; Linder, T.; Wawrosch, C.; Uhrin, P.; Temml, V.; Wang, L.; Schwaiger, S Heiss, E.H.; et al. Discovery and resupply of pharmacologically active plant-derived natural products: A review. *Biotechnol. Adv.* **2015**, *33*, 1582–1614. [CrossRef]
28. Zheng, X.; Baker, H.; Hancock, W.S.; Fawaz, F.; McCaman, M.; Pungor, E., Jr. Proteomic Analysis for the Assessment of Different Lots of Fetal Bovine Serum as a Raw Material for Cell Culture. Part IV. Application of Proteomics to the Manufacture of Biological Drugs. *Biotechnol. Prog.* **2006**, *22*, 1294–1300. [CrossRef]
29. Baker, M. Reproducibility: Respect your cells! *Nature* **2016**, *537*, 433–435. [CrossRef] [PubMed]
30. Wang, W.T.; Catto, J.W.F.; Meuth, M. Differential response of normal and malignant urothelial cells to CHK1 and ATM inhibitors. *Oncogene* **2015**, *34*, 2887–2896. [CrossRef] [PubMed]
31. Zhao, P.; Chen, X.; Wang, Q.; Zou, H.; Xie, Y.; Liu, H.; Zhou, Y.; Liu, P.; Dai, H. Differential toxicity mechanism of gold nanoparticles in HK-2 renal proximal tubular cells and 786-0 carcinoma cells. *Nanomedicine* **2020**, *15*, 1079–1096. [CrossRef]
32. Ohana, G.; Bar-Yehuda, S.; Barer, F.; Fishman, P. Differential effect of adenosine on tumor and normal cell growth: Focus on the A3 adenosine receptor. *J. Cell Physiol.* **2001**, *186*, 19–23. [CrossRef]
33. SeSeashore-Ludlow, B.; Rees, M.G.; Cheah, J.H.; Cokol, M.; Price, E.V.; Coletti, M.E.; Jones, V.; Bodycombe, N.E.; Soule, C.K. Gould, J.; et al. Harnessing Connectivity in a Large-Scale Small-Molecule Sensitivity Dataset. *Cancer Discov.* **2015**, *5*, 1210–1223 [CrossRef] [PubMed]
34. Yao, F.; Madani Tonekaboni, S.A.; Safikhani, Z.; Smirnov, P.; El-Hachem, N.; Freeman, M.; Manem, V.S.; Haibe-Kains, B. Tissue specificity of in vitro drug sensitivity. *J. Am. Med. Inform. Assoc.* **2017**, *25*, 158–166. [CrossRef] [PubMed]
35. Zhang, L.; Zhang, J.; Ye, Z.W.; Muhammad, A.; Li, L.; Culpepper, J.W.; Townsend, D.M.; Tew, K.D. Adaptive changes in tumor cells in response to reductive stress. *Biochem. Pharmacol.* **2024**, *219*, 115929. [CrossRef] [PubMed]

Mfotie Njoya, E.; Ndemangou, B.; Akinyelu, J.; Munvera, A.M.; Chukwuma, C.I.; Mkounga, P.; Mashele, S.S.; Makhafola, T.J.; McGaw, L.J. In Vitro Antiproliferative, Anti-Inflammatory Effects and Molecular Docking Studies of Natural Compounds Isolated from Sarcocephalus pobeguinii (Hua ex Pobég). *Front. Pharmacol.* **2023**, *14*, 1205414. Available online: https://www.frontiersin.org/journals/pharmacology/articles/10.3389/fphar.2023.1205414/full (accessed on 30 January 2025). [CrossRef]
Nakahara, R.; Maeda, K.; Aki, S.; Osawa, T. Metabolic adaptations of cancer in extreme tumor microenvironments. *Cancer Sci.* **2023**, *114*, 1200–1207. [CrossRef]
Ismail, M.; Hassan, M.H.; Mohamed, E.I.; Azmy, A.F.; Moawad, A.; Mohammed, R.; Zaki, M.A. New insights into the anti-inflammatory and anti-melanoma mechanisms of action of azelaic acid and other Fusarium solani metabolites via in vitro and in silico studies. *Sci. Rep.* **2024**, *14*, 14370. [CrossRef]
Breathnach, A.S. Azelaic acid: Potential as a general antitumoural agent. *Med. Hypotheses.* **1999**, *52*, 221–226. [CrossRef]
Dongdong, Z.; Jin, Y.; Yang, T.; Yang, Q.; Wu, B.; Chen, Y.; Luo, Z.; Liang, L.; Liu, Y.; Xu, A.; et al. Antiproliferative and Immunoregulatory Effects of Azelaic Acid Against Acute Myeloid Leukemia via the Activation of Notch Signaling Pathway. *Front. Pharmacol.* **2019**, *10*, 1396. [CrossRef]
Pan, Y.; Liu, D.; Wei, Y.; Su, D.; Lu, C.; Hu, Y.; Zhou, F. Azelaic Acid Exerts Antileukemic Activity in Acute Myeloid Leukemia. *Front. Pharmacol.* **2017**, *8*. Available online: https://www.frontiersin.org/journals/pharmacology/articles/10.3389/fphar.2017.00359/full (accessed on 2 December 2024). [CrossRef]
Passi, S.; Picardo, M.; Zompetta, C.; Luca C de Breathnach, A.S.; Nazzaro-porro, M. The Oxyradical-Scavenging Activity of Azelaic Acid in Biological Systems. *Free Radic. Res. Commun.* **1991**, *15*, 17–28. [CrossRef]
Perez-Dominguez, F.; Carrillo-Beltrán, D.; Blanco, R.; Muñoz, J.P.; Leon-Cruz, G.; Corvalan, A.H.; Urzua, U.; Calaf, G.M.; Aguayo, F. Role of Pirin, an Oxidative Stress Sensor Protein, in Epithelial Carcinogenesis. *Biology* **2021**, *10*, 116. [CrossRef] [PubMed]
Pirin is an Iron-Dependent Redox Regulator of NF-κB. Available online: https://www.pnas.org/doi/10.1073/pnas.1221743110 (accessed on 1 December 2024).
Jayab, N.A.; Abed, A.; Talaat, I.M.; Hamoudi, R. The molecular mechanism of NF-κB dysregulation across different subtypes of renal cell carcinoma. *J. Adv. Res.* 2024. Available online: https://www.sciencedirect.com/science/article/pii/S209012322400314X (accessed on 1 December 2024).
Artemov, A.V.; Zhigalova, N.; Zhenilo, S.; Mazur, A.M.; Prokhortchouk, E.B. VHL inactivation without hypoxia is sufficient to achieve genome hypermethylation. *Sci. Rep.* **2018**, *8*, 10667. [CrossRef] [PubMed]
Hu, J.; Tan, P.; Ishihara, M.; Bayley, N.A.; Schokrpur, S.; Reynoso, J.G.; Zhang, Y.; Lim, R.J.; Dumitras, C.; Yang, L.; et al. Tumor heterogeneity in VHL drives metastasis in clear cell renal cell carcinoma. *Signal Transduct. Target. Ther.* **2023**, *8*, 155. [CrossRef]
Jiao, M.; Hu, M.; Pan, D.; Liu, X.; Bao, X.; Kim, J.; Li, F.; Li, C.Y. VHL loss enhances antitumor immunity by activating the anti-viral DNA-sensing pathway. *iScience* **2024**, *27*, 110285. Available online: https://www.cell.com/iscience/abstract/S2589-0042(24)01510-4 (accessed on 2 December 2024). [CrossRef]
Xu, X.; Lu, W.J.; Shi, J.Y.; Su, Y.L.; Liu, Y.C.; Wang, L.; Xiao, C.X.; Chen, C.; Lu, Q. The gut microbial metabolite phenylacetylglycine protects against cardiac injury caused by ischemia/reperfusion through activating β2AR. *Arch. Biochem. Biophys.* **2021**, *697*, 108720. [CrossRef]
Hu, W.; Kuang, X.; Zhang, Y.; Luo, Y.; Zhang, L. Neuroprotective effects of phenylacetylglycine via β2AR on cerebral ischemia/reperfusion injury in rats. *Saudi Pharm. J.* **2024**, *32*, 102210. [CrossRef]
Arif, E.; Nihalani, D. Beta2-adrenergic receptor in kidney biology: A current prospective. *Nephrology* **2019**, *24*, 497–503. [CrossRef]
Albiñana, V.; Gallardo-Vara, E.; de Rojas-P, I.; Recio-Poveda, L.; Aguado, T.; Canto-Cano, A.; Aguirre, D.T.; Serra, M.M.; González-Peramato, P.; Martínez-Piñeiro, L.; et al. Targeting β2-Adrenergic Receptors Shows Therapeutical Benefits in Clear Cell Renal Cell Carcinoma from Von Hippel–Lindau Disease. *J. Clin. Med.* **2020**, *9*, 2740. [CrossRef]
Albiñana, V.; Recio-Poveda, L.; González-Peramato, P.; Martinez-Piñeiro, L.; Botella, L.M.; Cuesta, A.M. Blockade of β2-Adrenergic Receptor Reduces Inflammation and Oxidative Stress in Clear Cell Renal Cell Carcinoma. *Int. J. Mol. Sci.* **2022**, *23*, 1325. [CrossRef]
Kamiar, A.; Yousefi, K.; Dunkley, J.C.; Webster, K.A.; Shehadeh, L.A. β2-Adrenergic receptor agonism as a therapeutic strategy for kidney disease. *Am. J. Physiol. Regul. Integr. Comp. Physiol.* **2021**, *320*, R575–R587. [CrossRef]
Nakamura, A.; Niimi, R.; Yanagawa, Y. Renal β2-adrenoceptor Modulates the Lipopolysaccharide Transport System in Sepsis-induced Acute Renal Failure. *Inflammation* **2009**, *32*, 12–19. [CrossRef] [PubMed]
Emwas, A.H.; Roy, R.; McKay, R.T.; Ryan, D.; Brennan, L.; Tenori, L.; Luchinat, C.; Gao, X.; Zeri, A.C.; Gowda, G.N.; et al. Recommendations and Standardization of Biomarker Quantification Using NMR-Based Metabolomics with Particular Focus on Urinary Analysis. *J. Proteome Res.* **2016**, *15*, 360–373. [CrossRef] [PubMed]
Ryan, M.J.; Johnson, G.; Kirk, J.; Fuerstenberg, S.M.; Zager, R.A.; Torok-Storb, B. HK-2: An immortalized proximal tubule epithelial cell line from normal adult human kidney. *Kidney Int.* **1994**, *45*, 48–57. [CrossRef] [PubMed]
Glube, N.; Giessl, A.; Wolfrum, U.; Langguth, P. Caki-1 Cells Represent an in vitro Model System for Studying the Human Proximal Tubule Epithelium. *Nephron Exp. Nephrol.* **2007**, *107*, e47–e56. [CrossRef]

59. Caratozzolo, M.F.; Valletti, A.; Gigante, M.; Aiello, I.; Mastropasqua, F.; Marzano, F.; Ditonno, P.; Carrieri, G.; Simonnet, H.; D'Erchia, A.M.; et al. TRIM8 anti-proliferative action against chemo-resistant renal cell carcinoma. *Oncotarget* **2014**, *5*, 7446–745 [CrossRef]
60. Mevik, B.H.; Wehrens, R. The pls Package: Principal Component and Partial Least Squares Regression in R. *J. Stat. Softw.* **2007**, *1* 1–23. [CrossRef]
61. Kuhn, M. Building Predictive Models in R Using the caret Package. *J. Stat. Softw.* **2008**, *28*, 1–26. [CrossRef]
62. Bijlsma, S.; Bobeldijk, I.; Verheij, E.R.; Ramaker, R.; Kochhar, S.; Macdonald, I.A.; Van Ommen, B.; Smilde, A.K. Large-Scal Human Metabolomics Studies: A Strategy for Data (Pre-) Processing and Validation. *Anal. Chem.* **2006**, *78*, 567–574. [CrossRef]

molecules

MDPI

view

Natural Alkaloids as Antiviral Agents Against RNA Viruses: A Comprehensive and Mechanistic Review

risti Leka [1,2], Lucia Mamede [1], Elyn Vandeberg [1,2], Mutien-Marie Garigliany [1,2] and Allison Ledoux [1,*]

[1] Faculty of Medicine, University of Liège, CIRM, Building B36 Quartier Hôpital, 4000 Liege, Belgium; kristi.leka@uliege.be (K.L.); mmgarigliany@uliege.be (M.-M.G.)

[2] Faculty of Veterinary Medicine, University of Liège, FARAH-INDEEP, Building B42 Quartier Vallée 2 Avenue de Cureghem 6, 4000 Liege, Belgium

* Correspondence: allison.ledoux@uliege.be

Abstract

RNA viruses pose a persistent global threat due to their high mutation rates, zoonotic potential, and rapid adaptability. Emergence events have risen steadily, as demonstrated by major outbreaks caused by Influenza A, Ebola, Zika, and Chikungunya viruses, followed by the coronavirus epidemics of Severe Acute Respiratory Syndrome coronavirus (SARS-CoV-1) and Middle East Respiratory Syndrome Coronavirus (MERS-CoV) and culminating in the COVID-19 pandemic. These characteristics frequently compromise the durability of existing vaccines and antiviral therapies, highlighting the urgent need for new antiviral agents. Alkaloids, a structurally diverse class of nitrogen-containing natural compounds, have gained attention for their ability to interfere with multiple stages of the viral life cycle, including entry, replication, protein synthesis, and host immune modulation. To our knowledge, this review compiles all currently reported alkaloids with antiviral activity against RNA viruses and summarizes their proposed mechanisms of action, distinguishing evidence from in vitro, in vivo, and in silico studies. Quaternary alkaloids are discussed separately because their permanent ionic charge enables distinctive interactions with membranes and host pathways. Although many findings are promising, clinical translation remains limited by incomplete mechanistic validation, scarce in vivo data, suboptimal bioavailability, narrow therapeutic windows, and inconsistent experimental methodologies. To advance the field, future research should prioritize RT-qPCR–based antiviral evaluation to accurately quantify viral replication, incorporate mechanistic assays to clarify modes of action, apply structure–activity relationship (SAR) approaches for rational optimization, and expand in vivo pharmacokinetic and efficacy studies to assess therapeutic feasibility. Overall, alkaloids represent a promising yet underdeveloped reservoir for next-generation antiviral discovery against rapidly evolving RNA viruses.

Keywords: alkaloids; RNA viruses; broad-spectrum antivirals; mechanisms of action; quaternary alkaloids; pharmacokinetics and toxicity; antiviral drug discovery

1. Introduction

The objective of this review is to comprehensively compile naturally occurring alkaloids reported to exhibit antiviral activity against RNA viruses, a group characterized by frequent emergence, high mutation rates, and substantial pandemic potential. Beyond cataloging these compounds, the review critically examines their proposed mechanisms of action in relation to the experimental approaches used to characterize antiviral activity.

Because the strength and translational relevance of reported findings depend strongly on study design, it is essential to distinguish between predictive computational analyses, cell-based assays, and animal models. While in silico screening and early in vitro studies provide valuable insights for compound prioritization, mechanistic validation and in vivo evaluation are required to substantiate therapeutic potential. Accordingly, this review explicitly differentiates alkaloids supported by in silico, in vitro, and in vivo evidence and discusses computationally identified candidates in a dedicated section, with emphasis on their limitations and translational relevance.

To support this analysis, a comprehensive literature search was conducted between August 2025 and December 2025 to identify peer-reviewed studies reporting antiviral activity of natural alkaloids against RNA viruses. The search strategy combined broad academic literature databases and targeted keyword-based queries to ensure extensive coverage across virus families and experimental contexts. Search terms included combinations of "alkaloids", "quaternary alkaloids", "antiviral", and "RNA viruses", together with the names of individual RNA viruses searched separately, including SARS-CoV-1, SARS-CoV-2, MERS-CoV, influenza A virus (IAV), dengue virus (DENV), Zika virus (ZIKV), Ebola virus (EBOV), West Nile virus (WNV), Japanese encephalitis virus (JEV), chikungunya virus (CHIKV), respiratory syncytial virus (RSV), hepatitis C virus (HCV), foot-and-mouth disease virus (FMDV), and porcine epidemic diarrhea virus (PEDV). Searches were performed iteratively to capture studies spanning in silico, in vitro, and in vivo experimental approaches.

Only peer-reviewed articles published in English were considered. Studies were included if they reported antiviral activity, target engagement, or mechanistic effects of natural alkaloids against RNA viruses based on primary computational, cellular, or animal data. Reviews, editorials, patents, and studies lacking primary experimental or computational evidence were excluded, except when cited to provide contextual background. Data extraction focused on alkaloid identity, viral target or virus family, experimental model, and reported antiviral outcomes, with emphasis on quantitative parameters when available.

In contrast to previous reviews focusing primarily on compound listings or single virus families, the present review emphasizes mechanistic patterns across alkaloid classes, distinguishes classical from quaternary alkaloids, and critically evaluates antiviral evidence according to experimental level (in silico, in vitro, in vivo). Particular attention is given to translational bottlenecks, including toxicity, pharmacokinetics, and formulation challenges.

Overview of RNA Viruses and Their Global Health Impact

Viruses are microscopic infectious agents incapable of independent replication, as they lack the cellular machinery required for reproduction [1]. To propagate, they must invade a host cell and exploit its biosynthetic systems for their own replication. In 1971, David Baltimore introduced a classification system for viruses based on the type and replication strategy of their genetic material [2]. Within this framework, RNA viruses are distinguished by their ribonucleic acid (RNA) genomes, which confer unique structural and functional characteristics distinct from those of desoxyribonucleic acid (DNA) viruses.

A defining feature of RNA viruses is their exceptionally high mutation rate, primarily attributed to the low fidelity of RNA-dependent RNA polymerases, the enzymes responsible for genome replication [3]. This intrinsic genetic instability enables RNA viruses to adapt rapidly to new hosts and environmental pressures, often resulting in immune evasion and the reduced efficacy of antiviral drugs.

Many human RNA viruses are zoonotic in origin, arising in animal reservoirs before crossing the barrier into human populations. The likelihood of such "spillover" events is increasing due to anthropogenic factors such as deforestation, urbanization, climate change, and intensified human–wildlife interactions [4]. Notable zoonotic RNA viruses include

https://doi.org/10.3390/molecules31030539

the Human Immunodeficiency Virus (HIV), Influenza A virus subtypes H1N1 and H5N1, Severe Acute Respiratory Syndrome coronavirus (SARS-CoV), Middle East Respiratory Syndrome coronavirus (MERS-CoV), and Severe Acute Respiratory Syndrome coronavirus 2 (SARS-CoV-2).

The COVID-19 pandemic, caused by SARS-CoV-2 virus, underscored not only the profound global health threat posed by RNA viruses but also the degree to which the world remains unprepared for such challenges. The pandemic demonstrated how rapidly RNA viruses can evolve and disseminate across populations by producing numerous variants within a remarkably short period. Each of the variants was characterized by distinct levels of transmissibility, virulence, and clinical presentation [5–9]. The emergence and global spread of these variants placed extraordinary long-term strain on healthcare systems and triggered extensive social, economic and public health disruptions.

When a novel RNA virus emerges in the human population, pre-existing immunity is typically absent, rendering communities highly susceptible to infection. The continuous co-evolution of RNA viruses and their hosts further complicates outbreak prediction and as such it delays the development of effective vaccines and therapeutic solutions [10]. In summary the high mutability, zoonotic potential, and rapid adaptability of RNA viruses underscore their significance as persistent and evolving threats to global health.

2. Virus Classification in the Genomic Era

The classification of viruses is a dynamic and continuously evolving process, shaped largely by advances in genomic sequencing technologies. Historically, viruses were categorized by phenotypic traits, host range, and associated diseases [11]. Today, the International Committee on Taxonomy of Viruses (ICTV) serves as the authoritative body for viral taxonomy [12]. Modern systems rely heavily on phylogenetics and genomic data, frequently structured within the Baltimore classification framework [11].

Metagenomic expansion has produced large datasets that challenge existing taxonomic frameworks, especially given recombination and reassortment [12]. Integrative approaches combining sequence identity analyses with phylogenetic methods, for example in *Betaflexiviridae* as described by Silva et al. (2022), have been proposed [13]. These methods aim to refine viral classification to better account for the diversity revealed by high-throughput sequencing [13].

Despite advances, the Baltimore system remains a pillar in organizing viruses by genome type and replication mode although minor revisions have reflected newly identified behaviors and genome structures [12].

Baltimore divides the viruses into the following groups:

- I: dsDNA (e.g., Herpes simplex virus)
- II: ssDNA → dsDNA before transcription (e.g., Parvovirus)
- III: dsRNA; mRNA transcribed from RNA genome (e.g., Rotavirus)
- IV: (+)ssRNA; genome functions as mRNA (e.g., Picornavirus, Coronavirus, Flavivirus)
- V: (−)ssRNA; mRNA transcribed from RNA genome (e.g., Rabies virus, Influenza virus, Ebola virus)
- VI: ssRNA with reverse transcriptase (e.g., HIV)
- VII: dsDNA with reverse transcriptase (e.g., Hepatitis B virus)

This review aims its focus on Groups III, IV, V, and VI concerning only the RNA viruses responsible for major outbreaks in recent decades, including Influenza A virus, MERS-CoV, SARS-CoV, HIV, Ebola virus, Zika virus, Japanese Encephalitis virus and SARS-CoV-2.

https://doi.org/10.3390/molecules31030539

3. Alkaloid Biosynthetic Pathways and Structural Diversity

Plant alkaloids are structurally diverse nitrogen-containing compounds with broad biological activity and long medicinal use. Their complexity arises from enzyme-mediated modifications, glycosylation, acylation, reduction, oxidation, and methylation, which diversify structures and modulate bioactivity [14]. Cytochrome P450–mediated oxidations are particularly pivotal, enabling ring formation, cleavage, and expansion that greatly increase molecular diversity [15].

Advances in molecular biology and biotechnology have illuminated genes involved in alkaloid biosynthesis. For example, Kishimoto et al. 2016 used synthetic biology and microbial systems (*Escherichia coli*, *Saccharomyces cerevisiae*) to verify gene function, optimize production, and generate structural analogs, facilitating scalable pharmaceutical applications [15,16].

Obtaining useful amounts of alkaloids from plants often requires large quantities of raw material, considerable time, and specialized equipment to extract and identify each compound. Because of these demands, traditional isolation can be costly and environmentally taxing. Biosynthetic approaches, on the other hand, offer a more sustainable and practical alternative, allowing alkaloids to be produced without relying heavily on plant harvesting.

3.1. *Classification of Alkaloids*

Alkaloids are commonly classified by biosynthetic origin, emphasizing the presence and source of the nitrogen atom. Protoalkaloids are defined by the nitrogen outside the heterocyclic ring. On the contrary, for the true alkaloids the nitrogen is incorporated within a heterocyclic ring. Whereas for the pseudoalkaloids, the nitrogen is derived from non–amino acid precursors [17].

A structural classification based on the ring systems is also available into the following groups: quinoline, isoquinoline, pyrrole, pyrrolidine, pyridine, piperidine, pyrrolizidine, and indole-containing alkaloids [18].

3.2. *Host Hijacking and Viral Translation*

RNA viruses have evolved strategies to hijack host machinery for replication and protein synthesis. By manipulating ribosomes, they redirect translation toward viral proteins, often at the expense of host antiviral factors [19]. Some cleave key host factors or modulate PRR (pattern recognition receptor) activation, evading immune detection and suppression [20,21].

Because many RNA viruses lack a canonical 5′ cap and differ in other structural features, they employ non-canonical translation mechanisms to initiate protein synthesis, including internal ribosome entry, leaky scanning, and ribosomal frameshifting. Viral RNAs can also form circular structures that, similar to cellular mRNAs, recruit host initiation factors to drive protein production [22].

A major challenge for RNA viruses is coordinating translation and replication, which compete for shared host factors. To ensure efficient infection, viruses must precisely regulate these processes, maintaining a balance that allows genome replication to proceed while still producing the proteins required for virion assembly.

3.3. *Alkaloid Antiviral Modes of Action*

Alkaloids exhibit antiviral activity against RNA viruses and, in some cases, DNA viruses, demonstrating effects at multiple stages of the viral life cycle [23,24]. Their antiviral mechanisms include inhibition of viral entry, either by interfering with receptor binding or by inducing membrane perturbation. They may also suppress viral replication through

https://doi.org/10.3390/molecules3103053

direct inhibition of essential enzymes such as RNA-dependent RNA polymerase (RdRp) or viral proteases. Additionally, several alkaloids disrupt viral protein synthesis, further limiting viral proliferation.

Beyond direct antiviral effects, alkaloids can exert indirect activity by modulating immune responses, including the regulation of cytokines, reactive oxygen species (ROS), and interferons [25].

Examples that illustrate these mechanisms include 10-hydroxyusambarensine and cryptospirolepine, which show strong binding affinity in silico for multiple viral targets [26,27]. Isoquinoline alkaloids modulate key inflammatory pathways such as NF-κB and MAPK/ERK [28]. Meanwhile, compounds such as tetrandrine, oxymatrine, and berberine demonstrate potent immunomodulatory activity by suppressing excessive inflammation [29].

During the course of this review, clear distinctions emerged between the antiviral activities of general alkaloids and those of their specialized subclass, the quaternary alkaloids. These differences were not limited to their molecular structures, specifically the permanently charged quaternary ammonium group, but were also evident at the pharmacological level.

Because quaternary alkaloids carry a permanent positive charge, their limited membrane permeability often restricts their antiviral activity to early steps of the viral life cycle, such as viral attachment or entry. They often act through mechanisms such as membrane disruption, interference with viral entry, or modulation of lysosomal and host signaling pathways. In contrast, many non-quaternary alkaloids display activity across a broader range of viral processes, including replication, protein synthesis, and immune modulation.

Differences were also observed in potency, cytotoxicity, selectivity index, bioavailability, and overall potential for clinical translation. Given these mechanistic and pharmacological distinctions, we determined that it was more appropriate to present alkaloids and quaternary alkaloids in two separate tables (Tables 1 and 2) accordingly to ensure clarity and scientific accuracy.

In this review, antiviral evidence is discussed according to the experimental level at which it was generated, namely in silico, in vitro, and in vivo. While computational studies provide valuable hypotheses regarding potential targets and binding modes, they do not constitute proof of antiviral efficacy and must be interpreted with caution. In vitro and in vivo studies are therefore discussed separately where possible to reflect their respective levels of biological validation.

Figure 1 presents SARS-CoV-2, the most recent coronavirus responsible for a global pandemic, as an example to illustrate the viral replication cycle within the host and the key steps at which a known alkaloid can exert inhibitory activity.

https://doi.org/10.3390/molecules31030539

Table 1. Alkaloids displaying antiviral activity.

Compound & Origin	**References**	**Study Type**	**Virus**	**Mechanisms**	**Values**
1. ***1,2,3,4-Tetrahydroisoquinolines derivatives*** (From: e.g., *Papaveraceae* family) 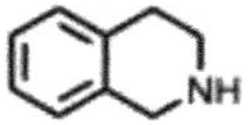	Y. Liao et al. 2023 [30]	In vitro, in silico and in vivo	IAV H1N1 H5N1 H3N2 Influenza B virus (IBV)	Mechanistic studies demonstrated that compound **35** could bind tightly to the PAN endonuclease of RNA-dependent RNA polymerase, thus blocking the viral replication to exert antiviral activity.	IC_{50} for compound **35** = 0.20 ± 0.01 μM EC_{50} = 0.88 μM SI = 113.1
	George, A. et al. (2018) [31]	In vitro	HIV-1	Inhibit the LEDGF/p75- IN	Compound **6d**: IC_{50} of ~10 μM
	Wang, X. et al. (2023) [32]	In vitro and in silico	SARS-CoV-2	Tt mainly inhibited the post-entry viral replication in both Vero E6 and Calu-3 cells.	Compound trans-1 EC_{50} = 2.78 μM SI > 71.94
2. **7-Methoxycryptopleurine** (From: *Menispermaceae* family) 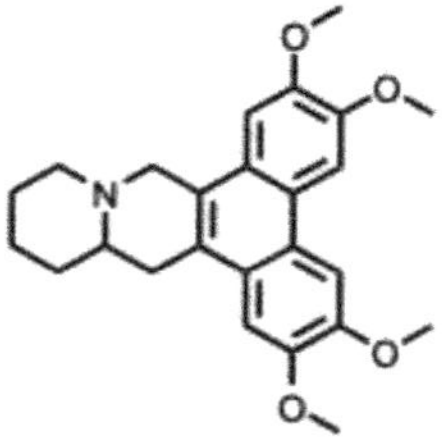	Majnooni, M. et al. (2001) [33]	*Review*	SARS-CoV-2	Blocking the S and N proteins, 3CLpro inhibitor	EC_{50} = 58 nM
3. ***10-Hydroxyusambarensine*** (From: *Strychnos usambarensis*, *Loganiaceae* family) 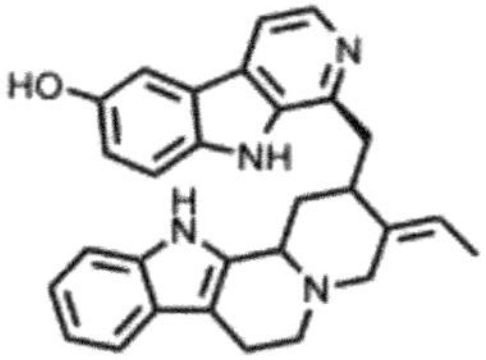	Ogunyemi, O.M. et al. (2020) [34]	In silico	SARS-CoV-2	Strong binding affinity to the RNA-dependent RNA polymerase (RdRp)	AutoDock version 4.2 programVina score: 10.1
4. ***Ajmaline*** (From: *Rauwolfia serpentina*, *Apocynaceae* family) 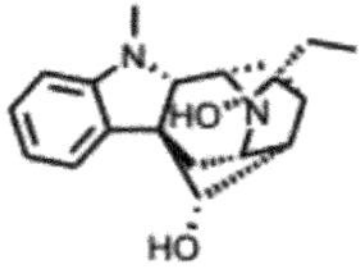	Cheng, F. et al. (2016) [35]	Computational biology	EBOV	Ajmaline predicted to up-regulates expression of several important Ebola-related genes, such as MERTK, FURIN, TYRO3, FURIN, and CTSB	q = 0.002

https://doi.org/10.3390/molecules31030539

Table 1. *Cont.*

Compound & Origin	References	Study Type	Virus	Mechanisms	Values
5. ***Aloperine and derivatives*** (From: *Sophora alopecuroides*/*Sophora flavescens*, *Fabaceae* family) 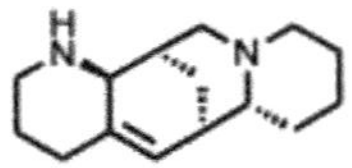	Cheng, F. et al. (2016) [35]	In vitro and in vivo	SARS-CoV-2	Inhibiting host cathepsin B activity and anti-cytokine effects	Compound **8a**: EC 50 = 39.1 µM SI > 6.8
	Zhou, P. et al. (2023) [36]	In vitro, in silico and in vivo	ZIKV	Targeting its RNA-dependent RNA polymerase (RdRp)	EC_{50}: from 2.98–6.036 µM depending on the cells line SI: 31.72–66.95 depending on the cells line Aloperine administration resulted in an improved survival rate in mice and reduced viremia
6. ***Amarbellisine*** (From: *Amaryllidaceae* family) 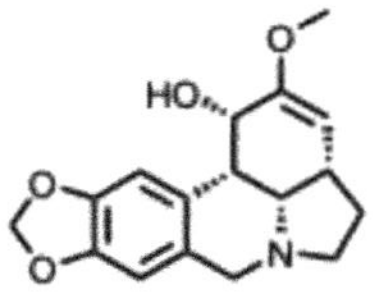	Merindol, N. et al. (2024) [37]	In vitro	HCoV−OC43 HCoV-OC43	Potential inhibition of viral replication	SI = 60 EC_{50} = 0.2 µM
7. ***Berbamine*** (From: *Berberis* spp., *Berberidaceae* family) 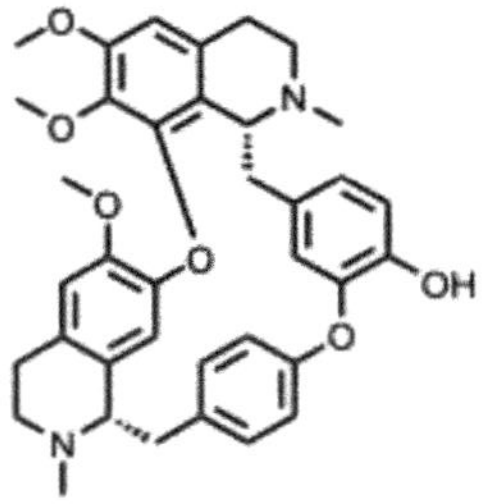	Huang, L. et al. (2021) [38]	In vitro	SARS-CoV-2	Compromising TRPMLs-mediated endolysosomal trafficking of ACE2	SARS-CoV-2: 2.35 ± 0.92 µM
8. ***Canthin-6-one and derivatives*** (From: *Rutaceae* family) 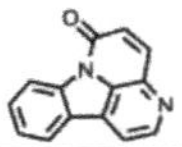	Wang, C. et al. (2024) [39]	In vitro	Newcastle disease virus (NDV)	Entry inhibition via Akt/ERK pathway; COX-2 induction	Analogue compounds: IC_{50} = 5.26–11.76 µM
	Verma, D. et al. (2020) [40]	In silico	SARS-CoV-2	Predicted binding to M^{pro} and PL^{pro}, potential protease inhibition	Canthin-6-one 9-O-β-glucopyranoside PL^{pro}: −9.4 kcal/mol M^{pro}: −8.5 kcal/mol

Table 1. *Cont.*

Compound & Origin	References	Study Type	Virus	Mechanisms	Values
9. ***Capsaicin*** (From: *Capsicum* spp., *Solanaceae* family) 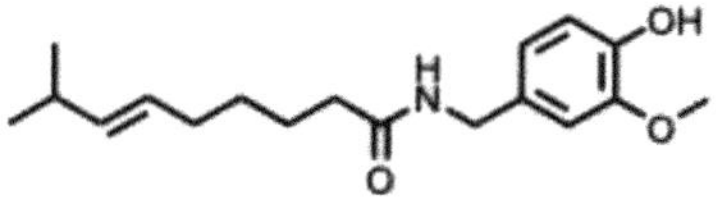	Zhang, M. et al. (2023) [41]	In vitro and in vivo	Encephalomyocarditis virus Vesicular stomatitis virus IAV H1N1	Capsaicin directly binds STAT3, promoting its lysosomal degradation	NA
	Marois, I. et al. (2014) [42]	In vitro	IAV	Capsaicin reduced influenza PA gene expression by ~48%	The EC_{50} was between 44.69–55.17 µM depending on the different strains of influenza virus
	Trischitta, P. et al. (2024) [43]	In vitro	Lassa Virus	Inhibits GP-mediated membrane fusion during viral entry	EC_{50} = 6.9–10.0 µmol/L
10. ***Cepharanthine*** (From: *Stephania* spp., *Menispermaceae* family) 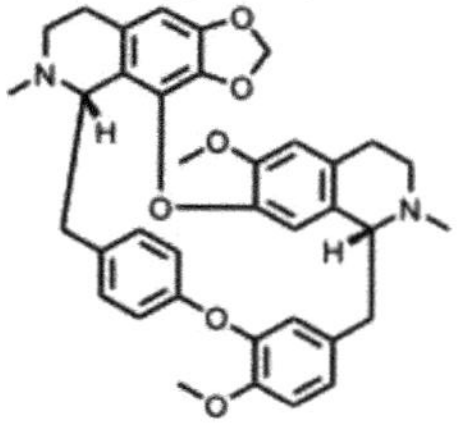	Xia, B. et al. (2023) [44]	Review	SARS-CoV-2	Inhibit viral entry and post-entry steps and attenuate the potential inflammatory effects	IC_{50} = 28.51 ng/mL
	Liu, K. et al. (2023) [45]	Review	1. SARS-CoV-2 2. SARS-CoV pseudovirus 3. MERS-CoV pseudovirus 4. HCoV-OC43 5. PEDV 6. SARS-CoV 7. EBOV 8. ZIKV 9. PRRSV 10. HIV-1	Multiple potential mechanisms of actions depending on the virus	1. EC_{50} = 0.15 M 2. EC_{50} ≈ 0.0417 µM 3. EC_{50} ≈ 0.14 µM 4. IC_{50} ≈ 0.83 µM 5. EC_{50} ≈ 2.53 µM; in vivo: 11.1 mg/kg oral dose reduced viral load 6. EC_{50} ≈ 0.79 µM 7. IC_{50} ≈ 0.42 µM 8. IC_{50} ≈ 2.19 µM 9. 10 µM reduced $TCID_{50}$ ~5.6-fold; NF-κB inhibition 10. EC_{50} = 0.026 M
11. ***Cephaeline*** (From: *Cephaelis ipecacuanha*, *Rubiaceae* family) 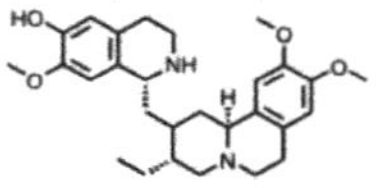	Ren, P. et al. (2022) [46]	In vitro and in silico	SARS-CoV-2	Targeting the host ribosome, and viral RNA, RdRp as well as N protein to interfere with the translating, propagating, replicating, and assembling process of the virus.	EC_{50} = 0.01 µM
	Yang, S. et al. (2018) [47]	In vitro, in silico and in vivo	ZIKV EBOV	Acts on the ZIKV RdRp NS5, host cell's lysosome, and 40S ribosomal subunit	IC_{50} values of less than 42 nM IC_{50} = 16.9 nM with 95% CI of 10.7–25.8 nM
12. ***Cherylline*** (From: *Amaryllidaceae* family) 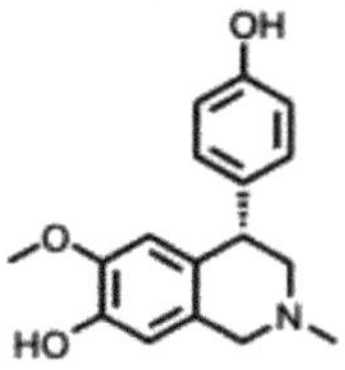	Ka, S. et al. (2021) [48]	In vitro and in silico	DENV ZIKV	Post-entry inhibition of RNA replication	EC_{50} = 8.8 µM SI = 28 EC_{50} = 20.3 µM SI = 12

https://doi.org/10.3390/molecules3103053

Table 1. *Cont.*

Compound & Origin	References	Study Type	Virus	Mechanisms	Values
13. *Cinchonine* (From: *Cinchona* spp., *Rubiaceae* family) 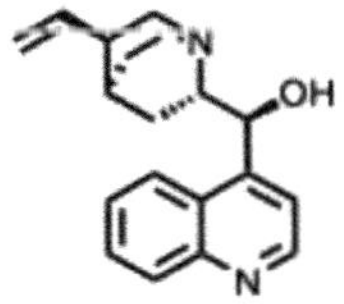	Ren, J. et al. (2022) [49]	In vitro	Porcine Epidemic Diarrhea Virus (PEDV)	Induction of autophagy, inhibiting early (adsorption/entry) and replication stages	Dose-dependent suppression of viral mRNA and N protein; ~100 μM nearly complete inhibition; RT-qPCR and $TCID_{50}$ assays show significant reduction
14. *Colchicine* (From: *Colchicaceae* family) 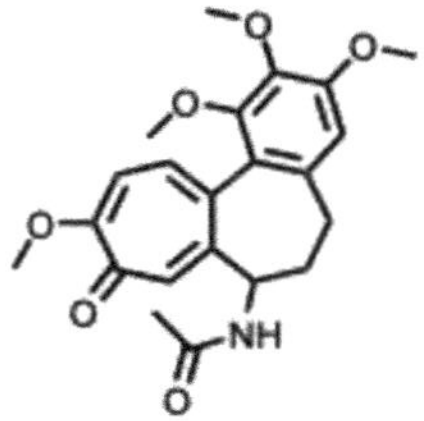	Hegazy, A. et al. (2024) [50]	In vitro and in silico	IAV H5N1 H1N1	Inhibition of viral adsorption and replication; docking supports NA/M2 binding	$IC_{50} = 0.111$ μg/mL $IC_{50} = 0.326$ μg/mL
15. *Conessine* (From: *Apocynaceae* family) 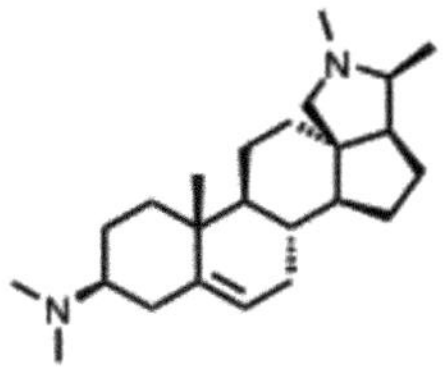	Majnooni, M. et al. (2021) [33]	*Review*	SARS-CoV-2	M^{pro} inhibitor	$EC_{50} = 2.34$ μM,
16. *Cyclopamine* (From: *Melanthiaceae* family) 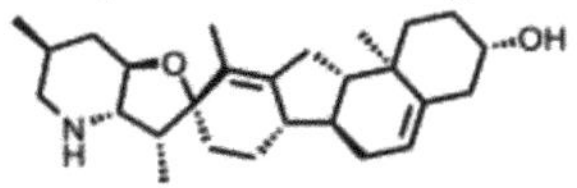	Bailly, C. et al. (2016) [51]	In vitro and In vivo	Human RSV (hRSV)	Disrupts inclusion bodies via M2-1 protein interaction, impairs RdRp complex	$IC_{50} \approx 380$ nM; $CC_{50} > 320$ μM
	Diot, C. et al. (2023) [52]	In vitro and in silico	hRSV Bovine RSV (BRSV)	Hardens inclusion bodies, interferes with M2-1–P–RNA dynamics	Dose-dependent inhibition
	Fix, J. et al. (2023) [53]	In vitro	BRSV	Similar M2-1-mediated mechanism to hRSV	$EC_{50} = 76$ nM

https://doi.org/10.3390/molecules31030539

Table 1. *Cont.*

Compound & Origin	References	Study Type	Virus	Mechanisms	Values
17. ***Dehydroevodiamine*** (From: *Rutaceae* family) 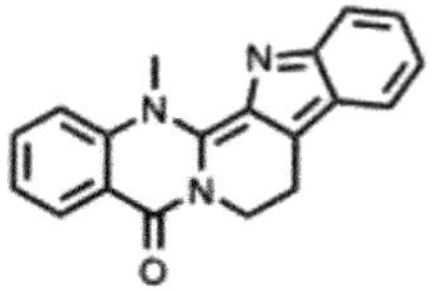	Li, K. et al. (2024) [54]	In vitro	PEDV	Inhibition of the PEDV replication stage, and its downregulation oft he ERK1/2 MAPH pathway	48 h PI: $IC_{50} = 3.574 \pm 0.566$ μg/mL SI = 3.503
18. ***Deoxynojirimycin and derivatives*** (From: e.g., *Morus alba*, *Moraceae* family) 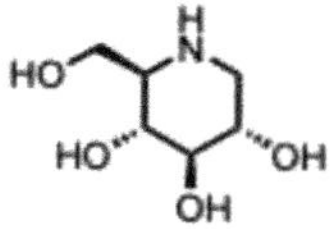	Hussain, M. et al. (2015) [55]	In vitro	IAV H3N2 strains	α-glucosidase inhibition, impaired viral glycoprotein folding	NN-DNJ $IC_{50} = 0.5$–2.5 μM depending on the viral strain
	Perera, N. et al. (2022) [56]	In vitro	DENV	Blocks viral release; inhibits glycoprotein maturation via α-glucosidase inhibition	IC_{50} of DNJ-iminosugars 48 h PI: 2THO-DNJ 1.6 ± 0.8 μM EOO-DNJ 3.1 ± 1.3 μM NN-DNJ 3.3 ± 1.5 μM
	Bhushan, G. et al. (2020) [57]	In vitro	ZIKV	Impairs viral replication by disrupting glycoprotein folding via ER glucosidase inhibition	At 1 μM, DNJ significantly reduced ZIKV RNA levels in supernatants compared to vehicle control ($p \leq 0.0318$)
19. ***Emetine*** (From: *Rubiaceae* family) 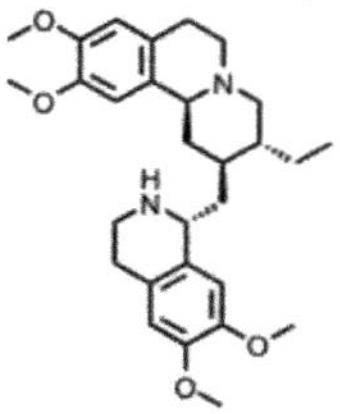	Yang, S. et al. (2018) [47]	In vitro and in vivo	ZIKV EBOV	Inhibits NS5 polymerase activity and disrupts lysosomal function; also inhibits EBOV entry	$IC_{50} = 52.9$ nM (95% CI: 35.4–73.2 nM) SJL mice; 1 mg/kg/day ~10-fold reduction in blood viremia at 7-day PI $IC_{50} = 16.9$ nM (95% CI:10.7–25.8 nM) 67% survival (4/6 mice) with IP emetine dosing
	Khandelwal, N. et al. (2017) [58]	In vitro and *in ovo*	PPRV NDV	Inhibits viral polymerase and entry; reduces viral RNA/protein synthesis	NDV: Reduced virus production 2-3 log in NDV infected cells
	Valipour, M. (2022) [59]	Review	SARS-CoV-2	Likely affects both virus- and host-based targets (translation machinery, NF-κB, etc.)	$EC_{50} = 0.007$ μM $EC_{50} = 0.46$ μM $IC_{50} = 0.52$ μM $EC_{50} = 0.000147$ μM $EC_{50} = 0.00771$ μM
	Bleasel, M. D. (2020) [60]	Commentary	SARS-CoV MERS-CoV	Broad-spectrum coronavirus inhibition suggested	EC_{50} SARS: 0.054 μM; EC_{50} MERS: 0.014 μM
20. ***Ephedrine and derivatives*** (From: *Ephedraceae* family) 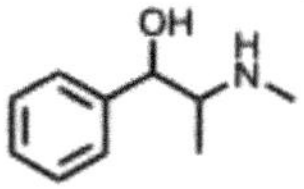	Wei, Y. et al. (2019) [61]	In vitro and in vivo	IAV (H1N1)	Modulation of TLR3/4/7 signaling, reducing TNF-α and increasing IFN-β	$EC_{50} = 5.66$–10.96 μg/mL depending on the drug delivery way

https://doi.org/10.3390/molecules31030539

Table 1. *Cont.*

Compound & Origin	References	Study Type	Virus	Mechanisms	Values
21. ***Fangchinoline*** (From: *Menispermaceae* family) 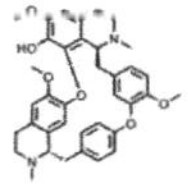	Yang, S. et al. (2024) [62]	In vitro and in vivo	ZIKA	Inhibits viral internalization	EC_{50}: 0.86 ± 0.47 μM
	Wan, Z. et al. (2012) [63]	In vitro	HIV-1	Inhibits gp160 proteolytic processing, blocking envelope maturation	EC_{50}: 0.8–1.7 μM depending on the HIV-1 strains
	Zhang, Q. Y. et al. (2024) [64]	In vitro	Enterovirus	Inhibits early-stage infection; VP1 mutations (E145G, V258I) reduce sensitivity	At concentration of 10 μmol/L FAN resulted in about 600-fold reduction in viral titers
22. ***Harmaline*** (From: *Nitrariaceae/Malpighiaceae* family) 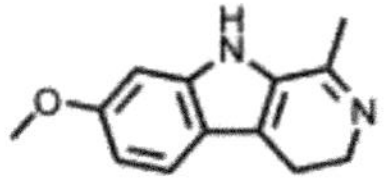	Hegazy, A. et al. (2023) [65]	In vitro	IAV H1N1 and H5N1	Not specified: measured via viral inhibition/cytopathic effect assay	H1N1: Harmaline: IC_{50} = 0.056 μg/mL H5N1: Harmaline: IC_{50} = 3.42 μg/mL
23. ***Harmine*** (From: *Nitrariaceae/Malpighiaceae* family) 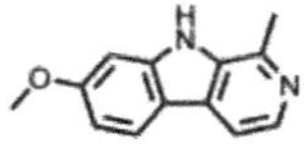	Hegazy, A. et al. (2023) [65]	In vitro	IAV H1N1 and H5N1	Not specified: measured via viral inhibition/cytopathic effect assay	H1N1: Harmine IC_{50} = 0.033 μg/mL H5N1: Harmine IC_{50} = 0.023 μg/mL
	Dahal, S. et al. (2023) [66]	In vitro	HCoV-229E SARS-CoV-2 variants HIV-1	Post-entry inhibition of viral replication via SR kinase inhibition; reduced viral protein expression and subgenomic RNAs.	Dose-dependent inhibition
24. ***Harringtonine*** (From: *Cephalotaxaceae* family) 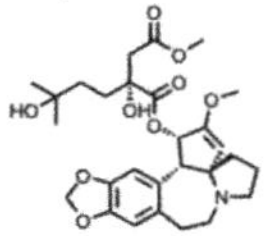	Kaur, P. et al. (2013) [67]	In vitro	CHIKV ZIKV	Inhibits viral protein synthesis by targeting host translation machinery	IC_{50} = 0.24 μM (plaque assay); EC_{50} = 0.29 μM (immunofluorescence assay)
	Lai, Z. et al. (2020) [68]	In vitro	ZIKV	Inhibits multiple stages: binding, entry, replication, release; also virucidal and prophylactic	~625 nM caused ~3-log reduction in viral RNA and titers
	Yang, Y. et al. (2023) [69]	In silico	SARS-CoV-2	Binds to spike RBD, TMPRSS2, and RBD–ACE2 complex to inhibit viral entry	High binding affinity in silico with RBD and TMPRSS2
25. ***Hernandezine*** (From: *Menispermaceae/Ranunculaceae* family) 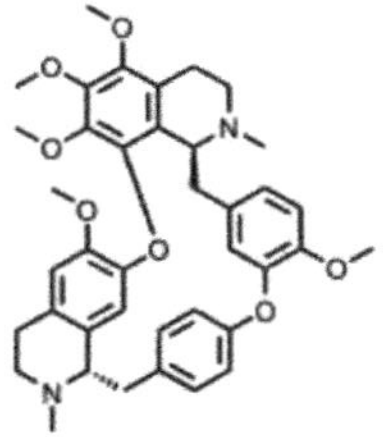	Majnooni, M. et al. (2021) [33]	*Review*	SARS-CoV-2	Blocking the calcium transition	EC_{50} = 10 μM,

https://doi.org/10.3390/molecules31030539

Table 1. *Cont.*

Compound & Origin	References	Study Type	Virus	Mechanisms	Values
26. ***Homoharringtonine*** (From: *Cephalotaxaceae* family) 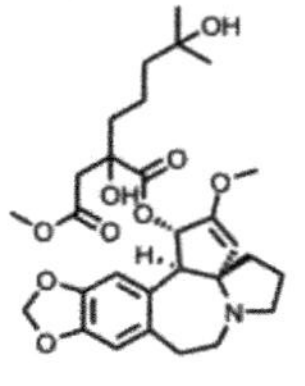	Dong. H. J. et al. (2018) [70]	In vitro	PEDV	Acts at early replication stages; additive effect with hydroxychloroquine (HCQ) or peptide tHR2	HHT (150 nM) reduced viral titers by ~3.5-fold; at 300 nM, approx. 40-fold reduction ($TCID_{50}$ assay)
	Gong, M. et al. (2019) [71]	In vitro	Foot-and-mouth disease virus (FMDV)	Inhibits early stages of replication	Dose-dependent inhibition
	Harisha, K. R. et al. (2025) [72]	In vitro	Rabies virus (RABV)	Post-entry and antiviral across both lab-adapted and clinical isolates	$EC_{50} \approx 0.3$ μM (BHK-21); $EC_{50} \approx 0.4$ μM (Neuro-2a); cell–cell spread inhibited at ≈ 1.0 μM
	Neerukonda, S. N. et al. (2020) [73]	Review	SARS-CoV-2	Likely suppresses viral replication by targeting phosphorylated eIF4E and inhibiting host protein translation	$EC_{50} \approx 2.10$ μM in Vero E6 cells
27. ***Indirubin*** (From: *Brassicaceae* family) 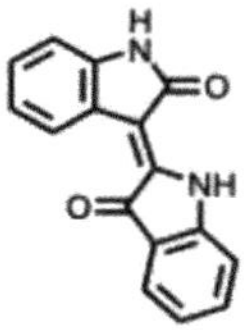	Mok, C. K. P. et al. (2014) [74]	In vitro	H5N1	Anti-inflammatory and antiviral via indirubin derivatives. Delays virus replication; reduces pro-inflammatory cytokines (IP-10)	Indirubin derivatives strongly suppress the pro-inflammatory cytokines including IP-10 (CXCL10), one of the key factors which contribute to the lung inflammation during H5N1 virus infection.
	Chang, S.-J. et al. (2012) [75]	In vitro and in vivo	JEV	Blocks viral attachment and has virucidal activity	$EC_{50} = 0.006–0.105$ mg/mL (concentration-dependent inhibition)
	Jie, C. et al. (2017) [76]	In vivo	IAV (H1N1)	Promotes MAVS-mediated IFN-β production and protects mitochondrial antiviral signaling.	Reduced lung NP levels, improved survival and lung pathology; enhanced IFN-β and IFITM3 signaling
	Medina-Moreno, S. et al. (2017) [77]	In vivo	HIV-1	Inhibits CDK9 to suppress HIV proviral transcription	Significant reduction in viremia at 5 mg/kg/day dosing
28. ***Lycorine and derivatives*** (From: *Amaryllidaceae* family) 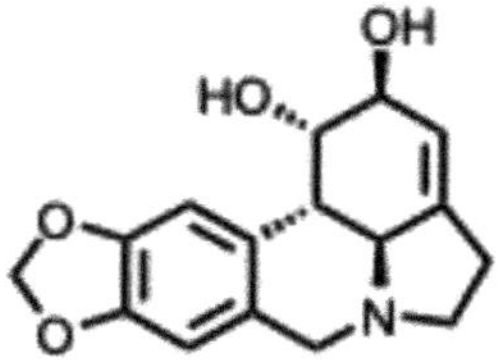	Jin, Y.-H. et al. (2021) [78]	In vitro	1. MERS-CoV	RdRp inhibition	$IC_{50} = 1.406 \pm 0.260$ μM
		In vitro	2. SARS-CoV	RdRp inhibition	$IC_{50} = 1.021 \pm 0.025$ μM
		In vitro	3. SARS-CoV-2	RdRp inhibition	$IC_{50} = 0.878 \pm 0.022$ μM
	Chen, H. et al. (2020) [79]	In vitro and in vivo	ZIKV	Post-entry inhibition of RdRp activity	$EC_{50} = 0.22–0.39$ μM in different cell lines; $CC_{50} = 4.4–21$ μM; RdRp $IC_{50} = 25$ μM; 10 mg/kg in mice protected against lethality (~83%)
	Li, N. et al. (2021) [80]	In vitro	CHIKV	Inhibits viral translation post-entry	$EC_{50} \approx 10$ μM
	Narayanan, A. et al. (2022) [81]	In vitro	SARS-CoV-2	M^{pro} inhibition; 88% reduction in viral spread in cell culture	$EC_{50} = 0.01$ μM (Lycorine HCl against M^{pro}; SI = 1878)
	Fielding, B. C. et al. (2020) [82]	In vitro	SARS-CoV-2	Likely host-targeted antiviral modulation	$EC_{50} = 300$ nM; SI ≈ 130

https://doi.org/10.3390/molecules31030539

Table 1. *Cont.*

Compound & Origin	References	Study Type	Virus	Mechanisms	Values
29. ***Matrine and derivatives*** (From: *Fabaceae* family) 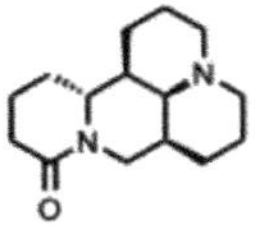	Pan, Q.-M. et al. (2015) [83]	In vitro	IAV H3N2	Inhibits viral replication	For Matrine type alkaloids: IC_{50} = 63.07–242.46 μM
	Qiao, W.-T. et al. (2024) [84]	In vivo	PEDV	Inhibits PEDV attachment and entry to cells.	EC_{50} = 0.09 μM (SI = 358.9)
30. ***Michellamine B*** (From: *Ancistrocladaceae* family) 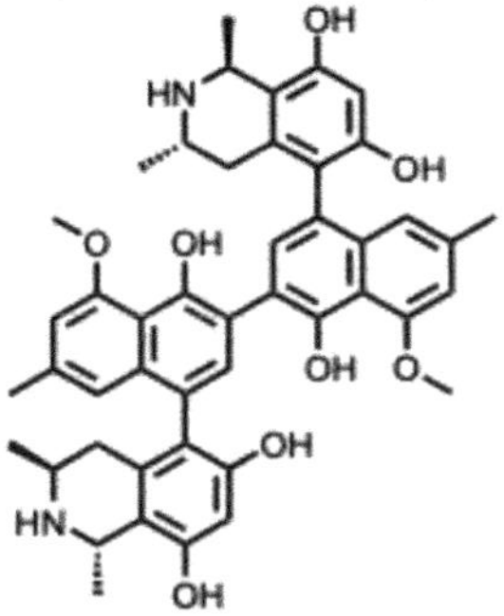	McMahon, J. B. et al. (1995) [85]	In vitro	HIV-1	Inhibits reverse transcriptase activity and viral-induced cellular fusion	EC_{50} = 1–20 μM Complete inhibition of cytopathic effects; RT enzymatic inhibition; $IC_{50} \approx 10$ μM for fusion
31. ***Neferine*** (From: *Nelumbonaceae* family) 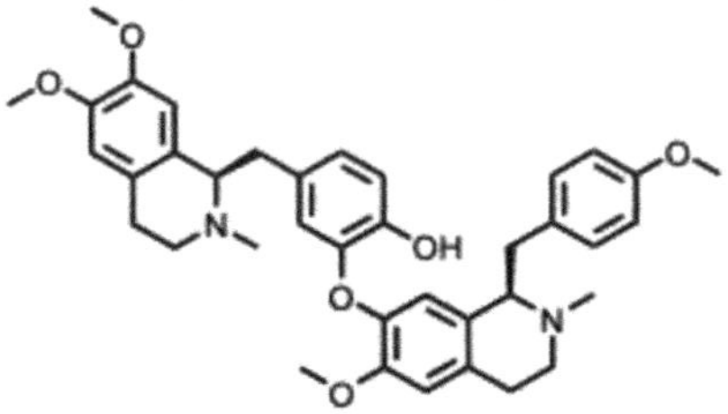	Yang, Y. et al. (2021) [86]	In vitro	SARS-CoV-2 Pseudovirus	Blocks host Ca^{2+}-dependent membrane fusion → viral entry	EC_{50} = 0.13–0.41 μM
	Yang, D. et al. (2025) [87]	In vitro	SARS-CoV-2	Enhanced virucidal activity by salt form	Neferine free base: EC_{50} = 36.01 μM; Neferine salt: EC_{50} = 4.78 μM

https://doi.org/10.3390/molecules31030539

Table 1. *Cont.*

Compound & Origin	References	Study Type	Virus	Mechanisms	Values
32. ***Panicutine*** (From: *Ranunculaceae* family) 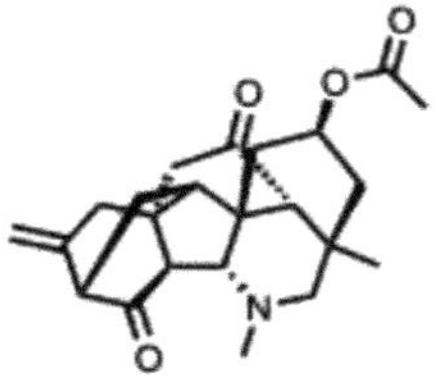	Kumar, A et al. (2022) [87]	In silico	SARS-CoV-2	Potential M^{pro} inhibition	Binding energy: −7.4 kcal/mol
33. ***Piperine*** (From: *Piperaceae* family) 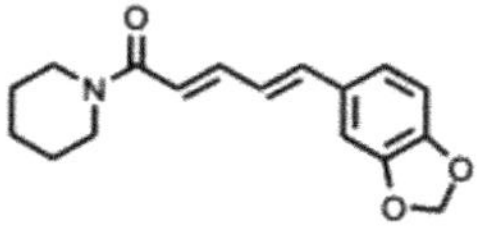	Nag, A. & Chowdhury, S. (2020) [88]	In silico	DENV EBOV	Potential inhibition of enzymatic targets	Stronger binding than ribavirin
	Pareek, A. et al. (2022) [89]	In vitro	CHIKV	RdRp inhibition, reduces viral replication	Kd = 0.08 μM (RdRp); EC_{50} = 6.68 μM
34. ***Pseudoephedrine*** (From: *Ephedraceae* family) 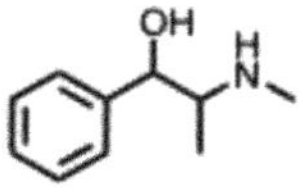	Deng, L et al. (2020) [90]	Review	IAV H1N1	Blunts cytokine storm, reduces lung inflammation, and inhibits virus replication	Increased life span in infected mice treated by PE
	Yu, S. et al. (2021) [91]	In vitro	SARS-CoV-2	Blocks viral entry and reduces inflammatory response	Inhibits spike pseudovirus entry; EC_{50} < 20 μM
35. ***Quinidine*** (From: *Rubiaceae* family) 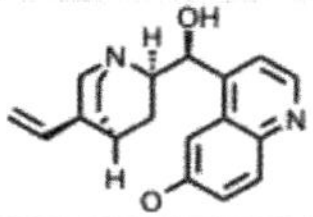	Yang et al. (2021) [92]	In vitro	SARS-CoV-2	Moderate antiviral activity among approved drugs in a high-throughput screen	IC_{50} = 0.42 μM (compound **6g**); IC_{50} = 1.41 μM (compound **7k**)
	Persoons, L. et al. (2021) [93]	In vitro	SARS-CoV-2 pseudovirus/HCoV-229E/OC43	Showed some broad-spectrum activity but only at relatively high doses	EC_{50} = 6 μM (SARS-CoV-2); EC_{50} = 0.2–9.4 μM (other HCoVs)
	Mamidala, E. et al. (2022) [94]	In silico	SARS-CoV-2 main protease (M^{pro})	Binds SARS-CoV-2 protease with predicted inhibitory affinity	Strong docking affinity

https://doi.org/10.3390/molecules31030539

Table 1. *Cont.*

Compound & Origin	References	Study Type	Virus	Mechanisms	Values
36. ***Quinine*** (Ттоııı. *Rubiaceae* family) 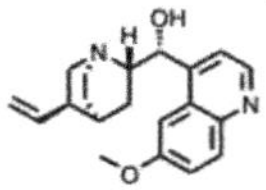	Malakar, S. et al. (2018) [95]	In vitro	DENV	Significant inhibition of DENV replication across multiple serotypes	~80% reduction vs. control
	D'Alessandro, S. et al. (2020) [96]	Review	IAV	Prophylactic plaque reduction at non-cytotoxic doses	NR
	Große, M. et al. (2021) [97]	In vitro	SARS-CoV-2	Complete viral inhibition at ≥50 µM; wider effective range depending on MOI and cell line	IC_{50} ≈ 25 µM (range: ~3.7–50 µM)
37. ***Rutaecarpine*** (From: *Rutaceae* family) 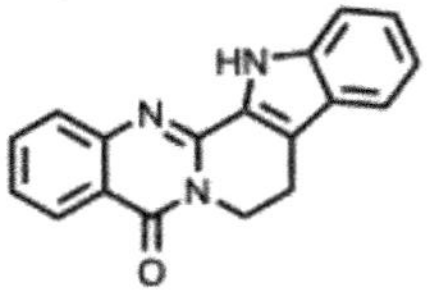	Lin, S. et al. (2023) [98]	In vitro and in silico	SARS-CoV-2 pseudovirus	Disrupts binding between spike protein and ACE2 receptor	IC_{50} ≈ 30 µM Omicron variant: IC_{50} ≈ 15 µM Binding energy ≈ −8.6 kJ/mol (wild-type), −10.2 kJ/mol (omicron)
38. ***Scopolamine*** (From: *Solanaceae* family) 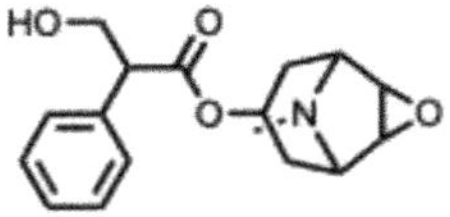	Bhattacharjee, A. et al. (2021) [99]	In ovo and in silico	JEV	Binds to the JEV NS5 protein and modulates TLR and IFN signaling pathways, potentiating antiviral innate immunity	Significant reduction in the viral load in CAM ($p < 0.0001$) and brain tissues ($p < 0.0001$) of the embryonated chick eggs when pre-treated with scopolamine hydrobromide
39. ***Sophocarpine*** (From: *Fabaceae* family) 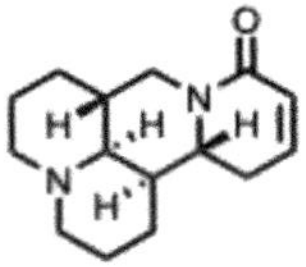	Jin, Z. et al. (2017) [100]	In vitro	Enterovirus 71 (E71)	Inhibits viral attachment, penetration, and RNA replication	IC_{50} = 350 µg/mL CC_{50} = 1346 µg/mL
40. ***Sophoridine*** (From: *Fabaceae* family) 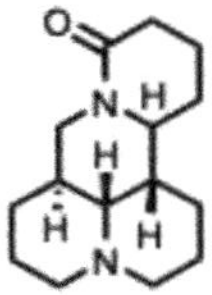	Ren, G. et al. (2019) [101]	In vitro	E71	Inhibits viral adsorption when added before infection	IC_{50} = 61.39 µg/mL

https://doi.org/10.3390/molecules31030539

Table 1. *Cont.*

Compound & Origin	References	Study Type	Virus	Mechanisms	Values
41. ***Strychnine*** (From: *Loganiaceae* family) 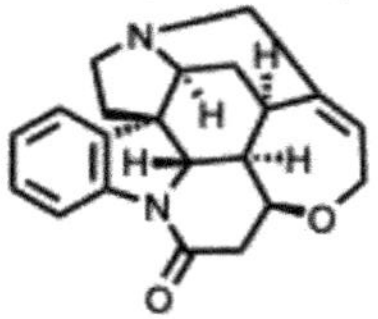	Hegazy, A. et al. (2023) [65]	In vitro	IAV H5N1 H1N1	Inhibits viral adsorption to host cells	IC_{50} = 11.85 μg/mL SI > 843 IC_{50} = 0.06 μg/mL; SI ≈ 167,000
42. ***Tetrahydropalmatine*** (From: *Papaveraceae* family) 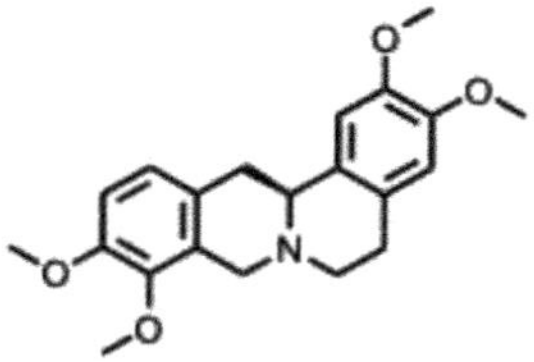	Lixia, H. et al. (2018) [102]	In vivo	JEV	Protects against neuronal apoptosis, reduces reactive oxygen/nitrogen species, lowers proinflammatory mediators, exhibits antiviral effects in brain tissues	Reduction in viral load in brain and CAM via neuroprotective and anti-inflammatory effects
43. ***Tetrandrine*** (From: *Menispermaceae* family) 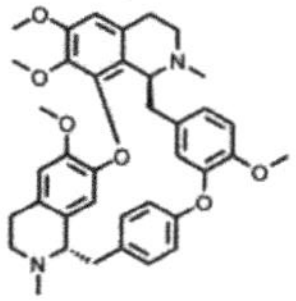	Kim, D. E. et al. (2019) [103]	In vitro	Human Coronavirus 0C43	Prevents viral replication and viral protein expression; effective mainly when administered pre- or during infection; also activates p38 MAPK signaling	IC_{50} ≈ 0.33 μM; SI > 40
	Liu, J. et al. (2023) [104]	In vitro and in vivo	SARS-CoV-2	Blocks viral entry at early stage by interfering with endosomal trafficking exhibits favorable lung biodistribution when inhaled	EC_{50} = 0.40–5.03 μM depending on a cell line
44. ***Tylophorine and analogues*** (From: *Apocynaceae* family) 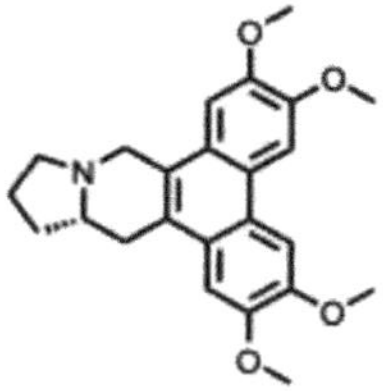	Fielding et al. (2020) [81]	In vitro	SARS-CoV	Blocks viral replication and cytopathic effects; potent inhibition of virus-induced apoptosis	EC_{50} = 5–340 nM (synthetic derivatives); 8–1468 nM (natural)
	T. I. M. et al. (2021) [105]	In vitro	SARS-CoV SARS-CoV-2	Inhibits viral RNA replication and NF-κB activation via JAK2 signaling pathway	IC_{50} = 58 nM (tylophorine), IC_{50} = 20 nM (7-methoxycryptopleurine)
	Yang et al. (2010) [106]	In vitro	Transmissible gastroenteritis virus (TGEV)	Targets viral RNA/RNP complex and inhibits NF-κB-mediated pro-inflammatory signaling	Substantial suppression of viral RNA replication (~4-log reduction)
	Wang, Y. et al. (2017) [107]	In vitro	Hepatitis C	Binds Hsc70 NBD, enhances ATPase activity → disrupts viral replication	Synthetic analogues of tylophorine DCB-3503 and rac-cryptopleurine EC_{50} = 30 nM EC_{50} =300 nM

https://doi.org/10.3390/molecules31030539

Table 1. *Cont.*

Compound & Origin	References	Study Type	Virus	Mechanisms	Values
45. ***Tryptanthrine*** (From: *Brassicaceae* family) 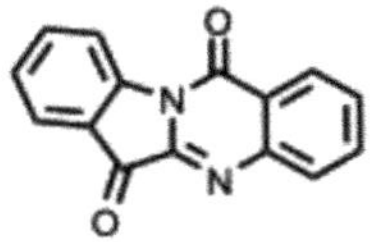	Mani, J. S. et al. (2020) [108]	Review	SARS-CoV-2 & Other HCoVs	Blocks coronaviral replication in early and late stages via inhibition of RdRp and papain-like protease	EC_{50} = 1.52 µM IC_{50} = 0.06 µM
46. ***Vilmorrianone*** (From: *Ranunculaceae* family) 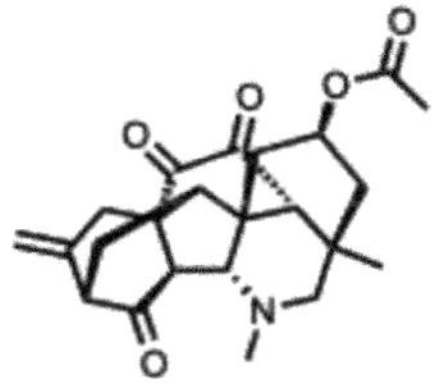	Kumar, A et al. (2022) [87]	In silico	SARS-CoV-2	Potential M^{pro} inhibition	Binding energy: −7.0 kcal/mol
47. ***Vinblastine*** (From: *Apocynaceae* family) 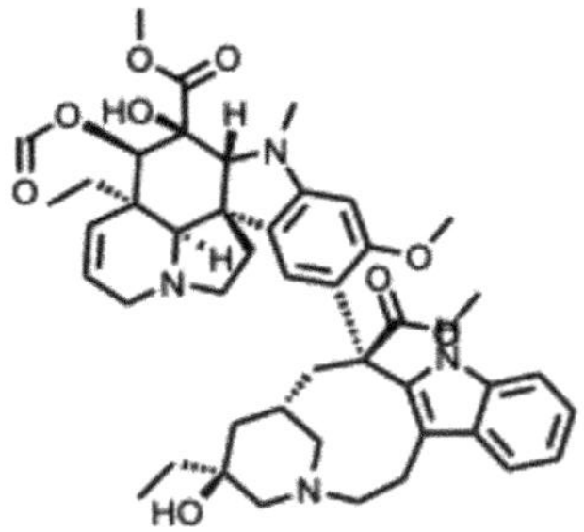	Akan, E. et al. (1997) [109]	In vitro	HIV-1	Vinblastine induced transcription through the HIV-1 long terminal repeat, suggesting modulation of NF-κB pathways, not antiviral action	~9–10-fold induction

When reported in µg/mL, IC_{50} or EC_{50} values were converted to µM using the molecular weight of the corresponding alkaloid when the chemical form was clearly specified. Values were left unconverted when compound form or purity was not explicitly stated. In the transversal summary table, viral families are reported using commonly accepted abbreviated forms to preserve readability. To ensure taxonomic clarity, a legend has been added below the table explicitly indicating the corresponding ICTV family names (e.g., CoV = Coronaviridae, Flavi = Flaviviridae, Alpha = Alphaviridae).

https://doi.org/10.3390/molecules31030539

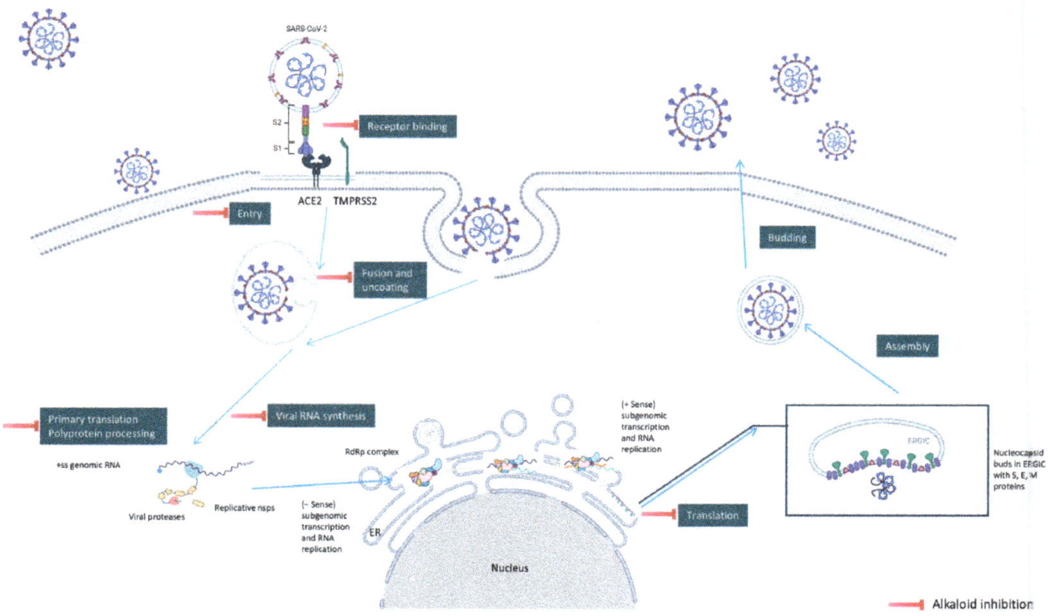

Figure 1. Overview of the SARS-CoV-2 replication cycle and the critical stages targeted by antiviral alkaloids.

4. Quaternary Alkaloids: Distinct Mechanisms and Pharmacology

Quaternary alkaloids constitute a distinct subclass of alkaloids defined by a permanently charged quaternary ammonium group. Because of this ionic structure, they exhibit physiochemical and pharmacological properties distinguishing them from other alkaloid types.

Quaternary alkaloids can influence several stages of the viral replication cycle, including entry, replication, protein synthesis, and modulation of host responses, similar to alkaloids in general. However, because of their permanent positive charge, these compounds have low membrane permeability and limited intracellular diffusion. As a result, their antiviral activity is often strongest during the early steps of infection, particularly at the level of viral attachment or entry. For example, tetrandrine, a known quaternary alkaloid, blocks viral entry by targeting NPC1, causing lysosomal cholesterol accumulation and triggering interferon responses via NPC1–STING [110].

Unfortunately, quaternary alkaloids display poor bioavailability, rapid clearance, and short half-lives [111]. However, emerging delivery technologies such as liposomes, nanoparticles, transdermal systems can improve absorption, extend circulation, and enhance pharmacological performance [111–113]. Another important issue with the quaternary alkaloids is that the activity often occurs near the cytotoxic thresholds [114]. The following (Table 2) summarizes the key quaternary alkaloids studied to date and their antiviral activities across RNA viruses.

https://doi.org/10.3390/molecules31030539

Table 2. Quaternary alkaloids displaying antiviral activity.

Compound & Origin	References	Study Type	Virus	Mechanisms	Antiviral Information Reported in the Original Studies (IC_{50}, EC_{50}, SI, Docking Scores, Qualitative Observations) *
1. ***Berberine and derivates*** (From: *Berberidaceae/Ranunculaceae* family) 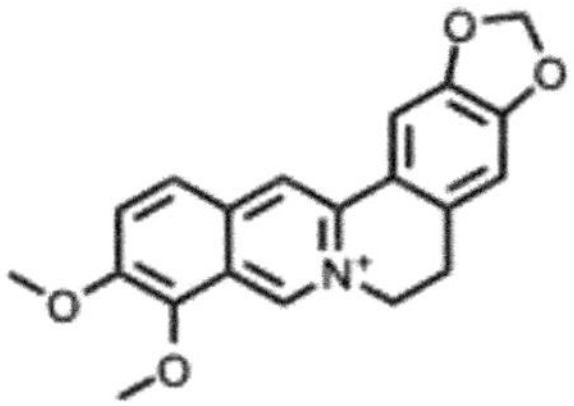	Wu et al. 2011 [115]	In vitro and in vivo	IAV	Berberine exerted strong inhibition on the inflammatory substances production	IC_{50} = 0.025 g/L. Decrease in mice mortality from 90% to 55%
	Wang et al. 2018 [116]	In vitro	EV71	Might inhibit MEK/ERK, suppresses autophagy (AKT, JNK, PI3KIII)	IC_{50}: 7.12–14.8 μM (Compound **2d**) and 7.43–10.25 μM (berberine)
	Shao et al. 2020 [117]	In vitro and in silico	HIV-1, clade B	It binds in the pocket of NHR and CHR of gp41.	IC_{50}: 5.5–10.25 μg/mL
	Ratanakomol et al. 2021 [118]	In vitro	DENV, ZIKV, CHIKV	Potential AMPK activation, lipid metabolism disruption, direct virucidal activity	IC_{50}: DENV: 42.87 μM; ZIKV: 11.42 μM; CHIKV: 14.21 μM
	Botwina et al. 2020 [119]	In vitro	IAV (H3N2)	Inhibits MAPK/ERK	IC_{50}: MDCK = 52 μM; A549 = 17 μM; LET1 = 4 μM; Human airway epithelial (HAE) = 16 μM
	Enkhtaivan et al. 2017 [120]	In vitro and in silico	IAV	Competitive neuraminidase inhibition	Berberine derivatives IC_{50} H1N1: 0.87–1.63 μg/mL H3N2: 1.15–2.98 μg/mL
	Varghese et al. 2016 [121]	In vitro	CHIKV	Might be affecting one or several host factors important for CHIKV replication	IC_{50}: 1.9 ± 0.9 μM
	Nguyen, C. Q. et al. (2021) [122]	In vitro and in silico	ZIKV	Potential candidate to inhibit NS2B-NS3 protease	Compound **4d**: Selectivity index (SI): 15.3 IC_{50} = 5.3 ± 1.9 μM
2. ***Chelerythrine*** (From: *Papaveraceae* family) 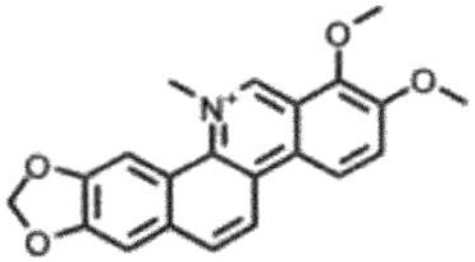	Españo, E. et al. (2022) [123]	In vitro, in silico	ZIKV	Potential entry/attachment inhibition	EC_{50} = 692.4 nM SI = 6.0
	Loe, M. et al. 2023 [124]	In vitro, in vivo	ZIKV	Potent inhibitor of ZIKV infection that targets the ZIKV NS4B protein	Chelerythrine chloride IC_{50} = 0.2513 μM. A 1.73 and 2.0 log 10 reduction in RNA copies/mL
	Guo, W. et al. 2020 [125]	In vitro (*plant*)	Tobacco Mosaic Virus (TMV)	Inactivation/proliferation inhibition	Chelerythrine at 0.5 mg/mL: 72.67% inactivation, (corresponding to 1.4 mM **)

https://doi.org/10.3390/molecules31030539

Table 2. *Cont.*

Compound & Origin	References	Study Type	Virus	Mechanisms	Antiviral Information Reported in the Original Studies (IC_{50}, EC_{50}, SI, Docking Scores, Qualitative Observations) *
3. ***Dehydrocorydaline*** (From: *Papaveraceae* family) 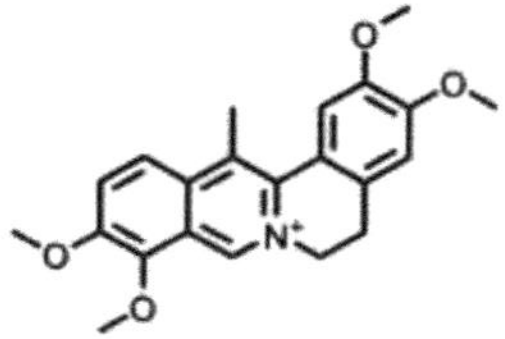	Orhan, I. et al. 2007 [126]	In vitro	Parainfluenza-3	N/A	CPE inhibitory concentration 16 μg/mL (≈ 40–44 μM ***)
4. ***Oxymatrine*** (From: *Fabaceae* family) 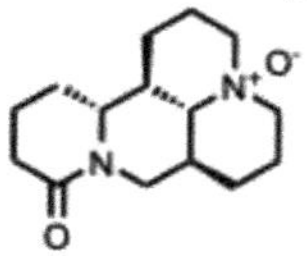	Dai, J. et al. 2018 [127]	In vitro and in vivo	IAV multiple strains (H1N1, H9N2, H5N1, H3N2)	Could significantly decrease the promoter activity of TLR signaling pathways TLR4, p38 MAPK, NF-κB	EC_{50}: 5.91–23.67 μg/mL (strain-dependent) (= 22–90 μM *)
	Zhi et al. 2024 [128]	In vitro	H9N2 Avian Influenza Virus (AIV)	TLR signaling pathways TLR3, NF-κB, IRF-3	Dose-dependent effects on the cell survival rate
	Chen, N. et al. (2016) [129]	In vitro	HCV	Proliferation inhibition	Inhibition of cell proliferation: up to 85.4% at 72 h, 12mg/mL; mRNA expression 0.59 ± 0.12 vs. control
5. ***Oxysophoridine*** (From: *Fabaceae* family) 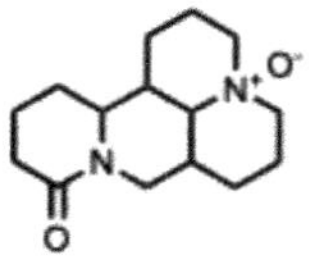	Majnooni, M. et al. (2021) [33]	*Review*	SARS-CoV-2	Nucleotide biosynthesis inhibitor	EC_{50} = 0.31 μM,

https://doi.org/10.3390/molecules3103053

Table 2. *Cont.*

Compound & Origin	References	Study Type	Virus	Mechanisms	Antiviral Information Reported in the Original Studies (IC_{50}, EC_{50}, SI, Docking Scores, Qualitative Observations) *
6. ***Palmatine*** (From: *Ranunculaceae; Rutaceae* family) 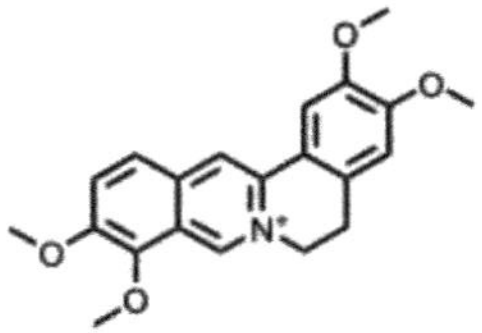	W. Zhang, 2024 [130]	In vitro	Infectious Bronchitis virus	Inactivating the virus, inhibiting its replication, modulating NF-κB/IRF7/JAK-STAT signaling pathways, and regulating apoptosis	IC_{50} = 7.76 μM Selection index (SI) was 86.74
	Fan Jia et al. 2010 [131]	In vitro	WNV	Palmatine could significantly inhibit the activity of NS2B-NS3 protease and that the inhibition was reversible	WNV: EC_{50}: 3.6 μM, IC_{50}: 96 μM
	Yi-Jung Ho et al. 2019 [132]	In vitro	ZIKV JEV	Inhibits Zika virus infection by disrupting virus binding, entry, and stability	Inhibited ZIKV binding by 95% and ZIKV entry by 69% Palmatine from 20–80 mM decreased JEV RNA levels.
7. ***Sanguinarine*** (From: *Papaveraceae* family) 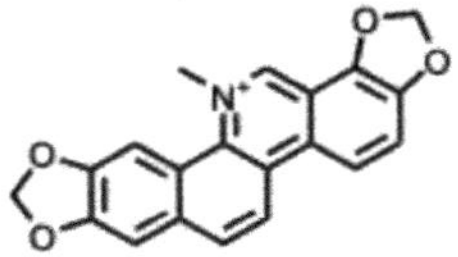	Qiyun Ke et al. 2023 [133]	In vitro and in silico	PRRSV	Targeting internalization, replication, and release stages of the viral life cycle	Sanguinarine inhibits the propagation of PRRSV in a dose-dependent manner

* When reported in μg/mL, IC_{50} or EC_{50} values were converted to μM using the molecular weight of the corresponding alkaloid when the chemical form was clearly specified. Values were left unconverted when compound form or purity was not explicitly stated. ** Converted assuming chelerythrine base; values were not converted when salt form was explicitly reported or not specified. *** Converted using the molecular weight of dehydrocorydaline (base) or dehydrocorydaline chloride when the salt form was specified; otherwise reported as an approximate range. In the transversal summary table, viral families are reported using commonly accepted abbreviated forms to preserve readability. To ensure taxonomic clarity, a legend has been added below the table explicitly indicating the corresponding ICTV family names (e.g., CoV = Coronaviridae, Flavi = Flaviviridae, Alpha = Alphaviridae).

https://doi.org/10.3390/molecules31030539

To improve clarity and methodological transparency, we compiled an exhaustive transversal overview summarizing the level of experimental evidence supporting the antiviral activity of each alkaloid discussed in this review. Alkaloids are traditionally presented alphabetically to facilitate compound identification; however, this organization may obscure differences in experimental validation levels. The present table (Table 3) addresses this limitation by distinguishing in silico, in vitro, and in vivo evidence without duplicating compound-specific antiviral data already detailed in Tables 1 and 2. Missing information is indicated by "–", reflecting the absence of reported data rather than negative results. This overview highlights the current imbalance between extensive computational or cell-based screening and the limited number of compounds validated in vivo, underscoring key translational gaps in alkaloid-based antiviral research.

Table 3. Experimental Evidence Levels for Alkaloid Antivirals against RNA Viruses.

Alkaloid	In Silico	In Vitro	In Vivo	Virus families	Main Targets
1,2,3,4-Tetrahydroisoquinoline derivatives	✓	✓	✓	Orthomyxo, Retro, CoV	RdRp (PAN), IN
7-Methoxycryptopleurine	–	✓	–	CoV	Blocking the S and N proteins, 3CLpro inhibitor
10-Hydroxyusambarensine	✓	–	–	CoV	RdRp
Ajmaline	✓	–	–	Filo	Host gene regulation
Aloperine	–	✓	✓	Flavi, CoV	RdRp/Cathepsin B
Amarbellisine	–	✓	–	CoV	Potential inhibition of viral replication
Berbamine	–	✓	–	CoV	Endolysosomal trafficking
Berberine	✓	✓	✓	Orthomyxo, Flavi, Alpha, Retro	Host pathways/Entry
Canthin-6-one derivatives	✓	✓	–	Paramyxo, CoV	Proteases/Entry
Capsaicin	–	✓	✓	Orthomyxo, Picorna	STAT3/Host immunity
Cepharanthine	✓	✓	✓	CoV, Flavi, Alpha, Retro	Entry/Host pathways
Cephaeline	✓	✓	✓	CoV, Flavi	Ribosome/RdRp
Cherylline	✓	✓	–	Flavi	RdRp
Cinchonine	–	✓	–	CoV	Autophagy
Colchicine	✓	✓	–	Orthomyxo	Adsorption/Replication
Conessine	–	✓	–	CoV	M^{pro} inhibitor
Cyclopamine	–	✓	✓	Pneumoviridae	Inclusion bodies (M2-1)
Dehydroevodiamine	–	✓	–	CoV	ERK/MAPK
Deoxynojirimycin derivatives	–	✓	–	Orthomyxo, Flavi	α-glucosidase
Emetine	✓	✓	✓	CoV, Flavi, Filo, Retro	Translation/Entry
Ephedrine	–	✓	✓	Orthomyxo	TLR signaling
Fangchinoline	–	✓	✓	Flavi, Retro, Picorna	Viral internalization
Harmaline	–	✓	–	Orthomyxo	–
Harmine	–	✓	–	Orthomyxo, CoV, Retro	SR kinase
Harringtonine	✓	✓	–	Alpha, Flavi, CoV	Translation
Hernandezine	–	✓	–	CoV	Blocking calcium transition
Homoharringtonine	–	✓	✓	CoV, Flavi, Rhabdo	Translation
Indirubin	–	✓	✓	Orthomyxo, Flavi, Retro	MAVS/Cytokines
Lycorine	✓	✓	✓	CoV, Flavi, Alpha	RdRp
Matrine	–	✓	✓	Orthomyxo, Arteri	Viral entry
Michellamine B	–	✓	–	Retro	Reverse transcriptase
Neferine	–	✓	–	CoV	Ca^{2+}-dependent fusion
Oxysophoridine	–	✓	–	CoV	Nucleotide biosynthesis inhibitor
Panicutine		✓			Protease (M^{pro})
Piperine	✓	✓	–	Flavi, Alpha	RdRp
Pseudoephedrine	–	✓	✓	Orthomyxo, CoV	Entry/Inflammation
Quinidine	✓	✓	–	CoV	Protease (M^{pro})

https://doi.org/10.3390/molecules31030539

Table 3. *Cont.*

Alkaloid	In Silico	In Vitro	In Vivo	Virus families	Main Targets
uinine	–	✓	–	Flavi, CoV	–
utaecarpine	✓	✓	–	CoV	Spike–ACE2
copolamine	✓	✓	–	Flavi	NS5/Innate immunity
ophocarpine	–	✓	–	Picorna	Viral attachment
ophoridine	–	✓	–	Picorna	Viral adsorption
trychnine	–	✓	–	Orthomyxo	Viral adsorption
etrahydropalmatine	–	–	✓	Flavi	Neuroprotection
etrandrine	✓	✓	✓	CoV, Orthomyxo	Entry/Endosomes
ylophorine & analogues	–	✓	–	CoV, Alpha	RdRp/NF-κB
ryptanthrine	–	✓	–	CoV	RdRp/PL^{pro}
ilmorrianone	✓	–	–	CoV	Protease (M^{pro})
inblastine	–	✓	–	Retro	Host transcription (No antiviral activity reported; host transcription modulation)

✓ = at least one primary study reported; – = no data reported in the reviewed literature. In the transversal summary table, viral families are reported using commonly accepted abbreviated forms to preserve readability. To ensure taxonomic clarity, a legend has been added below the table explicitly indicating the corresponding ICTV family names (e.g., CoV = Coronaviridae, Flavi = Flaviviridae, Alpha = Alphaviridae).

5. Discussion

Across both datasets, a coherent picture emerges, alkaloids represent a structurally diverse and mechanistically rich source of antiviral agents, with activity documented against a broad range of RNA viruses. Throughout this review, antiviral evidence is discussed according to the experimental level at which it was generated, namely in silico, in vitro, and in vivo, to reflect the degree of biological validation associated with each approach. In the first dataset, 42 compounds supported by 77 experimental entries show that approximately 45 percent of studies were performed exclusively in vitro, around 30 percent combined in vitro and in vivo approaches, and 25 percent relied mainly on in silico predictions. While computational studies provide valuable hypotheses and help prioritize candidate scaffolds, they do not constitute proof of antiviral efficacy and therefore require careful interpretation in the absence of experimental validation. This imbalance highlights extensive early screening but a shortage of deeper, mechanistic, and translational work. Another major bottleneck often occurs at the "in vitro–in vivo" transition, where a majority of compounds showing antiviral activity in vitro ultimately fail to demonstrate efficacy in vivo [23].

A central theme is the convergence of many alkaloids on viral replication machinery, particularly the RdRp. Compounds such as tetrahydroisoquinoline derivatives, lycorine, and aloperine consistently interact with RdRp or polymerase-associated subunits, often at low micromolar or nanomolar levels. Their rigid, polycyclic scaffolds appear intrinsically suited to binding nucleic acid–proximal enzymatic pockets, which may underlie their broad-spectrum effectiveness across coronaviruses, orthoflaviviruses, alphaviruses, and orthomyxoviruses.

A second mechanistic cluster involves entry inhibition. Neferine, fangchinoline, rutaecarpine, berbamine and capsaicin interfere with viral attachment, membrane fusion, or endosomal trafficking. These activities correlate with amphiphilic or cationic structural elements capable of perturbing lipid or acidic intracellular environments. Because many of these effects are host-directed, resistance may emerge more slowly, though this benefit must be balanced against increased toxicity risks.

Host-directed immunomodulation represents a third recurring pattern. Indirubin, emetine, ephedrine, and canthin-6-one derivatives modulate key pathways including NFκB, JAK, STAT, TLR, MAPK, and ERK. This dual antiviral and anti-inflammatory activity may

https://doi.org/10.3390/molecules31030539

be especially valuable in infections characterized by immunopathology such as influenza and SARS-CoV-2. At the same time, manipulating host signaling requires careful dosing and safety evaluation.

The second dataset, focused on berberine, chelerythrine, dehydrocorydaline, oxymatrine, palmatine, and sanguinarine, reveals similar mechanistic tendencies but with a clearer emphasis on innate immunity, inflammatory regulation, and endolysosomal physiology. Study distribution again skews toward in vitro work (≈50 percent), though about 35 percent include in vivo components. Berberine, palmatine, and oxymatrine frequently modulate TLR3, TLR4, NFκB, IRF3, AMPK, and autophagy-related pathways, producing broad-spectrum antiviral effects across influenza viruses, enteroviruses, flaviviruses, chikungunya virus, hepatitis C virus, and SARS-CoV-2. Meanwhile, quaternary or quaternary-like alkaloids such as chelerythrine and sanguinarine are particularly effective at disrupting viral entry and endosomal maturation through their permanent or semi-permanent cationic charge.

Cepharanthine (CEP) is one of the few alkaloids near clinical translation. Although an enhanced oral formulation (PD 001) received FDA clearance for Phase I and II testing in mild to moderate COVID 19, no PD 001 trials have begun, and the only active study is a Phase II placebo-controlled trial in mild COVID 19 (NCT05398705). Because intravenous CEP is impractical and standard oral dosing fails to reach antiviral levels, animal pharmacokinetic data indicating over 64 percent bioavailability with pulmonary delivery highlight inhaled formulations as the most promising route for clinical development [45].

Despite these strengths, significant limitations remain. Many alkaloids show narrow therapeutic windows, with active concentrations approaching cytotoxic levels. Depending on their chemical structure and administered dose, alkaloids may induce neurotoxicity, hepatotoxicity, nephrotoxicity, or cardiotoxicity [134]. Moreover, species-specific differences in metabolism and organ sensitivity further complicate the translation of preclinical findings to humans. A well-documented example is the class of pyrrolizidine alkaloids, which are strongly associated with hepatotoxic and carcinogenic effects due to their metabolic conversion into reactive intermediates [135]. Toxicity assessment is often limited to single- dose MTT assay (3-(4,5-dimethylthiazol-2-yl)-2,5-diphenyltetrazolium bromide assay) which measures cell metabolic activity or LDH assays (Lactate Dehydrogenase release assay) which measure cell membrane integrity in immortalized cell lines. However, these assays cannot predict organ-specific or long-term effects. Pharmacokinetic challenges such as poor oral bioavailability, limited solubility, extensive first-pass metabolism and short half-lives, further restrict their translational potential [136]. These issues highlight the need for optimized formulations, including nanoparticles, liposomes, prodrugs, and targeted delivery systems.

Beyond conventional molecular docking and virtual screening approaches, artificial intelligence (AI) and machine learning (ML)-based methods are increasingly being explored in antiviral drug discovery to prioritize natural compounds, including alkaloids. These approaches differ fundamentally from classical docking by integrating large multidimensional datasets to identify non-obvious structure–activity relationships through quantitative structure–activity relationship (QSAR) models, supervised learning algorithms, or deep neural networks. In principle, AI-driven screening may enable rapid prioritization of alkaloids with multi-target potential, optimization of scaffold–activity relationships, and early prediction of pharmacokinetic and ADMET properties. However, despite their growing visibility, the application of AI-based models to alkaloid antivirals against RNA viruses remains largely prospective. Model performance is strongly dependent on the quality, size, and chemical diversity of training datasets, which are often biased toward synthetic compounds and well-characterized viral targets. Moreover, AI predictions require

https://doi.org/10.3390/molecules3103053

experimental validation, as computational confidence does not necessarily translate into antiviral efficacy or acceptable toxicity profiles. At present, AI-based approaches should therefore be viewed as complementary hypothesis-generating tools rather than substitutes for experimental screening, and their integration into alkaloid-based antiviral research will depend on the generation of standardized, high-quality biological datasets.

In addition, natural source diversity remains underexplored. While terrestrial plants dominate the current research landscape, marine organisms and endophytic fungi represent rich reservoirs of chemically unique alkaloids featuring halogenation, sulfur bridges, or complex polycyclic structures [137,138]. Advances in metagenomics, genome mining, and synthetic biology offer powerful tools to identify and optimize these underused scaffolds, potentially unlocking new antiviral pharmacophores with mechanisms distinct from those found in traditional plant-derived compounds.

Complementary to these discovery-driven approaches, late-stage functionalization (LSF) strategies, including photochemical transformations, provide a powerful means to expand chemical diversity directly from complex natural alkaloid scaffolds. By enabling selective modification at advanced synthetic stages, LSF allows the generation of novel scaffold variants and unexplored chemical space that may not be accessible through biosynthesis or de novo synthesis alone, thereby broadening the pool of potential antiviral pharmacophores.

Resistance considerations also shape the translational landscape. Traditional antiviral development often suffers from the "one drug- one bug" limitation, where each compound is tailored to a single virus. Because many alkaloids act on conserved host pathways or multiple viral targets, they may avoid this bottleneck, reducing the likelihood of rapid resistance. Combination therapies represent another promising avenue [134]. Berberine's ability to enhance antibiotic activity via efflux inhibition and immunomodulation suggests that pairing alkaloids with established antivirals could yield synergistic effects [139]. Still, such strategies require thorough pharmacokinetic and toxicological evaluation before advancing to clinical stages.

Major methodological and infrastructural challenges persist. Many RNA viruses of public health concern require high-containment biosafety laboratories, which remain limited globally. As a result, many studies rely on pseudoviruses or computational models rather than live virus assays. RT-qPCR is used to quantify viral RNA, but it does not measure infectious virus. Viral titration assays which directly quantify infectious particles remain the gold standard for assessing antiviral efficacy. Pharmacokinetic analyses and tissue distribution studies are rare, and in vivo experiments often rely on short-term endpoints rather than detailed immunological or virological profiling. More robust preclinical and clinical work is needed to establish dosing, optimize structures, and generate high-quality efficacy data [135,140].

Vaccination remains a central pillar of RNA virus control. Yet rapid antigenic evolution, as seen in influenza A virus and SARS-CoV-2, limits vaccine durability [141]. Newer mRNA and vector-based platforms improve adaptability, but they cannot replace the need for broad-spectrum antivirals. Alkaloids, with their diverse mechanisms and host-targeted effects, may serve as complementary tools alongside vaccination to strengthen preparedness for future outbreaks.

These findings highlight both the promise and the challenges of alkaloid-based antiviral development. Their structural diversity, multi-target mechanisms, and activity across unrelated RNA viruses make them compelling candidates for next-generation therapeutics. Yet meaningful progress will require deeper mechanistic validation, expanded in vivo research, improved pharmacokinetic and toxicity profiling, and better access to biosafety level laboratories infrastructure. By combining modern drug development strategies with

https://doi.org/10.3390/molecules31030539

the chemical richness of alkaloids, it may be possible to generate broad-spectrum antivirals capable of responding quickly and effectively to future RNA virus threats.

6. Conclusions

Overall, the evidence shows that alkaloids form a rich and versatile source of antiviral candidates, acting through multiple conserved mechanisms including RdRp inhibition, blockade of viral entry, disruption of glycoprotein maturation, and modulation of host immune pathways. Several compounds, such as lycorine, emetine, homoharringtonine, cepharanthine, berberine, and palmatine, demonstrate broad spectrum activity across diverse RNA viruses, suggesting that alkaloid scaffolds are well suited for addressing the rapid evolution and cross species transmission typical of these pathogens.

Yet significant gaps hinder clinical translation. Most studies depend on in vitro assays without confirming true viral suppression, in vivo evidence remains limited, and toxicity, bioavailability, and pharmacokinetic profiles are poorly characterized. Many alkaloids also have narrow therapeutic windows, and the lack of high containment biosafety facilities restricts more rigorous evaluation. To advance these compounds meaningfully, future research must prioritize standardized antiviral testing, comprehensive toxicology, improved delivery strategies, and exploration of combination therapies. With such efforts, alkaloids could evolve from promising laboratory findings into antivirals for future RNA virus threats.

Author Contributions: The manuscript was written through contributions of all authors. All authors have read and agreed to the published version of the manuscript.

Funding: This research was supported by the National Fund for Scientific Research (FNRS) with a FRIA grant to K.L. [grant number: 40009257] and with a CDR grant to A.L. [grant number: 40021286] and as well as the Foundation Léon Frédéricq.

Institutional Review Board Statement: Not applicable.

Informed Consent Statement: Not applicable.

Data Availability Statement: No new data were created or analyzed in this study. Data sharing is not applicable to this article.

Acknowledgments: The authors warmly thank Michel Frédérich and Luc Angenot for their generous guidance and for sharing their extensive knowledge of alkaloids, which substantially contributed to the conceptual framing and refinement of this review.

Conflicts of Interest: The authors declare no conflicts of interest.

Abbreviations

The following abbreviations are used in this manuscript:

ACE2	Angiotensin Converting Enzyme 2
ADMET	Absorption, Distribution, Metabolism, Excretion, and Toxicity
AIV	Avian Influenza Virus
AMPK	AMP-activated Protein Kinase
CAM	Chorioallantoic Membrane
CC_{50}	50% Cytotoxic Concentration
CEB	Cepharanthine (context dependent, but in your text CEP is used)
CEP/PD-001	Cepharanthine/PharmaDrug oral formulation PD-001
CHIKV	Chikungunya Virus
CI	Confidence Interval
CPE	Cytopathic Effect
COVID-19	Coronavirus Disease 2019

https://doi.org/10.3390/molecules31030539

CYP450	Cytochrome P450 Enzyme System
DENV	Dengue Virus
DNJ	Deoxynojirimycin
EC_{50}	50% Effective Concentration
EBOV/Ebola	Ebola Virus
ELISA	Enzyme-Linked Immunosorbent Assay (appears implied in mechanistic work)
ER	Endoplasmic Reticulum
ERK	Extracellular Signal-Regulated Kinase
FDA	U.S. Food and Drug Administration
GP	Glycoprotein
HAE	Human Airway Epithelium
HBV	Hepatitis B Virus (if referenced)
HCV	Hepatitis C Virus
HCoV	Human Coronavirus
HIV	Human Immunodeficiency Virus
hRSV	Human Respiratory Syncytial Virus
IC_{50}	50% Inhibitory Concentration
IFN	Interferon
IL	Interleukin
IRF	Interferon Regulatory Factor
JAK	Janus Kinase
JEV	Japanese Encephalitis Virus
JNK	c-Jun N-terminal Kinase
LDH	Lactate Dehydrogenase
LEDGF/p75	Lens Epithelium Derived Growth Factor
MAPK	Mitogen-Activated Protein Kinase
MERS-CoV	Middle East Respiratory Syndrome Coronavirus
MOI	Multiplicity of Infection
Mpro (3CLpro)	Main Protease of SARS-CoV-2
MTT	3-(4,5-Dimethylthiazol-2-yl)-2,5-Diphenyltetrazolium Bromide
NA	Neuraminidase
NBD	Nucleotide Binding Domain
NDV	Newcastle Disease Virus
NF-κB	Nuclear Factor kappa-light-chain-enhancer of activated B cells
NS proteins	Non-structural viral proteins (e.g., NS2B, NS3, NS4B, NS5)
PA	Polymerase Acidic Protein (Influenza)
PEDV	Porcine Epidemic Diarrhea Virus
PK	Pharmacokinetics
qPCR/RT-qPCR	Quantitative Real-Time Polymerase Chain Reaction
RdRp	RNA-dependent RNA Polymerase
RNP	Ribonucleoprotein
ROS	Reactive Oxygen Species
RSV/BRSV	Respiratory Syncytial Virus/Bovine RSV
SI	Selectivity Index
SARS-CoV-2	Severe Acute Respiratory Syndrome Coronavirus 2
TLR	Toll-Like Receptor
TMPRSS2	Transmembrane Protease Serine 2
TRPML	Transient Receptor Potential Mucolipin
TGEV	Transmissible Gastroenteritis Virus
US NLM	United States National Library of Medicine
VSV	Vesicular Stomatitis Virus
WNV	West Nile Virus
ZIKV	Zika Virus

https://doi.org/10.3390/molecules31030539

References

1. Forterre, P.; Prangishvili, D. The origin of viruses. *Res. Microbiol.* **2009**, *160*, 466–472. [CrossRef] [PubMed]
2. Baltimore, D. Expression of animal virus genomes. *Bacteriol. Rev.* **1971**, *35*, 235–241. [CrossRef]
3. Gebreyes, W.A.; Dupouy-Camet, J.; Newport, M.J.; Oliveira, C.J.B.; Schlesinger, L.S.; Saif, Y.M.; Kariuki, S.; Saif, L.J.; Saville, W Wittum, T.; et al. The Global One Health Paradigm: Challenges and Opportunities for Tackling Infectious Diseases at the Huma Animal, and Environment Interface in Low-Resource Settings. *PLoS Neglected Trop. Dis.* **2014**, *8*, e3257. [CrossRef]
4. Ludwig, B.; Kraus, F.B.; Allwinn, R.; Doerr, H.W.; Preiser, W. Viral zoonoses—A threat under control? *Intervirology* **2003**, *46*, 71–7 [CrossRef]
5. Tao, K.; Tzou, P.L.; Nouhin, J.; Gupta, R.K.; de Oliveira, T.; Kosakovsky Pond, S.L.; Fera, D.; Shafer, R.W. The biological an clinical significance of emerging SARS-CoV-2 variants. *Nat. Rev. Genet.* **2021**, *22*, 757–773. [CrossRef]
6. V'kovski, P.; Kratzel, A.; Steiner, S.; Stalder, H.; Thiel, V. Coronavirus biology and replication: Implications for SARS-CoV-2. *Na Rev. Microbiol.* **2021**, *19*, 155–170. [CrossRef]
7. Mlcochova, P.; Kemp, S.A.; Dhar, M.S.; Papa, G.; Meng, B.; Ferreira, I.A.T.M.; Datir, R.; Collier, D.A.; Albecka, A.; Singh, S.; et a SARS-CoV-2 B.1.617.2 Delta variant replication and immune evasion. *Nature* **2021**, *599*, 114–119. [CrossRef]
8. Karim, S.S.A.; Karim, Q.A. Omicron SARS-CoV-2 variant: A new chapter in the COVID-19 pandemic. *Lancet* **2021**, *398*, 2126–212 [CrossRef] [PubMed]
9. Ewen, C. Omicron Variant Puts. *Nature* **2021**, *600*, 21. [CrossRef] [PubMed]
10. Carrasco-Hernandez, R.; Jácome, R.; Vidal, Y.L.; de León, S.P. Are RNA viruses candidate agents for the next global pandemic? review. *ILAR J.* **2017**, *58*, 343–358. [CrossRef] [PubMed]
11. Bimrew, S.; Abera, M. Review on Classification and Nomenclature of Viruses. *Am. J. Life Sci.* **2023**, *11*, 11–23. [CrossRef]
12. Simmonds, P.; Adams, M.J.; Benkő, M.; Breitbart, M.; Brister, J.R.; Carstens, E.B.; Davison, A.J.; Delwart, E.; Gorbalenya, A.E Harrach, B.; et al. Consensus statement: Virus taxonomy in the age of metagenomics. *Nat. Rev. Microbiol.* **2017**, *15*, 161–16 [CrossRef]
13. Silva, J.M.F.; Melo, F.L.; Elena, S.F.; Candresse, T.; Sabanadzovic, S.; Tzanetakis, I.E.; Blouin, A.G.; Villamor, D.E.V.; Mollov, D Constable, F.; et al. Virus classification based on in-depth sequence analyses and development of demarcation criteria using th Betaflexiviridae as a case study. *J. Gen. Virol.* **2022**, *103*, 001806. [CrossRef]
14. Bhambhani, S.; Kondhare, K.R.; Giri, A.P. Diversity in Chemical Structures and Biological Properties of Plant Alkaloids. *Molecule* **2021**, *26*, 3374. [CrossRef]
15. Nguyen, T.D.; Dang, T.T.T. Cytochrome P450 Enzymes as Key Drivers of Alkaloid Chemical Diversification in Plants. *Front. Plan Sci.* **2021**, *12*, 682181. [CrossRef] [PubMed]
16. Kishimoto, S.; Sato, M.; Tsunematsu, Y.; Watanabe, K. Evaluation of biosynthetic pathway and engineered biosynthesis o alkaloids. *Molecules* **2016**, *21*, 1078. [CrossRef] [PubMed]
17. Masci, V.L.; Bernardini, S.; Modesti, L.; Ovidi, E.; Tiezzi, A. Medicinal Plants as a Source of Alkaloids. *Microorg. Sustain.* **201** *15*, 85–113. [CrossRef]
18. Dey, P.; Kundu, A.; Kumar, A.; Gupta, M.; Lee, B.M.; Bhakta, T.; Dash, S.; Kim, H.S. *Analysis of Alkaloids (Indole Alkaloid Isoquinoline Alkaloids, Tropane Alkaloids)*; Elsevier Inc.: Amsterdam, The Netherlands, 2020. [CrossRef]
19. Wang, X.; Zhu, J.; Zhang, D.; Liu, G. Ribosomal control in RNA virus-infected cells. *Front. Microbiol.* **2022**, *13*, 1026887. [CrossRef
20. Cesaro, T.; Michiels, T. Inhibition of PKR by Viruses. *Front. Microbiol.* **2021**, *12*, 757238. [CrossRef]
21. Lee, H.R.; Choi, U.Y.; Hwang, S.W.; Kim, S.; Jung, J.U. Viral inhibition of PRR-Mediated innate immune response: Learning fron KSHV evasion strategies. *Mol. Cells* **2016**, *39*, 777–782. [CrossRef] [PubMed]
22. Firth, A.E.; Brierley, I. Non-canonical translation in RNA viruses. *J. Gen. Virol.* **2012**, *93*, 1385–1409. [CrossRef] [PubMed]
23. Faisal, S.; Badshah, S.L.; Kubra, B.; Emwas, A.H.; Jaremko, M. Alkaloids as potential antivirals. A comprehensive review. *Na Prod. Bioprospecting* **2023**, *13*, 4. [CrossRef] [PubMed]
24. Abookleesh, F.L.; Al-Anzi, B.S.; Ullah, A. Potential Antiviral Action of Alkaloids. *Molecules* **2022**, *27*, 903. [CrossRef] [PubMed]
25. Seteyen, A.L.S.; Girard-Valenciennes, E.; Septembre-Malaterre, A.; Gasque, P.; Guiraud, P.; Sélambarom, J. Anti-Alphavira Alkaloids: Focus on Some Isoquinolines, Indoles and Quinolizidines. *Molecules* **2022**, *27*, 5080. [CrossRef]
26. Gonzalez, B.L.; de Oliveira, N.C.; Ritter, M.R.; Tonin, F.S.; Melo, E.B.; Sanches, A.C.C.; Fernandez-Llimos, F.; Petruco, M.V de Mello, J.C.P.; Chierrito, D.; et al. The naturally-derived alkaloids as a potential treatment for COVID-19: A scoping review *Phytother. Res.* **2022**, *36*, 2686–2709. [CrossRef]
27. Kittakoop, P.; Darshana, D.; Sangsuwan, R.; Mahidol, C. Alkaloids and Alkaloid-Like Compounds are Potential Scaffolds o Antiviral Agents against SARS-CoV-2 (COVID-19) Virus. *Heterocycles* **2022**, *105*, 115. [CrossRef]
28. Sharma, D.; Sharma, N.; Manchanda, N.; Prasad, S.K.; Sharma, P.C.; Thakur, V.K.; Rahman, M.M.; Dhobi, M. Bioactivity anc In Silico Studies of Isoquinoline and Related Alkaloids as Promising Antiviral Agents: An Insight. *Biomolecules* **2022**, *13*, 17 [CrossRef]

https://doi.org/10.3390/molecules3103053

Xian, Y.; Zhang, J.; Bian, Z.; Zhou, H.; Zhang, Z.; Lin, Z.; Xu, H. Bioactive natural compounds against human coronaviruses: A review and perspective. *Acta Pharm. Sin. B* **2020**, *10*, 1163–1174. [CrossRef]

Liao, Y.; Ye, Y.; Liu, M.; Liu, Z.; Wang, J.; Li, B.; Huo, L.; Zhuang, Y.; Chen, L.; Chen, J.; et al. Identification of N- and C-3-Modified Laudanosoline Derivatives as Novel Influenza PAN Endonuclease Inhibitors. *J. Med. Chem.* **2023**, *66*, 188–219. [CrossRef]

George, A.; Reddy, A.G.K.; Satyanarayana, G.; Raghavendra, N.K. 1,2,3,4-Tetrahydroisoquinolines as inhibitors of HIV-1 integrase and human LEDGF/p75 interaction. *Chem. Biol. Drug Des.* **2018**, *91*, 1133–1140. [CrossRef]

Wang, X.; Burdzhiev, N.T.; Hu, H.; Li, Y.; Li, J.; Lozanova, V.V.; Kandinska, M.I.; Wang, M. Novel Tetrahydroisoquinoline-Based Heterocyclic Compounds Efficiently Inhibit SARS-CoV-2 Infection In Vitro. *Viruses* **2023**, *15*, 502. [CrossRef] [PubMed]

Majnooni, M.B.; Fakhri, S.; Bahrami, G.; Naseri, M.; Farzaei, M.H.; Echeverr, J. Alkaloids as Potential Phytochemicals against SARS-CoV-2: Approaches to the Associated Pivotal Mechanisms. *Evid.-Based Complement. Altern. Med.* **2021**, *2021*, 6632623. [CrossRef] [PubMed]

Ogunyemi, O.M.; A Gyebi, G.; A Elfiky, A.; O Afolabi, S.; Ogunro, O.B.; Adegunloye, A.P.; Ibrahim, I.M. Alkaloids and flavonoids from African phytochemicals as potential inhibitors of SARS-CoV-2 RNA-dependent RNA polymerase: An in silico perspective. *Antivir. Chem. Chemother.* **2020**, *28*, 2040206620984076. [CrossRef] [PubMed]

Cheng, F.; Murray, J.L.; Zhao, J.; Sheng, J.; Zhao, Z.; Rubin, D.H. Systems Biology-Based Investigation of Cellular Antiviral Drug Targets Identified by Gene-Trap Insertional Mutagenesis. *PLoS Comput. Biol.* **2016**, *12*, e1005074. [CrossRef] [PubMed]

Zhou, P.; Lao, Z.; Long, H.; Pan, P.; Liao, F.; Zheng, W.; Li, Z.; Dai, J.; Liu, H.; Jiang, Y.; et al. The in vitro and in vivo antiviral effects of aloperine against Zika virus infection. *J. Med. Virol.* **2023**, *95*, e28913. [CrossRef]

Merindol, N.; Martins, L.L.B.; Elfayres, G.; Custeau, A.; Berthoux, L.; Evidente, A.; Desgagné-Penix, I. Amaryllidaceae Alkaloids Screen Unveils Potent Anticoronaviral Compounds and Associated Structural Determinants. *ACS Pharmacol. Transl. Sci.* **2024**, *7*, 3527–3539. [CrossRef]

Huang, L.; Yuen, T.T.-T.; Ye, Z.; Liu, S.; Zhang, G.; Chu, H.; Yue, J. Berbamine inhibits SARS-CoV-2 infection by compromising TRPMLs-mediated endolysosomal trafficking of ACE2. *Signal Transduct. Target. Ther.* **2021**, *6*, 2–4. [CrossRef]

Wang, C.; Wang, T.; Dai, J.; Han, Y.; Hu, R.; Li, N.; Yang, Z.; Wang, J. Canthin-6-one analogs block Newcastle disease virus proliferation via suppressing the Akt and ERK pathways. *Poult. Sci.* **2024**, *103*, 103944. [CrossRef]

Verma, D.; Mitra, D.; Kamboj, A.; Mahakur, B.; Chaudhary, P.; Shrivastav, R.; Janmeda, P.; Pant, K.; Mohapatra, P.K.D. Canthin-6-One 9-O-Beta-Glucopyranoside: A natural inhibitor of SARS-CoV-2 Canthin-6-One 9-O-Beta-Glucopyranoside: An inhibitor of SARS-CoV-2 (COVID 19) proteases PLpro and Mpro/3CLpro. *Res. Sq.* **2020**, *2*, 1–16. [CrossRef]

Zhang, M.-Q.; Jia, X.; Cheng, C.-Q.; Wang, Y.-X.; Li, Y.-Y.; Kong, L.-D.; Li, Q.-Q.; Xie, F.; Yu, Y.-L.; He, Y.-T.; et al. Capsaicin functions as a selective degrader of STAT3 to enhance host resistance to viral infection. *Acta Pharmacol. Sin.* **2023**, *44*, 2253–2264. [CrossRef]

Marois, I.; Cloutier, A.; Meunier, I.; Weingartl, H.M.; Cantin, A.M.; Richter, M.V. Inhibition of influenza virus replication by targeting broad host cell pathways. *PLoS ONE* **2014**, *9*, e110631. [CrossRef]

Trischitta, P.; Tamburello, M.P.; Venuti, A.; Pennisi, R. Pseudovirus-Based Systems for Screening Natural Antiviral Agents: A Comprehensive Review. *Int. J. Mol. Sci.* **2024**, *25*, 5188. [CrossRef]

Xia, B.; Zheng, L.; Li, Y.; Sun, W.; Liu, Y.; Li, L.; Pang, J.; Chen, J.; Li, J.; Cheng, H. The brief overview, antivirus and anti-SARS-CoV-2 activity, quantitative methods, and pharmacokinetics of cepharanthine: A potential small-molecule drug against COVID-19. *Front. Pharmacol.* **2023**, *14*, 1098972. [CrossRef]

Liu, K.; Hong, B.; Wang, S.; Lou, F.; You, Y.; Hu, R.; Shafqat, A.; Fan, H.; Tong, Y. Pharmacological Activity of Cepharanthine. *Molecules* **2023**, *28*, 5019. [CrossRef]

Ren, P.-X.; Shang, W.-J.; Yin, W.-C.; Ge, H.; Wang, L.; Zhang, X.-L.; Li, B.-Q.; Li, H.-L.; Xu, Y.-C.; Xu, E.H.; et al. A multi-targeting drug design strategy for identifying potent anti-SARS-CoV-2 inhibitors. *Acta Pharmacol. Sin.* **2022**, *43*, 483–493. [CrossRef] [PubMed]

Yang, S.; Xu, M.; Lee, E.M.; Gorshkov, K.; Shiryaev, S.A.; He, S.; Sun, W.; Cheng, Y.-S.; Hu, X.; Tharappel, A.M.; et al. Emetine inhibits Zika and Ebola virus infections through two molecular mechanisms: Inhibiting viral replication and decreasing viral entry. *Cell Discov.* **2018**, *4*, 31. [CrossRef]

Ka, S.; Merindol, N.; Sow, A.A.; Singh, A.; Landelouci, K.; Plourde, M.B.; Pépin, G.; Masi, M.; Di Lecce, R.; Evidente, A.; et al. Amaryllidaceae Alkaloid Cherylline Inhibits the Replication of Dengue and Zika Viruses. *Antimicrob. Agents Chemother.* **2021**, *65*, e0039821. [CrossRef] [PubMed]

Ren, J.; Zeng, W.; Jiang, C.; Li, C.; Zhang, C.; Cao, H.; Li, W.; He, Q. Inhibition of Porcine Epidemic Diarrhea Virus by Cinchonine via Inducing Cellular Autophagy. *Front. Cell. Infect. Microbiol.* **2022**, *12*, 856711. [CrossRef]

Hegazy, A.; Soltane, R.; Alasiri, A.; Mostafa, I.; Metwaly, A.M.; Eissa, I.H.; Mahmoud, S.H.; Allayeh, A.K.; Shama, N.M.A.; Khalil, A.A.; et al. Anti-rheumatic colchicine phytochemical exhibits potent antiviral activities against avian and seasonal Influenza A viruses (IAVs) via targeting different stages of IAV replication cycle. *BMC Complement. Med. Ther.* **2024**, *24*, 49. [CrossRef] [PubMed]

51. Bailly, B.; Richard, C.-A.; Sharma, G.; Wang, L.; Johansen, L.; Cao, J.; Pendharkar, V.; Sharma, D.-C.; Galloux, M.; Wang, Y.; et a Targeting human respiratory syncytial virus transcription anti-termination factor M2-1 to inhibit in vivo viral replication. *Sc Rep.* **2016**, *6*, 25806. [CrossRef]
52. Diot, C.; Richard, C.-A.; Risso-Ballester, J.; Martin, D.; Fix, J.; Eléouët, J.-F.; Sizun, C.; Rameix-Welti, M.-A.; Galloux, M. Hardenin of Respiratory Syncytial Virus Inclusion Bodies by Cyclopamine Proceeds through Perturbation of the Interactions of the M2- Protein with RNA and the P Protein. *Int. J. Mol. Sci.* **2023**, *24*, 13862. [CrossRef]
53. Fix, J.; Descamps, D.; Galloux, M.; Ferret, C.; Bouguyon, E.; Zohari, S.; Näslund, K.; Hägglund, S.; Altmeyer, R.; Valarcher, J.-F et al. Screening antivirals with a mCherry-expressing recombinant bovine respiratory syncytial virus: A proof of concept usin cyclopamine. *Vet. Res.* **2023**, *54*, 36. [CrossRef]
54. Li, K.; Wang, H.; Chen, T.; Wang, X.; Wang, X.; Zhong, M.; Gao, X.; Hao, Z. Dehydroevodiamine inhibits PEDV throug regulateing ERK1/2 MAPK pathway in Vero cells. *Virology* **2024**, *598*, 110166. [CrossRef]
55. Hussain, S.; Miller, J.L.; Harvey, D.J.; Gu, Y.; Rosenthal, P.B.; Zitzmann, N.; McCauley, J.W. Strain-specific antiviral activity o iminosugars against human influenza A viruses. *J. Antimicrob. Chemother.* **2015**, *70*, 136–152. [CrossRef]
56. Perera, N.; Brun, J.; Alonzi, D.S.; Tyrrell, B.E.; Miller, J.L.; Zitzmann, N. Antiviral effects of deoxynojirimycin (DNJ)-base iminosugars in dengue virus-infected primary dendritic cells. *Antivir. Res.* **2022**, *199*, 105269. [CrossRef]
57. Bhushan, G.; Lim, L.; Bird, I.; Chothe, S.K.; Nissly, R.H.; Kuchipudi, S.V. Iminosugars With Endoplasmic Reticulum α-Glucosidas Inhibitor Activity Inhibit ZIKV Replication and Reverse Cytopathogenicity in vitro. *Front. Microbiol.* **2020**, *11*, 531. [CrossRef [PubMed]
58. Khandelwal, N.; Chander, Y.; Rawat, K.D.; Riyesh, T.; Nishanth, C.; Sharma, S.; Jindal, N.; Tripathi, B.N.; Barua, S.; Kumar, N Emetine inhibits replication of RNA and DNA viruses without generating drug-resistant virus variants. *Antivir. Res.* **2017**, *14* 196–204. [CrossRef] [PubMed]
59. Valipour, M. Different Aspects of Emetine's Capabilities as a Highly Potent SARS-CoV-2 Inhibitor against COVID-19. *AC Pharmacol. Transl. Sci.* **2022**, *5*, 387–399. [CrossRef] [PubMed]
60. Bleasel, M.D.; Peterson, G.M. Emetine, ipecac, ipecac alkaloids and analogues as potential antiviral agents for coronaviruse *Pharmaceuticals* **2020**, *13*, 51. [CrossRef]
61. Wei, W.; Du, H.; Shao, C.; Zhou, H.; Lu, Y.; Yu, L.; Wan, H.; He, Y. Screening of antiviral components of Ma Huang Tang an investigation on the ephedra alkaloids efficacy on influenza virus type A. *Front. Pharmacol.* **2019**, *10*, 961. [CrossRef]
62. Yang, S.; Yang, X.; Wang, Z.; Li, W.; Cao, R.; Zhong, W. Fangchinoline Inhibits Zika Virus by Disrupting Virus Internalizatio *ACS Infect. Dis.* **2024**, *10*, 4066–4072. [CrossRef] [PubMed]
63. Wan, Z.; Lu, Y.; Liao, Q.; Wu, Y.; Chen, X. Fangchinoline inhibits human immunodeficiency virus type 1 replication by interferin with gp160 proteolytic processing. *PLoS ONE* **2012**, *7*, e39225. [CrossRef]
64. Zhang, Q.Y.; Li, J.-Q.; Li, Q.; Zhang, Y.; Zhang, Z.-R.; Li, X.-D.; Zhang, H.-Q.; Deng, C.-L.; Yang, F.-X.; Xu, Y.; et al. Identification o fangchinoline as a broad-spectrum enterovirus inhibitor through reporter virus based high-content screening. *Virol. Sin.* **2024** *39*, 301–308. [CrossRef]
65. Hegazy, A.; Mahmoud, S.H.; Elshaier, Y.A.M.M.; Shama, N.M.A.; Nasr, N.F.; Ali, M.A.; El-Shazly, A.M.; Mostafa, I.; Mostafa, A Antiviral activities of plant-derived indole and β-carboline alkaloids against human and avian influenza viruses. *Sci. Rep.* **202** *13*, 1612. [CrossRef] [PubMed]
66. Dahal, S.; Clayton, K.; Cabral, T.; Cheng, R.; Jahanshahi, S.; Ahmed, C.; Koirala, A.; Ocando, A.V.; Malty, R.; Been, T.; et al. On path toward a broad-spectrum anti-viral: Inhibition of HIV-1 and coronavirus replication by SR kinase inhibitor harmine. *J. Viro* **2023**, *97*, e0039623. [CrossRef]
67. Kaur, P.; Thiruchelvan, M.; Lee, R.C.H.; Chen, H.; Chen, K.C.; Ng, M.L.; Chu, J.J.H. Inhibition of Chikungunya virus replicatio by harringtonine, a novel antiviral that suppresses viral protein expression. *Antimicrob. Agents Chemother.* **2013**, *57*, 155–167 [CrossRef] [PubMed]
68. Lai, Z.Z.; Ho, Y.J.; Lu, J.W. Harringtonine inhibits zika virus infection through multiple mechanisms. *Molecules* **2020**, *25*, 4082 [CrossRef]
69. Yang, Z.; Fu, X.; Zhao, Y.; Li, X.; Long, J.; Zhang, L. Molecular insights into the inhibition mechanism of harringtonine agains essential proteins associated with SARS-CoV-2 entry. *Int. J. Biol. Macromol.* **2023**, *240*, 124352. [CrossRef]
70. Dong, H.J.; Wang, Z.-H.; Meng, W.; Li, C.-C.; Hu, Y.-X.; Zhou, L.; Wang, X.-J. The natural compound homoharringtonine presents broad antiviral activity in vitro and in vivo. *Viruses* **2018**, *10*, 601. [CrossRef]
71. Gong, M.; Li, S.; Xie, Y.; Zhao, F.; Shao, J.; Zhang, Y.; Wang, W.; Chang, H. Inhibitory effects of homoharringtonine on foot and mouth disease virus in vitro. *J. Med. Virol.* **2019**, *91*, 1595–1601. [CrossRef]
72. Harisha, K.R.; Kailaje, V.; Kondreddi, R.R.; Gudla, C.S.; Singh, S.; Ramakrishnaiah, S.; Isloor, S.; Narayanan, S.; Shandil R.K.; Rudramurthy, G.R. Homoharringtonine Inhibits CVS-11 and Clinical Isolates of Rabies Virus In Vitro: Identified via High-Throughput Screening of an FDA-Approved Drug Library. *Viruses* **2025**, *17*, 945. [CrossRef]
73. Neerukonda, S.N. A Review on SARS-CoV-2 Virology, Pathophysiology. *Pathogens* **2020**, *9*, 426. [CrossRef]

https://doi.org/10.3390/molecules3103053

Jie, C.; Luo, Z.; Chen, H.; Wang, M.; Yan, C.; Mao, Z.-F.; Xiao, G.-K.; Kurihara, H.; Li, Y.-F.; He, R.-R. Indirubin, a bisindole alkaloid from Isatis indigotica, reduces H1N1 susceptibility in stressed mice by regulating MAVS signaling. *Oncotarget* **2017**, *8*, 105615–105629. [CrossRef]
Chang, S.J.; Chang, Y.C.; Lu, K.Z.; Tsou, Y.Y.; Lin, C.W. Antiviral activity of Isatis indigotica extract and its derived indirubin against Japanese encephalitis virus. *Evid.-Based Complement. Altern. Med.* **2012**, *2012*, 925830. [CrossRef]
Medina-Moreno, S.; Dowling, T.C.; Zapata, J.C.; Le, N.M.; Sausville, E.; Bryant, J.; Redfield, R.R.; Heredia, A. Targeting of CDK9 with indirubin 3′-monoxime safely and durably reduces HIV viremia in chronically infected humanized mice. *PLoS ONE* **2017**, *12*, e0183425. [CrossRef] [PubMed]
Jin, Y.H.; Min, J.S.; Jeon, S.; Lee, J.; Kim, S.; Park, T.; Park, D.; Jang, M.S.; Park, C.M.; Song, J.H.; et al. Lycorine, a non-nucleoside RNA dependent RNA polymerase inhibitor, as potential treatment for emerging coronavirus infections. *Phytomedicine* **2021**, *86*, 153440. [CrossRef] [PubMed]
Chen, H.; Lao, Z.; Xu, J.; Li, Z.; Long, H.; Li, D.; Lin, L.; Liu, X.; Yu, L.; Liu, W.; et al. Antiviral activity of lycorine against Zika virus in vivo and in vitro. *Virology* **2020**, *546*, 88–97. [CrossRef] [PubMed]
Li, N.; Wang, Z.; Wang, R.; Zhang, Z.-R.; Zhang, Y.-N.; Deng, C.-L.; Zhang, B.; Shang, L.-Q.; Ye, H.-Q. In Vitro Inhibition of Alphaviruses by Lycorine. *Virol. Sin.* **2021**, *36*, 1465–1474. [CrossRef]
Narayanan, A.; Narwal, M.; Majowicz, S.A.; Varricchio, C.; Toner, S.A.; Ballatore, C.; Brancale, A.; Murakami, K.S.; Jose, J. Identification of SARS-CoV-2 inhibitors targeting Mpro and PLpro using in-cell-protease assay. *Commun. Biol.* **2022**, *5*, 169. [CrossRef]
Fielding, B.C.; da Silva Maia Bezerra Filho, C.; Ismail, N.S.M.; de Sousa, D.P. Alkaloids: Therapeutic potential against human coronaviruses *Molecules* **2020**, *25*, 5496. [CrossRef]
Pan, Q.M.; Li, Y.H.; Hua, J.; Huang, F.P.; Wang, H.S.; Liang, D. Antiviral Matrine-Type Alkaloids from the Rhizomes of Sophora tonkinensis. *J. Nat. Prod.* **2015**, *78*, 1683–1688. [CrossRef]
Qiao, W.T.; Yao, X.; Lu, W.-H.; Zhang, Y.-Q.; Malhi, K.K.; Li, H.-X.; Li, J.-L. Matrine exhibits antiviral activities against PEDV by directly targeting Spike protein of the virus and inducing apoptosis via the MAPK signaling pathway. *Int. J. Biol. Macromol.* **2024**, *270*, 132408. [CrossRef] [PubMed]
McMahon, J.B.; Currens, M.J.; Gulakowski, R.J.; Buckheit, R.W.; Lackman-Smith, C.; Hallock, Y.F.; Boyd, M.R. Michellamine B, a novel plant alkaloid, inhibits human immunodeficiency virus-induced cell killing by at least two distinct mechanism. *Antimicrob. Agents Chemother.* **1995**, *39*, 484–488. [CrossRef]
Yang, Y.; Yang, P.; Huang, C.; Wu, Y.; Zhou, Z.; Wang, X.; Wang, S. Inhibitory effect on SARS-CoV-2 infection of neferine by blocking Ca^{2+}-dependent membrane fusion. *J. Med. Virol.* **2021**, *93*, 5825–5832. [CrossRef]
Yang, D.D.; Chutiwitoonchai, N.; Wang, F.; Tian, P.; Sureram, S.; Lei, X.; Mahidol, C.; Ruchirawat, S.; Kittakoop, P. Effects of organic salts of virucidal and antiviral compounds from Nelumbo nucifera and Kaempferia parviflora against SARS-CoV-2. *Sci. Rep.* **2025**, *15*, 6380. [CrossRef]
Kumar, A.; Sharma, M.; Richardson, C.D.; Kelvin, D.J. Potential of Natural Alkaloids From Jadwar (*Delphinium denudatum*) as Inhibitors Against Main Protease of COVID-19: A Molecular Modeling Approach. *Front. Mol. Biosci.* **2022**, *9*, 898874. [CrossRef]
Nag, A.; Chowdhury, R.R. Piperine, an alkaloid of black pepper seeds can effectively inhibit the antiviral enzymes of Dengue and Ebola viruses, an in silico molecular docking study. *Virusdisease* **2020**, *31*, 308–315. [CrossRef] [PubMed]
Pareek, A.; Kumar, R.; Mudgal, R.; Neetu, N.; Sharma, M.; Kumar, P.; Tomar, S. Alphavirus antivirals targeting RNA-dependent RNA polymerase domain of nsP4 divulged using surface plasmon resonance. *FEBS J.* **2022**, *289*, 4901–4924. [CrossRef]
Li, D.; Chengzhi, C.; Xiaoyin, C.; Zhongping, W. Pseudoephedrine protects mice from infection of H1N1 virus. *Int. J. Clin. Virol.* **2020**, *4*, 14–20. [CrossRef]
Yu, S.; Chen, Y.; Xiang, Y.; Lin, H.; Wang, M.; Ye, W.; Zhang, P.; Chen, H.; Lin, G.; Zhu, Y.; et al. Pseudoephedrine and its derivatives antagonize wild and mutated severe acute respiratory syndrome-CoV-2 viruses through blocking virus invasion and antiinflammatory effect. *Phytother. Res.* **2021**, *35*, 5847–5860. [CrossRef] [PubMed]
Yang, L.; Pei, R.-J.; Li, H.; Ma, X.-N.; Zhou, Y.; Zhu, F.-H.; He, P.-L.; Tang, W.; Zhang, Y.-C.; Xiong, J.; et al. Identification of SARS-CoV-2 entry inhibitors among already approved drugs. *Acta Pharmacol. Sin.* **2021**, *42*, 1347–1353. [CrossRef]
Persoons, L.; Vanderlinden, E.; Vangeel, L.; Wang, X.; Do, N.D.T.; Foo, S.-Y.C.; Leyssen, P.; Neyts, J.; Jochmans, D.; Schols, D.; et al. Broad spectrum anti-coronavirus activity of a series of anti-malaria quinoline analogues. *Antivir. Res.* **2021**, *193*, 105127. [CrossRef]
Mamidala, E.; Davella, R.; Gurrapu, S.; Shivakrishna, P. In Silico Identification of Clinically Approved Medicines Against the main Protease of SARS-CoV-2—A Causative Agent of COVID-19. *Int. J. Life Sci. Pharma Res.* **2022**, *11*, 107–122. [CrossRef]
Malakar, S.; Sreelatha, L.; Dechtawewat, T.; Noisakran, S.; Yenchitsomanus, P.-T.; Chu, J.J.H.; Limjindaporn, T. Drug repurposing of quinine as antiviral against dengue virus infection. *Virus Res.* **2018**, *255*, 171–178. [CrossRef] [PubMed]
D'alessandro, S.; Scaccabarozzi, D.; Signorini, L.; Perego, F.; Ilboudo, D.P.; Ferrante, P.; Delbue, S. The use of antimalarial drugs against viral infection. *Microorganisms* **2020**, *8*, 85. [CrossRef]

https://doi.org/10.3390/molecules31030539

97. Große, M.; Ruetalo, N.; Layer, M.; Hu, D.; Businger, R.; Rheber, S.; Setz, C.; Rauch, P.; Auth, J.; Fröba, M.; et al. Quinine inhibi infection of human cell lines with SARS-CoV-2. *Viruses* **2021**, *13*, 647. [CrossRef]
98. Lin, S.; Wang, X.; Guo, H.; Dai, N.; Tang, R.W.-L.; Lee, H.C.; Leung, K.W.; Dong, T.T.-X.; Webb, S.E.; Miller, A.L.; et al. The Ethan Extract of Evodiae Fructus and Its Ingredient, Rutaecarpine, Inhibit Infection of SARS-CoV-2 and Inflammatory Responses. *Int. Mol. Sci.* **2023**, *24*, 762. [CrossRef] [PubMed]
99. Bhattacharjee, A.; Chaudhuri, R.; Dash, J.J.; Saha, M.; Choudhury, L.; Roy, S. Pre-treatment with Scopolamine Naturally Suppress Japanese Encephalitis Viral Load in Embryonated Chick Through Regulation of Multiple Signaling Pathways. *Appl. Biocher Biotechnol.* **2021**, *193*, 1654–1674. [CrossRef] [PubMed]
100. Jin, Z.; Yang, L.; Ding, G.; Yang, G.; Han, Y.; Zhang, X.; Li, W. Sophocarpine against enterovirus 71 in vitro. *Exp. Ther. Med.* **201** *14*, 3792–3797. [CrossRef]
101. Ren, G.; Ding, G.; Zhang, H.; Wang, H.; Jin, Z.; Yang, G.; Han, Y.; Zhang, X.; Li, G.; Li, W. Antiviral activity of sophoridine again enterovirus 71 in vitro. *J. Ethnopharmacol.* **2019**, *236*, 124–128. [CrossRef]
102. Lixia, H.; Jun, C.; Song, H.; FaHu, Y.; Jinwen, T. Neuroprotective effect of (-)-tetrahydropalmatine in Japanese encephalitis viru strain GP-78 infected mouse model. *Microb. Pathog.* **2018**, *114*, 197–203. [CrossRef]
103. Kim, D.E.; Min, J.S.; Jang, M.S.; Lee, J.Y.; Shin, Y.S.; Park, C.M.; Song, J.H.; Kim, H.R.; Kim, S.; Jin, Y.-H.; et al. Natural bi benzylisoquinoline alkaloids-tetrandrine, fangchinoline, and cepharanthine, inhibit human coronavirus oc43 infection of mrc- human lung cells. *Biomolecules* **2019**, *9*, 696. [CrossRef]
104. Liu, J.; Wang, F.; Wang, X.; Fan, S.; Li, Y.; Xu, M.; Hu, H.; Liu, K.; Zheng, B.; Wang, L.; et al. Antiviral effects and tissue exposure c tetrandrine against SARS-CoV-2 infection and COVID-19. *MedComm* **2023**, *4*, e206. [CrossRef]
105. Islam, M.T.; Sarkar, C.; El-Kersh, D.M.; Jamaddar, S.; Uddin, S.J.; Shilpi, J.A.; Mubarak, M.S. Natural products and their derivative against coronavirus: A review of the non-clinical and pre-clinical data. *Phytother. Res.* **2020**, *34*, 2471–2492. [CrossRef]
106. Yang, C.W.; Lee, Y.-Z.; Kang, I.-J.; Barnard, D.L.; Jan, J.-T.; Lin, D.; Huang, C.-W.; Yeh, T.-K.; Chao, Y.-S.; Lee, S.-J. Identificatio of phenanthroindolizines and phenanthroquinolizidines as novel potent anti-coronaviral agents for porcine enteropathogeni coronavirus transmissible gastroenteritis virus and human severe acute respiratory syndrome coronavirus. *Antivir. Res.* **201** *88*, 160–168. [CrossRef]
107. Wang, Y.; Lee, S.; Ha, Y.; Lam, W.; Chen, S.-R.; Dutschman, G.E.; Gullen, E.A.; Grill, S.P.; Cheng, Y.; Fürstner, A.; et al. Tylophorin Analogs Allosterically Regulates Heat Shock Cognate Protein 70 And Inhibits Hepatitis C Virus Replication. *Sci. Rep.* **201** *7*, 10037. [CrossRef]
108. Mani, J.S.; Johnson, J.B.; Steel, J.C.; Broszczak, D.A.; Neilsen, P.M.; Walsh, K.B.; Naiker, M. Natural product-derived phytochem cals as potential agents against coronaviruses: A review. *Virus Res.* **2020**, *284*, 197989. [CrossRef] [PubMed]
109. Akan, E.; Chang-Liu, C.M.; Watanabe, J.; Ishizawa, K.; Woloschak, G.E. The effects of vinblastine on the expression of huma immunodeficiency virus type 1 long terminal repeat. *Leuk. Res.* **1997**, *21*, 459–464. [CrossRef] [PubMed]
110. Qing, Z.-X.; Yang, P.; Tang, Q.; Cheng, P.; Liu, X.-B.; Zheng, Y.-J.; Liu, Y.-S.; Zeng, J.-G. Isoquinoline Alkaloids and Their Antivira Antibacterial, and Antifungal Activities and Structure-activity Relationship. *Curr. Org. Chem.* **2017**, *21*, 1920–1934. [CrossRef]
111. Zheng, X.; Wu, F.; Lin, X.; Shen, L.; Feng, Y. Developments in drug delivery of bioactive alkaloids derived from traditiona Chinese medicine. *Drug Deliv.* **2018**, *25*, 398–416. [CrossRef]
112. Gidwani, B.; Bhairam, M.; Shukla, S.S.; Verma, H.; Pandey, R.K. Herbal Bioenhancers in Pharmaceutical Drug Delivery Mechanisms, Challenges, and Future Innovations. *Chem. Biodivers.* **2025**, *22*, e202500760. [CrossRef] [PubMed]
113. Onoue, S.; Yamada, S.; Chan, H.K. Nanodrugs: Pharmacokinetics and safety. *Int. J. Nanomed.* **2014**, *9*, 1025–1037. [CrossRef]
114. Gabrielsen, B.; Monath, T.P.; Huggins, J.W.; Kefauver, D.F.; Pettit, G.R.; Groszek, G.; Hollingshead, M.; Kirsi, J.J.; Shannon, W.M Schubert, E.M.; et al. Antiviral (RNA) activity of selected Amaryllidaceae isoquinoline const ituents and synthesis of relate substances. *J. Nat. Prod.* **1992**, *55*, 1569–1581. [CrossRef] [PubMed]
115. Wu, Y.; Li, J.Q.; Kim, Y.J.; Wu, J.; Wang, Q.; Hao, Y. In vivo and in vitro antiviral effects of berberine on influenza virus. *Chin. Integr. Med.* **2011**, *17*, 444–452. [CrossRef]
116. Wang, Y.X.; Yang, L.; Wang, H.-Q.; Zhao, X.-Q.; Liu, T.; Li, Y.-H.; Zeng, Q.-X.; Li, Y.-H.; Song, D.-Q. synthesis and evolution o berberine derivatives as a new class of antiviral agents against enterovirus 71 through the MEK/ERK pathway and autophag *Molecules* **2018**, *23*, 2084. [CrossRef] [PubMed]
117. Shao, J.; Zeng, D.; Tian, S.; Liu, G.; Fu, J. Identification of the natural product berberine as an antiviral drug. *AMB Express* **202** *10*, 164. [CrossRef]
118. Ratanakomol, T.; Roytrakul, S.; Wikan, N.; Smith, D.R. Berberine inhibits dengue virus through dual mechanisms. *Molecules* **202** *26*, 5501. [CrossRef]
119. Botwina, P.; Owczarek, K.; Rajfur, Z.; Ochman, M.; Urlik, M.; Nowakowska, M.; Szczubiałka, K.; Pyrc, K. Berberine Hamper Influenza A Replication through. *Viruses* **2020**, *12*, 344. [CrossRef]
120. Enkhtaivan, G.; Muthuraman, P.; Kim, D.H.; Mistry, B. Discovery of berberine based derivatives as anti-influenza agent throug blocking of neuraminidase. *Bioorg. Med. Chem.* **2017**, *25*, 5185–5193. [CrossRef]

https://doi.org/10.3390/molecules3103053

1. Varghese, F.S.; Kaukinen, P.; Gläsker, S.; Bespalov, M.; Hanski, L.; Wennerberg, K.; Kümmerer, B.M.; Ahola, T. Discovery of berberine, abamectin and ivermectin as antivirals against chikungunya and other alphaviruses. *Antivir. Res.* **2016**, *126*, 117–124. [CrossRef]
2. Nguyen, C.Q.; Nguyen, T.H.M.; Nguyen, T.T.T.; Bui, T.B.H.; Nguyen, T.T.; Huynh, N.T.; Le, T.D.; Nguyen, T.M.P.; Nguyen, D.T.; Nguyen, M.T.; et al. Designs, Synthesis, Docking Studies, and Biological Evaluation of Novel Berberine Derivatives Targeting Zika Virus. *J. Chem.* **2021**, *2021*, 5567111. [CrossRef]
3. Españo, E.; Kim, J.; Lee, C.; Webster, R.G.; Webby, R.J. Chelerythrine as an anti-Zika virus agent: Therapeutic potential and mode of action. *bioRxiv* **2022**. [CrossRef]
4. Loe, M.W.C.; Lee, R.C.H.; Chin, W.-X.; Min, N.; Teo, Z.Y.; Ho, S.X.; Yi, B.; Chu, J.J.H. Chelerythrine chloride inhibits Zika virus infection by targeting the viral NS4B protein. *Antivir. Res.* **2023**, *219*, 105732. [CrossRef] [PubMed]
5. Guo, W.; Lu, X.; Liu, B.; Yan, H.; Feng, J. Anti-TMV activity and mode of action of three alkaloids isolated from Chelidonium majus. *Pest Manag. Sci.* **2021**, *77*, 510–517. [CrossRef]
6. Orhan, I.; Özçelik, B.; Karaoğlu, T.; Şener, B. Antiviral and antimicrobial profiles of selected isoquinoline alkaloids from Fumaria and Corydalis species. *Z. Naturforschung Sect. C J. Biosci.* **2007**, *62*, 19–26. [CrossRef] [PubMed]
7. Dai, J.P.; Wang, Q.-W.; Su, Y.; Gu, L.-M.; Deng, H.-X.; Chen, X.-X.; Li, W.-Z.; Li, K.-S. Oxymatrine inhibits influenza a virus replication and inflammation via TLR4, p38 MAPK and NF-κB pathways. *Int. J. Mol. Sci.* **2018**, *19*, 965. [CrossRef]
8. Regulation, I. Oxymatrine Modulation of TLR3 Signaling: A Dual-Action. *Molecules* **2024**, *29*, 1945. [CrossRef]
9. Chen, N.; Liu, Y.H.; Liu, X.J.; Chen, Y.R.; Guo, Y.H.; Liu, M. Oxymatrine inhibits target cell infection in the HCVcc system. *Zhonghua Gan Zang Bing Za Zhi* **2016**, *24*, 40–45. [CrossRef]
0. Zhang, W.; Chen, Y.; Yang, F.; Zhang, H.; Su, T.; Wang, J.; Zhang, Y.; Song, X. Antiviral effect of palmatine against infectious bronchitis virus through regulation of NF-κB/IRF7/JAK-STAT signalling pathway and apoptosis. *Br. Poult. Sci.* **2024**, *65*, 119–128. [CrossRef]
1. Jia, F.; Zou, G.; Fan, J.; Yuan, Z. Identification of palmatine as an inhibitor of West Nile virus. *Arch. Virol.* **2010**, *155*, 1325–1329. [CrossRef]
2. Ho, Y.J.; Lu, J W.; Huang, Y.L.; Lai, Z.Z. Palmatine inhibits Zika virus infection by disrupting virus binding, entry, and stability. *Biochem. Biophys. Res. Commun.* **2019**, *518*, 732–738. [CrossRef] [PubMed]
3. Ke, Q.; Duan, K.; Cheng, Y.; Xu, S.; Xiao, S.; Fang, L. Sanguinarine Exhibits Antiviral Activity against Porcine Reproductive and Respiratory Syndrome Virus via Multisite Inhibition Mechanisms. *Viruses* **2023**, *15*, 688. [CrossRef]
4. Bonincontro, G.; Scuderi, S.A.; Marino, A.; Simonetti, G. Synergistic Effect of Plant Compounds in Combination with Conventional Antimicrobials against Biofilm of *Staphylococcus aureus*, *Pseudomonas aeruginosa*, and *Candida* spp. *Pharmaceuticals* **2023**, *16*, 1531. [CrossRef]
5. Ng, Y.P.; Or, T.C.T.; Ip, N.Y. Plant alkaloids as drug leads for Alzheimer's disease. *Neurochem. Int.* **2015**, *89*, 260–270. [CrossRef]
6. Hukkanen, J.; Jacob, P.; Benowitz, N.L. Metabolism and disposition kinetics of nicotine. *Pharmacol. Rev.* **2005**, *57*, 79–115. [CrossRef]
7. Sweilam, S.H.; Alqarni, M.H.; Youssef, F.S. Antimicrobial Alkaloids from Marine-Derived Fungi as Drug Leads versus COVID-19 Infection: A Computational Approach to Explore their Anti-COVID-19 Activity and ADMET Properties. *Evid.-Based Complement. Altern. Med.* **2022**, *2022*, 5403757. [CrossRef] [PubMed]
8. Linnakoski, R.; Reshamwala, D.; Veteli, P.; Cortina-Escribano, M.; Vanhanen, H.; Marjomäki, V. Antiviral agents from fungi: Diversity, mechanisms and potential applications. *Front. Microbiol.* **2018**, *9*, 2325. [CrossRef]
9. Zhou, H.; Wang, W.; Cai, L.; Yang, T. Potentiation and Mechanism of Berberine as an Antibiotic Adjuvant Against Multidrug-Resistant Bacteria. *Infect. Drug Resist.* **2023**, *16*, 7313–7326. [CrossRef]
0. Tilaoui, M.; Mouse, H.A.; Zyad, A. Update and New Insights on Future Cancer Drug Candidates From Plant-Based Alkaloids. *Front. Pharmacol.* **2021**, *12*, 719694. [CrossRef] [PubMed]
1. Sandor, A.M.; Sturdivant, M.S.; Ting, J.P.Y. Influenza Virus and SARS-CoV-2 Vaccines. *J. Immunol.* **2021**, *206*, 2509–2520. [CrossRef]

https://doi.org/10.3390/molecules31030539

molecules

MDPI

Review

Bioactive Plant Peptides: Physicochemical Features, Structure-Function Insights and Mechanism of Action

Sara Avilés-Gaxiola [1], Israel García-Aguiar [1], Luis Alfonso Jiménez-Ortega [2], Erick Paul Gutiérrez-Grijalva [3] and José Basilio Heredia [2,*]

[1] Health Sciences Department, Universidad Autónoma de Occidente, Blvd. Lola Beltrán and Blvd. Rolando Arjona, Culiacán 80020, Mexico; sara.aviles@uadeo.mx (S.A.-G.); israel.garcia@uadeo.mx (I.G.-A.)
[2] Centro de Investigación en Alimentación y Desarrollo, AC Nutraceutical and Functional Foods, Carretera a Eldorado Km 5.5, Col. Campo el Diez, Culiacán 80110, Mexico; ljimenez222@estudiantes.ciad.mx
[3] Cátedras CONACYT—Centro de Investigación y Desarrollo, AC Carretera a Eldorado Km 5.5, Col. Campo el Diez, Culiacán 80110, Mexico; erick.gutierrez@ciad.mx
* Correspondence: jbheredia@ciad.mx

Abstract

Different cultures worldwide have attributed particular healing abilities to various plants for a long time. After decades of studies, research has demonstrated that their bioactivity is associated mainly with the presence of natural products, including short protein fragments known as peptides. These molecules may occur naturally in plants or be generated from plant protein through enzyme hydrolysis. In recent years, a growing body of evidence has linked plant-derived peptides to diverse biological activities, underscoring the importance of their structural and physicochemical features in determining functionality. Compared with peptides of animal or microbial origin, plant peptides stand out for their high abundance in sustainable sources, low allergenic potential, and distinctive structural traits- such as enrichment in hydrophobic and aromatic residues- that influence their stability, mechanisms of action, and biological functions. This review compiles and analyzes current literature to provide insights into how amino acid composition, secondary structure, net charge, and hydrophobicity influence peptide bioactivity. In addition, the review highlights the mechanisms of action most frequently described for plant peptides. Finally, the article discusses the current landscape and prospects of peptide-based drugs.

Keywords: plant bioactive peptides; structure; function; diseases; health

1. Introduction

For a long time, popular culture has established that certain foods and plants help alleviate various chronic ailments. However, in recent years, the role of specific dietary compounds in preventing diseases and improving human quality of life has become evident. In recent decades, the search for bioactive molecules has intensified, linking their activity with specific mechanisms of action [1]. Peptides are one of the most reported natural compounds with nutraceutical potential. They are linear chains composed of fewer than 40 amino acids joined by covalent bonds and released from a more complex protein structure called the parent protein [2]. Due to their size, peptides may be absorbed in the intestine, accessing the bloodstream, allowing these molecules to exert a physiological effect in vivo. They have been found to impact body functions and influence health positively, boosting the quality of human health, and therefore have been considered bioactive

https://doi.org/10.3390/molecules30183683

molecules, i.e., nutraceuticals [3]. These molecules are sometimes called cryptides, bioactive peptide fragments encrypted within larger protein sequences and released through enzymatic hydrolysis [4]. Furthermore, peptides have minimal side effects since they are not accumulated in organs and are absent from immunoreactions [5]. Over 5300 bioactive peptides have been cataloged in the BIOPEP-UWM^{tm} database [6].

Protein-rich foods are considered optimal sources for the extraction of peptides. Within this category, animal-derived foods are particularly prominent, encompassing milk, eggs, cheese, meat, fish industry by-products, and even bovine blood [7]. In recent years, bioactive peptide research has focused on alternative sources, such as those derived from protein-rich plants. This shift is mainly because producing plant-based proteins is more economical and sustainable than animal sources. Additionally, plants are plentiful, and the demand for plant-derived products continues to rise, as their consumption is linked to healthier dietary choices [8]. Moreover, agri-food byproducts have emerged as promising reservoirs of bioactive peptides, offering an opportunity to valorize waste streams while supporting circular economy approaches [9]. Beyond these advantages, plant peptides display distinctive features compared to animal- or microbe-derived peptides. They are generally associated with lower allergenic potential, which enhances their safety in nutraceutical and functional food applications [10]. Structurally, they often show a higher prevalence of hydrophobic and aromatic residues—particularly at the C-terminal—that improve their stability, antioxidant potential, and interactions with molecular targets [11]. These differences, combined with their sustainability and accessibility, highlight plant peptides as a unique and promising class of bioactive molecules. In this context, plant peptides have been found to display anti-thrombotic, antimicrobial, antihypertensive, immunomodulatory, antioxidative, and anticancer activities, among many others [12].

Bioactive peptides are mainly produced from plant proteins through enzymatic hydrolysis, gastrointestinal digestion, and microbial fermentation [13–15]. Enzymatic hydrolysis of proteins is the most common and effective method for producing bioactive peptides. This process involves using either a single enzyme or a combination of different enzymes. The most frequently used proteinases are digestive enzymes, such as pepsin, trypsin, chymotrypsin, and pancreatic enzymes. Additionally, enzymes derived from fungal and bacterial sources, such as alcalase and plant-based enzymes like papain, are also utilized in this process [16]. Each enzyme has a particular active site within the protein; consequently, the degree of hydrolysis of a specific enzyme depends on controllable and optimizable factors such as hydrolysis time, enzyme concentration, and enzyme combination [17]. Once produced, peptides require purification and characterization. Ultrafiltration and chromatographic methods are the principal strategies for isolation, whereas advanced techniques such as liquid chromatography coupled with UV detection or tandem mass spectrometry (MS/MS) enable structural identification [13–17]. These technological improvements have greatly expanded the discovery of new bioactive peptides and insights regarding their structure-function [18–20]. Research has shown that the activity of peptides is closely related to their chemical structure and the conformation of their amino acids. This review seeks to analyze the structure–activity relationship of plant peptides based on our current knowledge and analyses. Understanding this relationship is essential for developing therapeutic peptides, which offer an innovative, effective, and promising alternative for treating prevalent diseases in today's population.

2. Structure-Function Relationship of Bioactive Plant Peptides

2.1. Relation Between Plant Peptide Structure and Its Antioxidant Activity

Oxidative stress is an imbalance caused by a higher concentration of oxidants than antioxidants. This phenomenon breaks redox signaling, promoting cellular damage [21].

Reactive oxygen species (ROS), such as O_2^-, HO_2, H_2O_2, and OH^-, damage biomolecules, including DNA, proteins, and lipids; the damage caused by oxidative stress can both promote the onset of chronic diseases as well as their complications [22]. Over the past few decades, considerable interest has been focused on antioxidants, particularly those of natural origin [23]. Plant peptides are among the most extensively studied natural antioxidant molecules, emerging as a trending and prominent research area. Since 2010, 98 research articles have been published evaluating the antioxidant potential of plant peptides, with around 40 of these papers identifying the plant peptide sequences. Note that the primary methods used to determine the antioxidant activity of peptides are chemical-based, iron chelating activity, and radical scavenging (ABTS and DPPH). Cell-based methods have also been developed, such as cellular antioxidant activity assays using various cell lines like HepG2 and Caco-2 [24]. In Table 1, the antioxidant activity of plant peptides is presented. After our analysis, we determined that the peptides with higher antioxidant activity repeat key amino acids such as glutamic acid, aspartic acid, glycine, alanine, leucine, and phenylalanine. Glutamic acid and aspartic acid are negatively charged amino acids; since they have excess electrons, they have free radical quenching activity. Also, due to their charged residues, they chelate metals and inhibit metal-mediated oxidation [25,26]. On the other hand, these peptides show the most potent lipid peroxidation activity [27].

Table 1. Summarization of plant peptides with antioxidant activity.

Plant Source (Common Name)	Obtaining Method	Peptides Sequences	Activity	Reference
Apricot	Alcalase	SHNLPILR and SEAGVTE	Hydroxyl radical and superoxide anion radical scavenging capacities of 60 and 50% (1 mg/mL) and DPPH of 25% (1 mg/mL).	[28]
Asparagus	Alcalase	FAPVPFDF, MLLFPM * and FIARNFLLGW	DPPH scavenging Activity (EC_{50} of 4.14 μmol/L).	[29]
Basul	Alcalase	10 peptides (CCGDYY * among them)	ABTS and ORAC scavenging Activity (IC_{50} of 1.18 and 3.61 μmol TE/μmol).	[30]
Bitter bean	Alcalase	29 peptides	DPPH and reducing power of 2.9 mg GAE/g and 11.7 mM, respectively.	[31]
Cress	Alkaline protease and neutral protease	GRVGSSSC, GRAGGSYM, GHPNFKLNCSGG, GTKSCKA, ASSNARDMI, TAGGCYIPI, and KNCALQ	DPPH, OH^-, O_2^-, and ABTS scavenging activities of 89.2, 26.3, 40.6, and 42.9%, respectively (0.1 mg/mL).	[32]
Cherrie	Alcalase	14 peptides	ABTS, FRAP, and Hydroxyl scavenging activity of 40, 35, and 20%, respectively.	[33]
Cherries	Thermolysin	22 peptides	ABTS, FRAP, and Hydroxyl scavenging activity of 55, 45, and 30%, respectively.	[33]
Chickpea	Neutrase	LTEIIP	DPPH and hydroxyl radicals scavenging activity (IC_{50} of 0.24 and 0.57 mg/mL, respectively).	[34]
Chickpea	Alcalase	NRYHE	DPPH, Hydroxyl, and superoxide free radicals scavenging activity of 50, 80, and 60%, respectively.	[35]
Chickpea	Pepsin and pancreatin	ALEPDHR *, TETWNPNHPEL, FVPH, and SAEHGSLH	50 CAA units.	[26]
Chickpea	Alcalase	DHG and VGDI *	DPPH scavenging activity of 67% at 200 μg/mL	[36]
Chickpea	Alcalase and flavorzyme	RQSHFANAQP	DPPH and Hydroxyl radical scavenging activity of 64.9 and 95.8%, respectively (5 mg/mL)	[37]
Corn	Alcalase and flavouryzme	CSQAPLA *, YPKLAPNE, and YPQLLPNE	DPPH radical and superoxide anion radical (IC_{50} of 0.116 and 0.39 mg/mL, respectively).	[38]
Corn	Alcalase	MGGN, MNN * and MEN	CAA of 1213.79 μmol of QE/100 mmol	[39]

Table 1. *Cont.*

Plant Source (Common Name)	Obtaining Method	Peptides Sequences	Activity	Reference
Corn	Alcalase and trypsin	MI/LPP	DPPH scavenging activity (IC_{50} of 220 µg/mL).	[40]
Corn	Alcalase	AGI/LPM * and HAI/LGA	Scavenges hydroxyl radicals by 79.41% (10 mg/mL)	[41]
Corn	Alkaline protease and flavourzyme	LPF, LLPF, and FLPF *	DPPH and ABTS radical scavenging activity (IC_{50} of 1.51 and 2.83 mM, respectively).	[42]
Corn	Alcalase	YA and LMCH *	ABTS, DPPH, and O_2 of 25, 15, and 30%, respectively (1, 10, and 10 mg/mL, respectively)	[43]
Cottonseed	Microbial fermentation	YSNQNGRF	DPPH, ABTS, and hydroxy radical scavenging activity (EC_{50} of 0.49, 2.05, and 2.21 mg/mL).	[44]
Hemp	Pepsin and pancreatin	WVYY * and PSLPA	DPPH scavenging activity of 67% (0.5 mg/mL).	[45]
Hemp	Alcalase	NHAV and HVRETALV	Protective effects against cell death and oxidative apoptosis.	[46]
Hemp	Protamex	YGRDEISV and LDLVKPQ	ABTS, FE^{2+} chelating activity, and hydroxy radical scavenging activity of 52.3, 52.9, and 50.9%, respectively (0.4 mg/mL).	[47]
Hemp	Alcalase	LLY, LLR, IR, TY, VY, LH, EL, LK, AY, H, R and F.	ABTS, DPPH, FCA, and FRAP (IC_{50} of 0.5, 1, 1.2, and 1 mg/mL, respectively).	[48]
Jackfruit	Trypsin	VGPWQK	ABTS EC_{50} of 1 mg/mL	[49]
Jujube	Trypsin	VGQHTR * and GWLK	ABTS, DPPH, and chelating activity of 30, 20, and 15%, respectively (0.3 mg/mL).	[50]
Korean Pine	Alcalase	KWFCT and QWFCT *	ABTS scavenging activity of 74.9% (2 mg/mL), CAA of 916.32 µmol of QE/100 g.	[51]
Lentil	Savinase	LLSGTQNQPSFLSGF, NSLTLPILRYL * and TLEPNSVFLPVLLH	ORAC 1.432 µmol TE/µmol.	[52]
Lotus	Flavourzyme	16 peptides	DPPH and H_2O_2 scavenging (EC_{50} of 2.9 and 16.1 mg/mL, and reducing power activity of 8.0 mg/mL, respectively)-	[27]
Moringa	Pepsin and pancreatin	20 peptides	DPPH and ABTS of 45.70 and 93.09%, respectively (1.33 mg/mL), ORAC 1.27 mM TE/g.	[53]
Mulberry	Neutrase	SVL, EAVQ, and RDY *	DPPH and ABTS scavenging activity of 45 and 100%, respectively (0.4 mg/mL), CAA of 2204 µM QE/100 g.	[54]
Mung bean	Pepsin and pancreatin	MD, QSA, EW, LGW, KK, SVP, and DVAF	ABTS, ORAC, and iron chelating activity of 581.3 ascorbic acid, 127 TE, and 4843.5 EDTA equivalent, respectively (1 µM).	[55]
Mung Bean	Thermolysin	KK, DM, S, Y and W	ABTS, ORAC, and iron chelating activity of 646.2 ascorbic acid, 127 TE, and 5161.7 EDTA equivalent, respectively (1 µM).	[55]
Mung bean	Pancreatin	10 peptides (LLGIL * among them)	DPPH and hydroxyl radical neutralizers of 81.27% (4 mg/mL) and EC_{50} of 0.37 mM, respectively.	[56]
Mung Bean	Bromelain	CTN, HC, CGN, and CSGD	DPPH radical scavengers (EC_{50} values of 0.30, 0.29, 0.28, and 0.30 mg/mL, respectively).	[57]
Mungbean	Neutral protease	WN, WGN, LY, AW, RGWYE *, FW, GVPFW, and WLF	DPPH, ABTS, and OH^- radical scavenging activities of 60%, 35%, and 30%, respectively.	[58]
Palm	Papain	YLLLK, YGIKVGYAIP, GGIF *, GIFE, WAFS, GVQEGAGHYALL, WAF, AWFS, and LPWRPATNVF	DPPH scavenging Activity (IC_{50} of 0.02 µM).	[59]

Table 1. *Cont.*

Plant Source (Common Name)	Obtaining Method	Peptides Sequences	Activity	Reference
Pearl millet	Trypsin	SDRDLLGPNNQYLPK	DPPH, ABTS, Fe^{2+} chelating ability, and hydroxyl assay of 67.66, 78.81, 51.20, and 60.95%, respectively.	[60]
Peony	Alcalase	FSAP, PVETVR, QEPLLR, EAAY * and VLRPPLS	ABTS and hydroxyl radical-scavenging activity of 98.5 and 61.9%, respectively (0.5 mg/mL).	[61]
Perilla	Alkaline protease	YL and FY *	Inhibition of lipid peroxidation in rat liver (IC_{50} of 6.81 mg/mL), 0.05 mg/mL elevated H_2O_2 exposed cell ratios from 44.3 to 57.8%.	[62]
Perilla	Trypsin	ISPRILSYNLR	DPPH and ABTS scavenging activity of 48.13 and 78.14%, respectively (0.1 mg/mL).	[63]
Rice	Papain, flavourzyme, and protamex	RPNYTDA, TSQLLSDQ, TRTGDPFF * and NFHPQ	DPPH, ABTS, and FRAP scavenging activity (IC_{50} of 0.446, 0.658, and 0.235 mg/mL, respectively).	[64]
Rice	Trypsin	YSK	DPPH and reducing power (IC_{50} of 0.15 and 0.125 mg/mL, respectively).	[65]
Rice	Trypsin	91 peptides	ORAC 4.07 μmol TE/mg.	[66]
Rice	Alcalase	SGDWSDIGGR *, DFGSEILPR, GEPFPSDPKKQLQ, and GEKGGIPIGIGK	ORAC and DPPH of 41.57 μmol TE/g and 29.41%.	[67]
Sorghum	Protease-Purazyme	11 peptides	ABTS of 6.10%	[68]
Soybean	Alcalase	FDPAL	The scavenging rate on superoxide anion and hydroxyl radical was 75% and 80%, respectively. (2 mM)·500 μM elevated H_2O_2 exposed HeLa cells' survival rate.	[69]
Soybean	Pepsin and pancreatin	28 peptides (FSREEGQQQGEQ * among them)	ROS production by Caco-2 was lowered by half.	[70]
Sweet potato	Alcalase	TYQTF, SGQYFL, YMVSAIWG, YYIVS * and YYDPL	OH^- scavenging activity of 74.52% (1.0 mg/mL).	[71]
Tomato	Alcalase	29 peptides	DPPH scavenging activity of 60.36% (1 mg/mL).	[72]
Walnut	Alcalase	ALWPF and PLRWPF	Zebrafish embryo oxidative damage model values of 50 μg/mL and 30 μg/mL, respectively.	[73]
Wheat	Alcalase	CGFPGHC, QAC, RNF, SSC, and WF	TEAC and superoxide anion radical scavenging activity of 4760 μmol TE/g and 43.72%, respectively (5 mg/mL).	[74]

* This symbol was placed when the reported activity is associated with a particular peptide sequence.

Glycine, alanine, and leucine are aliphatic amino acids that provide structural flexibility and contribute to peptide–lipid interactions, enhancing cellular uptake [11,75]. In contrast, antioxidant capacity is primarily attributed to residues with aromatic or sulfur-containing side chains—such as tyrosine, tryptophan, phenylalanine, cysteine, and methionine—which can donate electrons or hydrogen atoms to neutralize free radicals. Histidine, with its imidazole group, also plays a key role in radical scavenging. Importantly, the position of these within the peptide sequence is critical: antioxidant activity is enhanced when hydrophobic or aromatic amino acids are located at the C-or N-terminal nds, improving solubility in lipid systems and protecting membranes from peroxyl radical–mediated oxidation [76,77].

For example, the hydrophobic amino acid leucine reduces $Fe_3{}^+$, mainly when located at the N and C-terminal, increasing DPPH radical scavenging activity. Leucine also contributes to antioxidant activity through its long aliphatic side chain, which interacts with the acyl chains of fatty acids [56]. Regarding phenylalanine, it has been reported that this aromatic amino acid provides protons from its benzene group, acting directly as a radical scavenger and promoting high values on the assays ABTS and ORAC [78]. Tyrosine contains a phenolic group, which acts as a hydrogen atom donor, providing protons to electron-deficient free radicals and quenching them in *in vitro* methods such as the DPPH assay. Also, the most effective position for tyrosine has been reported in the C-terminal. This amino acid also scavenges the peroxyl radicals generated during the AAPH assay. It has also been associated with high CAA values since it can remove peroxyl radicals more easily than other amino acids [29]. It has been reported that peptides containing tyrosine have twice the antioxidant activity as those without it in their structure [62].

The sulfur-containing amino acids, methionine and cysteine, have a nucleophilic character and have been found to contribute to the scavenging activity of peptides by sulfur-hydrogen donation [71]. Cysteine is another important amino acid since it directly interacts with free radicals because its thiol group protects cellular molecules from oxidation with high TEAC assay values [30,38,62]. Also, peptides containing cysteine possess higher ferric antioxidant power [57].

Nevertheless, since plant protein is deficient in sulfur-containing amino acids, they are not commonly found in all plant-derived peptides [41]. Another amino acid recognized for its antioxidant activity is histidine, a radical scavenger due to its imidazole group proton-donating ability [41]. Histidine's contribution to peptides' antioxidant activity is higher when located in the C-terminus position. Finally, valine at the N-terminal has been related to enhancing the activity of antioxidant peptides in oil systems [36].

Besides particular amino acids interacting with free radicals, particular peptides may influence gene expression. One is SOD-3, a crucial endogenous antioxidant enzyme, which is the case for the peptide FDPAL obtained from soybean [69]. In addition, other soy peptides were able to activate the Keap1/Nrf2 pathway, also increasing the expression of antioxidant and phase II enzymes such as SOD1, TrxR1, NQO1, and GR [70].

A relationship between peptides having a secondary structure and their antioxidant activity has been established, being higher with a lower random coil content [51]. Also, lower α-helix content has been reported to be linked with increased antioxidant capacity [41]. As for the length, antioxidant plant peptides have been reported to have 2 to 21 amino acids. We found that up to 66% of the identified antioxidant plant peptides have from 2 to 10 amino acids. This is very convenient because smaller peptides can easily pass through cell membranes. Also, short peptides may pass through the intestinal barrier and exert their effect in the body. The reported peptides have been found not to cause damage to healthy cells; cytotoxic and hemolytic assays have seen this. Plant peptides should be utilized as medicinal antioxidants and as nutraceuticals for functional foods to increase antioxidant activities and prevent food oxidation reactions, increasing shelf life, for example, in high lipid foods. Antioxidant plant peptides are an alternative to synthetic antioxidants considered harmful, such as BHA, BHT, and TBHQ, which affect the spleen, lung, and liver [74].

2.2. Relation Between Plant Peptides Structure and Its Antiproliferative Activity

Cancer is a process involving an uncontrolled division of the body's cells. It is one of the most relevant health problems globally, one of the leading causes of death. The short size of peptides allows them to penetrate cell membranes, where they can interact with oncogenic proteins inside the cell [79]. Also, they are efficacious signaling molecules that bind to specific cell surface receptors, such as ion channels. These actions trigger cell cycle arrest, suppress cancer cell growth, and inhibit invasion through various mechanisms, including cytoplasmic membrane disruption, mitochondrial membrane depolarization, DNA damage, and autophagy-like cell death [80,81].

Bowman-Birk trypsin inhibitor is an 8 kDa naturally occurring peptide isolated from various legumes and highlighted for its antiproliferative activity, which is attributed to its ability to inhibit serine proteases. This function is due to the presence of two inhibitory domains in its structure: CTKSNPPTC and CAYSNPPKC. The specific residues responsible for this activity are lysine (in the first domain) and tyrosine (in the second domain). Additionally, the high abundance of cysteine in this peptide facilitates the formation of disulfide bonds, enhancing its stability and resistance [82].

It has been established that the amino acids that repeat the most among plant peptides reported with anticancer activity are glutamic acid, leucine, serine, phenylalanine, and alanine (Table 2). As reported in the literature, associations between these amino acids and anticancer activity in peptides are detailed below. As for glutamic acid, it is a negatively charged amino acid and has been recognized to have antiproliferative activity on tumor cells [83,84]. Leucine and alanine are hydrophobic amino acids, improving peptide cell penetration [85]. Although the serine anticancer mechanism is not fully understood, Chatupheeraphat et al. [79] reported that removing serine from anticancer peptide sequences significantly reduced bioactivity. Peptides containing phenylalanine have a higher affinity for cancer cell membranes because phenylalanine contributes to the hydrophobicity of these peptides [83]. Hydrophobic and hydrophilic amino acids in the C or N-terminal promote an amphiphilic character, which has been reported to be beneficial. While hydrophobic amino acids are necessary to invade cancer cell membranes, hydrophilic ones give the peptide stability and a high possibility of interaction inside the cell [86]. Peptides containing hydrophobic amino acids (>30%) and also amino acids that give the peptide a positive charge (greater than or equal to 1), for example, lysine and arginine, bind to the membranes of cancer cells, which have a net negative charge associated with the presence of the phosphatidylserine phospholipid, which accounts for 9% of the total amount of phospholipids in the human cell membrane. Usually, this phospholipid faces the cytoplasm, but it goes to the outer part of the membrane in cancer cells, giving them a net negative charge. Once they attach to the membrane, peptides with these characteristics create pores through electrostatic interactions. This process leads to the leakage of intracellular components, resulting in necrosis. In the case of non-cationic peptides, the presence of hydrophobic amino acids enables them to penetrate the cell interior and interact with cytoplasmic regulatory proteins [79,87].

Table 2. Summarization of plant peptides with antiproliferative activity.

Plant Source (Common Name)	Obtaining Method	Peptides Sequences	Activity	Reference
Bean	Pepsin and pancreatin hydrolysis	GLTSK, LSGNK, GEGSGA *, MPACGSS and MTEEY	Cell cycle arrest (G1) and Apoptosis induction (loss of mitochondrial membrane potential) in HCT116 cells (IC_{50} of 134.6 μM).	[88]
Chickpea	Alcalase hydrolysis	ADLPGLK	Cytotoxicity to Ishikawa cells (IC_{50} of 101.5 μg/mL), inhibition of Ishikawa cells proliferation by 68.9% (500 μg/mL), and activation of caspase-3 and cell cycle arrest at S and G2.	[87]
Common bean	Naturally occurring	KTYENLADTY KGPYFTTGSH DHYKNKEHLRSGRMRDDFF	Cytotoxicity to L1210 and MBL2 cells (IC_{50} of 4.0 and 9.0 μM, respectively).	[89]
Ginger	Pepsin hydrolysis	RALGWSCL	Cytotoxicity to NB4 cells (IC_{50} of 600 μg/mL) and upregulation of caspase 3, 8, and 9.	[79]

Table 2. *Cont.*

Plant Source (Common Name)	Obtaining Method	Peptides Sequences	Activity	Reference
Lentil	Trypsin hydrolysis	27 peptides	Cytotoxicity to PC3 cells (IC_{50} of 0.96 mg/mL).	[90]
Lentil	Naturally occurring	Bowman-Birk trypsin inhibitor	Cytotoxicity to HT29 cells (IC_{50} of 32 μM).	[91]
Lima Bean	Naturally occurring	KTCENLATY and RGPCF	Antiproliferative activity BEL-7402 and SHSY5Ycells	[92]
Maize	Alcalase hydrolysis	EDLPGCPRE, LTKYNIT, DFGGGHPDPN, EDIIPF, MMAAAAGAG, DAVLQTLA, AAFVVAGT and MEVAMAM	Cytotoxicity to HepG2 cells (IC_{50} of 8.9 μg/mL).	[93]
Narcissus	Naturally occurring	GNNNLTSSQQQQIQFILQQI	Inhibited leukemia cells' proliferation by 100% (80 μM)	[94]
Nut	Trypsin hydrolysis	AYRNRYRRQY, RYEQRPRT, and LPTSEAAKY	Inhibited HeLa cell proliferation by 19% (50 μg/L).	[86]
Pea	Naturally occurring	Bowman-Birk trypsin inhibitor	Cytotoxicity to HT29 cells with an IC_{50} of 31 μM.	[82]
Pombalia calceolaria	Naturally occurring	GLPCAESCVFIPC TITAILGCSCRDRVCYD *, GIPCAESCVFIPCVTAIL GCSCKDRVCYN and GIPCGECVWIPCISSAIGC-SCKNKVCYRN	Cytotoxicity to MDA-MB-231 cells (IC_{50} of 1.8 μM).	[95]
Quinoa	Pepsin and pancreatin hydrolysis	SENIDDPS, DVYSPEAG, EAGRLTTL, IRAMPV, IFQEYI, SFFVFL, RELGEWGI GGLGDVLGGLP	Cytotoxicity to Caco-2, HT-29, and HCT-116 cells (IC_{50} of 0.256, 0.195, and 0.193 mg/mL, respectively).	[96]
Quinoa	Trypsin hydrolysis	FHPFPR, NWFPLPR, and HYNPFPG	Inhibited Ishikawa cells proliferation by 53.93% (8 g/L) and upregulation of caspase-3.	[96]
Rapeseed	Fermentation with *Bacillus subtilis* and *A. elegans* Plus, neutral protease hydrolysis	WTP	Inhibited HepG2 cells proliferation by 58.86% by increasing p53 and BAX expression and decreasing BCL-2 expression (800 μg/mL)	[97]
Rapeseed	Naturally occurring	NDGNQPL	Cytotoxicity to HepG2 cells (IC_{50} of 1.56 mmol/L) and cell cycle arrest at G0/G1.	[98]
Rice	Alcalase and Gastrointestinal juices hydrolysis	EQRPR	Inhibited Caco-2, MCF-7, and HepG-2 cells growth by 84%, 80%, and 84%, respectively.	[99]
Sacha inchi	Alkaline and neutral proteases	LLEPDVR, ALVEKAKAS and TGDGSLRPY	Cytotoxicity to HepG2 cells (IC_{50} of 109 μg/m).	[84]
Soybean	Alcalase hydrolysis	L/I-VPK	Cytotoxicity to HepG2, MCF-7 and HeLa cells (IC_{50} of 0.22, 0.15, and 0.32 μM, respectively).	[100]
Soybean	Naturally occurring	Vglycin	Promoted apoptosis and cell cycle arrest at G1/S in CT-26, SW480, and NCL-H716 (IC_{50} values of 4.21, 3.68, and 3.62 μmol/L, respectively). 38% inhibition of colon cancer in mice at 20 mg/kg/Day/21 days.	[101]
Soybean	Pepsin and pancreatin hydrolysis	15 peptides	Cytotoxicity to Caco-2, HT-29, and HCT-116 cells (IC_{50} of 10.3, 14.9, and 15.2 mg/mL, respectively).	[102]
Soybean	Alcalase, pepsin, and pancreatin hydrolysis	158 amino acid residues peptide	Inhibited Caco-2 and HCT-116 cells proliferation by 80% (700 μg/mL).	[103]
Soybean	Pepsin, trypsin, and chymotrypsin hydrolysis	CSDMRLNSCHSA, LSYPAQC, CYEPCKPSEDDKEN and PCKPSEDDKEN	Cytotoxicity to HT-29 cells (IC_{50} of 0.29 mg/mL).	[104]

Table 2. *Cont.*

Plant Source (Common Name)	Obtaining Method	Peptides Sequences	Activity	Reference
Sweet potato	Naturally occurring	AASTPVGGGR and RLDRGQ	Inhibited Panc-1 cells' proliferation by 55% by promoting apoptosis through the mitochondrial apoptotic pathway (100 μM)	[105]
Walnut	Papain hydrolysis	CTLEW	Cytotoxicity to Caco-2 and HeLa cells (IC_{50} of 0.65 and 0.60 mg/mL, respectively).	[86]
Walnut	Papain hydrolysis	CTLEW	Cytotoxicity to Caco-2, HeLa, and MCF-7 cells (IC_{50} of 0.65, 0.60, and 0.449 mg/mL, respectively).	[86]

* This symbol was placed when the reported activity is associated with a particular peptide sequence.

Other important peptide amino acids with anticancer activity include proline, histidine, tryptophan, and glycine. The presence of proline increases the flexibility of peptides [99]. Histidine enables peptides to induce cancer cytotoxicity by increasing membrane permeability. Tryptophan has been found to enter cancer cells by the endocytic pathway and bind to the major groove of nuclear DNA [83]. On the other hand, glycine-rich peptides have been found to stimulate NK cell activation, enhancing antitumor activity in both animal and human models, although the exact mechanism has not been established [106].

Amino acids described in the antioxidant section are also essential in the structure of anticancer peptides, as they prevent cells from being in an oxidative stress state and play a role in anti-tumor activity [97,107]. It has been previously reported that antioxidant peptides can potentially prevent and treat ROS-related diseases, particularly certain types of cancer. These peptides can act as anticancer compounds by inhibiting oxidative stress and reducing genetic alterations, such as mutations and chromosomal rearrangements [96]. As for the secondary structure, β-pleated sheet peptides usually have two or even more disulfide bonds and therefore are more stable than α-helical peptides [106]. As for the length of the identified anticancer plant peptides, they range from 3 to 158 amino acids.

Nevertheless, we have determined that approximately 63% of the identified anticancer plant peptides contain between 3 and 10 amino acids. Small peptides exhibit greater molecular mobility compared to those of longer length, and they have high diffusivity across membranes and a higher probability of interacting with cancer cell molecules within the cells. Peptides longer than 20 amino acids have been found to possess less cytotoxic effect on cancer cells. Molecular docking studies have shown that specific peptides interact with the active site of enzymes involved in regulating gene expression, such as HDAC1 and MDM2. In this context, the quinoa-derived peptide HYNPYFPG forms hydrogen bonds with the active site of HDAC1, thereby inhibiting the enzyme's activity. Histidine and aspartic acid were identified as the key amino acids responsible for this inhibitory effect [96]. Regarding MDM2, the rapeseed-derived peptide NDGNQPL exhibited high affinity for the enzyme's active site, promoting its inhibition primarily through establishing hydrophobic interactions [98].

Different mechanisms have been associated with plant peptides, such as the mitochondrial apoptotic pathway (rapeseed, sweet potato, pecan, soybean, and maize) and autophagic cell death (walnut peptides). On the other hand, bean peptides specifically affected gene expression in colon cancer cells HCT116 and RKO. They mainly upregulated transcriptionally activated genes that encode for antioxidant enzymes related to NRF-2, associated with cancer prevention. Also, it affected genes involved in MAPK signaling, having a role in proliferation and apoptosis [88]. Bean peptides also promote DNA damage through PARP cleavage and cell cycle arrest through nuclear translocation of p53 [81].

Another reported mechanism is for *Pombalia calceolaria* peptides, which inhibited cancer cells' migration [95].

2.3. Relation Between Plant Peptides Structure and Its Angiotensin-Converting Enzyme Inhibitory Activity

The feasibility of using bioactive peptides as potential hypotensive drugs depends on their bioavailability and bioactivity, Plant bioactive peptides can act as hypotensive drugs by known mechanisms. However, one of the most studied is their capacity to inhibit the angiotensin-converting enzyme (ACE). The renin-angiotensin and the kinin-nitric oxide systems are the hypotensive activity mechanisms that have been studied the most. However, the most well-studied bioactive peptides are the renin-angiotensin system, where the angiotensin-converting enzyme is one of the main protagonists [108,109]. The angiotensin-converting-enzyme, a glycosylated membrane-bound zinc metalloprotease, is the main protagonist of the renin-angiotensin system; in normal conditions, its primary metabolic function begins after renin acts on angiotensinogen to angiotensin I, an inactive peptide which is later catalyzed by the ACE to angiotensin II and then binds to vascular wall receptors to cause the contraction of the blood vessels. However, during hypertensive conditions, the abnormal conditions cause the renin-angiotensin system to function excessively, leading to high levels of angiotensin II, thus causing hypertension [110,111]. Therefore, the angiotensin-converting enzyme inhibition can help lower the blood pressure in hypertensive patients [112,113]. Moreover, most recent studies have shown that bioinformatic predictions can help to model peptides with specific bioactivities like antihypertensive biopeptides [114].

Previous studies have elucidated a correlation between the amino acid sequences and their ACE-inhibitory capacity (Table 3) [109]. In another related subject, one mode of action of bioactive peptides in the ACE inhibition is competition of the peptides with the ACE substrate for the union with the enzyme's catalytic site. This site contains three specific pockets, and the interaction between peptides and ACE involves various types of molecular forces, such as van der Waals interactions, hydrophobic contacts, and electrostatic attractions. Among these, hydrogen bonds are predominant in stabilizing the peptide–ACE complex [115]. According to studies employing molecular docking tools, amino acids such as isoleucine, leucine, and proline enhance the antihypertensive activity of peptides. This effect is attributed to their ability to establish specific interactions with residues like histidine, alanine, and arginine, respectively, which are located within the binding pockets of the ACE. These interactions hinder the enzyme's catalytic activity, thereby contributing to its inhibition [116,117]. On the other hand, A net negative charge resulting from the incorporation of strongly acidic amino acids, such as aspartic acid and glutamic acid, can potentially interfere with ACE activity by binding to the zinc ion essential for its catalytic function [118].

The above is not always the case, for some bioactive peptides like LW and IY, a noncompetitive inhibition has been reported. In the case of the peptides IW and FY, they have been cataloged as uncompetitive inhibitors [109,118]. It is worth noting that competitive ACE inhibitors have a practical limitation, as they must compete with the high physiological concentration of the natural substrate, angiotensin I. In this regard, noncompetitive and uncompetitive peptides like those mentioned above, which do not rely on substrate competition, may offer greater effectiveness [111]. The different mechanisms by which plant-derived peptides inhibit ACE are illustrated in a schematic representation in Figure 1.

Table 3. Summarization of plant peptides with ACE-inhibitory activity.

Plant Source (Common Name)	Obtaining Method	Peptides	ACE Inhibitor Activity	Reference
Hempseed	Pepsin, trypsin, and pancreatin	SHLNWVCIFLGFHSFGLYI, QIQFEGFCRF, and IPDKANLGFRFP	Determined as an ACE-inhibitor using the PeptideRanker and BIOPEP	[119]
Huangqi root	Aqueous extraction	LVPPHA	IC_{50} of 414.88 μM	[116]
Longan seeds	Pepsin and pancreatin	ETSGMKPTEL and ISSMGILVCL	IC_{50} of 0.003 mg/mL	[120]
Moringa	Flavourzyme,	IPPAYSK, ILVDR *, FFFPK and LLDPR	IC_{50} of 81.25 μM	[111]
Moringa	Alcalase	LGFF and GLFF	IC_{50} of 0.29 and 1.88 mM, respectively	[121]
Olive	Alcalase, thermolysin, neutrase, PTN, and flavourzyme	LTPTSN, LVVDGEGY, and FDAVGVK	IC_{50} of 0.029 mg/mL	[122]
Stink bean seeds	Alcalase	ASPAPAGLSYCVPEV DPLLLRAEDKLDLSDL	51.7–80.2% inhibition	[123]
Plum	Alcalase, thermolysin, flavourzyme, and protease P.	MLPSLPK, HLPLL, NLPLL, HNLPLL, KGVL, and HLPLLR	IC_{50} of 0.5 mg/mL	[124]
Rice	Neutrase	FDGSPVGY * and VFDGVLRPGQ	IC_{50} of 0.079 mg/mL	[125]
Soybean	Alkaline proteinase, papain, trypsin, pepsin, and pancreatin	LLPLPVLK, WLRL, SWLRL and MLPVMR	51.43% inhibition	[126]
Soybean	Alkaline protease, flavored protease, papain, trypsin, and chymotrypsin	GKGLW and GDGLKW *	IC_{50} of 33.98 μM	[115]
Soybean	Alcalase	LY *, YVVF, LVF, WMY, LVLL, and FF	IC_{50} of 0.53 μM	[127]
Soybean	Alcalase and Flavourzyme	ALKPDNR, VVPD, NDRP, and NDTP	IC_{50} of 148.28 μg/ml	[118]

* This symbol was placed when the reported activity is associated with a particular peptide sequence.

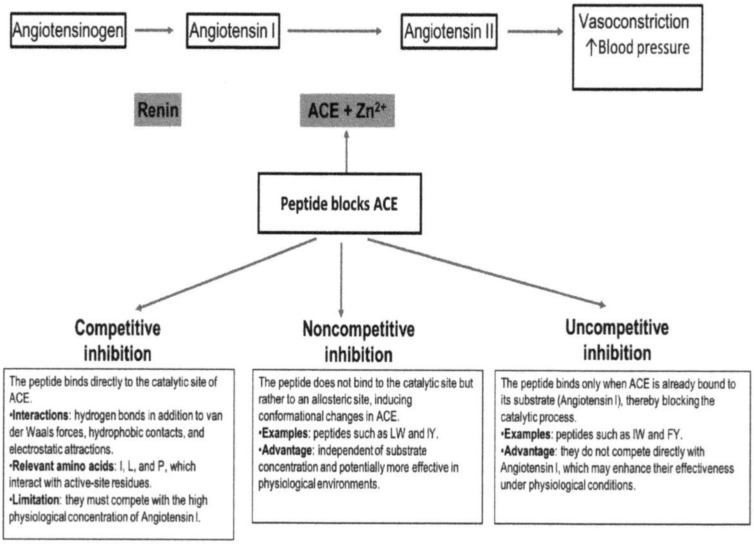

Figure 1. Schematic representation of the inhibitory mechanisms of plant-derived peptides on ACE.

Hydrophobic and aliphatic amino acids are also necessary for peptides' hypotensive activity [109]. Most ACE-inhibitory peptides are characterized by the presence of hydrophobic or branched-chain amino acids at the N-terminus. At the same time, residues with cyclic or aromatic rings—such as tyrosine, phenylalanine, tryptophan, and proline—are commonly found at the C-terminus. In fact, structural analyses of hypotensive peptides have shown that approximately 40% contain tryptophan at the C-terminal end [115]. One of the main reasons for the prevalence of hydrophobic amino acids in these bioactive peptides is the hydrophobic nature of the binding pockets in the ACE, which enhances interactions and binding affinity with peptides possessing such characteristics [112,117,118].

Peptide size also plays a critical role in determining bioactivity. In this regard, it has been documented that smaller peptides, particularly those composed of fewer than ten amino acids, exhibit a more potent antihypertensive effect compared to larger peptides. One contributing factor is that shorter peptides—especially dipeptides and tripeptides—can be absorbed intact through the gastrointestinal tract and reach the blood circulatory system. Additionally, specific amino acid residues, such as proline, phenylalanine at the C-terminal, and leucine at the N-terminal, have been shown to confer resistance to peptidase degradation. These residues also enhance affinity for peptide transporters, further facilitating absorption. Moreover, the smaller size of these peptides increases their likelihood of effectively interacting with the active site of ACE [121].

Moreover, it is essential to mention that unless there is sufficient and significant data regarding the pharmacological potential of bioactive peptides from plant origin, absorption, distribution, metabolism, and excretion studies are important. Unfortunately, this approach is often neglected, whether by instrumental or budget restrictions. Also, studies should focus on mimicking physiological conditions and concentrations of peptides, because the aleatory use of peptide concentrations in *in vitro* studies might lead to an overestimation of their antihypertensive potential [128,129]. Thus, an experimental design considering the abovementioned factors is needed for more physiologically relevant results.

2.4. Relation Between Plant Peptides Structure and Its Hypolipidemic Activity

Dyslipidemia is a metabolic disorder characterized by abnormal concentrations of lipids and lipoproteins in the bloodstream, most commonly involving elevated levels of total cholesterol, low-density lipoprotein cholesterol (LDL-C), triglycerides, or decreased levels of high-density lipoprotein cholesterol (HDL-C). It represents a significant risk factor for cardiovascular diseases and is influenced by genetic, dietary, and lifestyle factors, as well as enzymatic activity involved in lipid metabolism and various underlying pathological conditions.

Peptides isolated from soybean and lupine have been shown to affect 3-hydroxy-3-methylglutaryl coenzyme A reductase (HMGCoAR), since they act as competitive inhibitors of the enzyme. This effect is due to characteristics such as the number, position, and type of residues of amino acids, as well as peptide hydrophobicity. For example, proline and valine have been reported as crucial, and the inhibitory activity is higher when they are located within the first and the fourth N-terminal positions and when they are flanked by leucine, phenylalanine, alanine, and glycine [130,131].

By Molecular docking and *in vitro* studies, lupine and rapeseed peptides (Table 4) were found to inhibit HMGCoAR, low-density lipoprotein receptor (LDLR), and proprotein convertase subtilisin/kexin type-9 (PCSK9).

Table 4. Summarization of plant peptides with hypolipidemic activity.

Plant Source (Common Name)	Obtaining	Peptides	Activity	Reference
Amaranth	Pepsin, trypsin, and pancreatin hydrolysis	GGV, IVG/LVG, VGVI/VGVL	4 μg/mL inhibited HMG-CoA reductase by 45%.	[132]
Bean	Pepsin and pancreatin hydrolysis	LVTTTVDL, QTSTPLFS, VELVGPK, TRGVLV	700 mg/kg/day for nine weeks in BALB/c mice decreased lipid profile, total cholesterol, triglycerides, and HDL-c levels.	[133]

Table 4. *Cont.*

Plant Source (Common Name)	Obtaining	Peptides	Activity	Reference
Brewer's spent grain	Purazyme and flavourzyme hydrolysis	WNIHMEHQDLTTME, DFGIASF, LAAVEALSTNG	0.4 mg/mL inhibited cholesterol esterase by 35% and 1.4 mg /mL inhibited pancreatic lipase by 80%.	[134]
Coffee	Alcalase, pepsin, pancreatin, and thermolysin hydrolysis	79 peptides	Reduced *in vitro* micellar cholesterol solubility by 32%.	[135]
Cowpea	Pepsin and pancreatin hydrolysis	11 peptides	35 μg/mL reduced cholesterol micellar solubilization by 71.7% and 50 μg/mL inhibited HMGCoAR by 57.1%.	[136]
Hemp seed	Pepsin hydrolysis	90 peptides	1.0 mg/mL inhibited HMGCoAR by 80%, up-regulated the LDLR protein levels by 63% and increased the LDL-uptake in HepG2 cells.	[137]
Hemp seed	Trypsin	20 peptides	Reduced serum total cholesterol, triglycerides, and low-density lipoprotein cholesterol by 28, 34, and 40% in high-fat-induced mouse.	[138]
Lupin	Pepsin hydrolysis	YDFYPSSTKDQQS	100 μM inhibited HMGCoAR by 87.4%, modulated the cholesterol metabolism in HepG2 cells, improved the low-density lipoprotein levels by 50.5% and increased the receptor protein levels by 37.8% via SREBP-1 activation.	[139]
Lupin	Trypsin hydrolysis	GQEQSHQDEGVIVR	Inhibited HMGCoAR, with an IC_{50} of 99.5 μM, 00 μM, enhanced the LDLR protein levels by 43.4%, reduced the $PCSK9^{D374Y}$ protein level by 41.2% and decreased the HNF-1α level by 19.8% in HepG2 cells.	[140]
Lupin	Trypsin and pepsin hydrolysis Synthesized	YDFYPSSTKDQQS, LILPKHSDAD, LTFPGSAED, LTFPG	Produced a fast and efficient decrease in systolic blood pressure (−37 mmHg after 2 h).	[141]
Lupin	Trypsin and pepsin hydrolysis	LILPKHSDAD, LPKHSDAD, YDFYPSSTKDQQS, LTFPGSAED	Inhibited HMGCoAR (IC_{50} of 175.3 μM), increased LDLR expression, and decreased the PCSK9 production of HepG2 cells.	[142]
Olive seed	Alcalase hydrolysis	130 peptides	Reduced cholesterol micellar solubility by 49%, bound to 10% of total bile acid, and inhibited cholesterol esterase and HMGCoA enzymes by 30% and 15%, respectively.	[143]
Pinto bean	Protamex®	PPHMLP, PLPTGAGP, PPHMGGP, PLPLHMLP, LSSLGMGSLGALPVCM	4.59 μM inhibited pancreatic lipase by 87%.	[144]
Rapeseed	Alcalase	EFLELL	Decreased total cholesterol and triglyceride by 46% and 32%, respectively, at a concentration of 2 mM	[145]
Rice	Pepsin and trypsin hydrolysis	GEQQQQPGM	100 mg/body weight for 30 days reduced the atherogenic index.	[146]
Rice	N/A	IIAEK	300 mg/kg body weight/day reduced total serum cholesterol, LDL-cholesterol, and the atherogenic index.	[147]
Soybean	Trypsin and pepsin hydrolysis	IAVPGEVA, IAVPTGVA, LPYP	10 μM inhibited the activity of HMGCoAR and modulated cholesterol metabolism in HepG2 cells.	[148]
Soybean	Pepsin and trypsin hydrolysis	140 peptides	0.5 mg/mL inhibited HMGCoAR by 76.9% and Increased LDLR protein levels by 63.0% in HepG2 cells.	[149]

Table 4. *Cont.*

Plant Source (Common Name)	Obtaining	Peptides	Activity	Reference
Soybean	Trypsin and pepsin hydrolysis	YVVNPDNDEN, YVVNPDNNEN	250 µM inhibited HMGCoAR by 80% and increased the LDL uptake by 2-fold in HepG2 cells.	[150]
Tea	Pepsin	FLF, IYF, and QIF	Inhibited pancreatic lipase and cholesterol esterase (IC_{50} values of 0.153 and 0.549 mg/mL, respectively).	[131]

N/A = Not available.

As for PCSK9 inhibition, it was reported that the amino acid leucine within the peptide LPKHSDAD (Table 4) was inserted in the hydrophobic pocket of PCSK9, forming a stable peptide-PCSK9 union stabilized by hydrogen bonds [142,145]. Plant peptides containing at least four hydrophobic amino acids have a hypocholesterolemic effect [151]. Mainly if one is located at the C- or N-terminal [152–155]. Hydrophobic amino acids establish hydrophobic interactions with lipids and with the non-polar molecules of bile acids, which, although weak, promote the reduction in blood cholesterol levels [131,156–158]. On the other hand, peptides with cationic amino acids can also capture bile acids because the cationic residues interact with the carboxyl groups of bile acids. Some plant peptide sequences have been modified to increase their hypocholesterolemic activity by adding polar amino acids like serine residues to the C-terminal, improving their solubility in aqueous systems [155,159]. We determined that the amino acids that repeat the most among peptides with hypolipidemic activity are lysine, threonine, valine, glutamic acid, and isoleucine. These peptides are mainly made up of 2 to 20 amino acids.

2.5. Relation Between Plant Peptides Structure and Its Hypoglycemic Activity

Diabetes is a degenerative pathology that causes chronic hyperglycemia due to apoptosis of the pancreatic β-cells. If it is not controlled, cardiovascular diseases and other health complications can develop [160,161]. Peptides with hypoglycemic activity are characterized by inhibiting carbohydrate metabolism enzymes such as α-amylase, α-glucosidase, and DPP-IV. In addition, they can also act by inhibiting the glucose transporter system and acting as insulin-like molecules [162,163]. Table 5 shows plant sources of hypoglycemic peptides.

Table 5. Summarization of plant peptides with hypoglycemic activity.

Plant Source	Obtaining	Peptides	Activity	Reference
Bean	α-amylase, pepsin, and pancreatin hydrolysis	INEGSLLLPH, FVVAEQAGNEEGFE, INEGSLLLPH, SGGGGGGVAGAATASR, GSGGGGGGGFGGPRR, GGYQGGGYGGNSGGGYGNRG, GGSGGGGGSSSGRRP, GDTVTVEFDTFLSR,	α-amylase inhibitory activity (IC_{50} of 1.94 mg/mL)	[164]
Bean	Pepsin and pancreatin hydrolysis	22 peptides	Inhibited DPP-IV (IC_{50} 0.03 mg/mL)	[165]
Bean	Alcalase and bromelain hydrolysis	40 peptides	Inhibited α-amylase (50%), α-glucosidase (76.4%), and DPP-IV (55.3%)	[166]
Bittermelon	N/A	RVRVWVTERGIVARPPTIG.	600 mg/day/ 3 Regulated the blood glucose of diabetic patients and reduced the glycated hemoglobin by 7.4%.	[167]

Table 5. *Cont.*

Plant Source	Obtaining	Peptides	Activity	Reference
Bittermelon	Naturally occurring	GHPYYSIKKS	1 mg/kg significantly reduced blood glucose levels (7.68 mmol/L) in an in vivo model	[168]
Bittermelon	Naturally occurring	H-GHPYYSIKKS-OH	In a murine model, 1 mg/kg showed anti-hyperglycemic effects and lowered blood glucose levels.	[169]
Bittermelon	Pepsin and pancreatin hydrolysis	SRCQGKSSWPQLVGSTGAAAK AWIERENPRVRAVIIKV, GSGATKDFRCDRVRVWVTERGI-VARPPTIG, SRCQGKSSWPQLVGSTGA, GAAAKAWIERENPRVRAVI, VIIKVGSGATKDFRCDRVR, RVRVWVTERGIVARPPTIG, DFR-CDRVRVWVTERGIVARPPTIG, VTERGIVARPPTIG	5 nmol/kg displayed a hypoglycemic activity *in vitro* and in vivo	[170]
Black bean	Alcalase, Trypsin, proteinase K, Flavourzyme, Thermolysin, Pepsin, Papain, and Chymotrypsin hydrolysis	34 peptides	1 mg/mL inhibited DPP-IV, α-amylase, and α-glucosidase by 96.7, 53.4, and 66.1%, respectively	[171]
Brewer's spent grain	Alcalase, ultraFlo, shearzyme, and flavourzyme hydrolysis	IPY, LPY, IPLQP, LPLQP, APLP, VPIP, IPVP *, PLVP	DPP-IV inhibitory activity with an IC_{50} of 38.67 μM	[172]
Chickpea	Pepsin/pancreatin and bromelain hydrolysis	32 peptides	Had affinity for DPPV-IV, α-amylase, and α-glucosidase active sites.	[173]
Chickpea	Enzymatic hydrolysis (not specified)	LLR, FH, RQLPR, KGF, and NFQ	Inhibited α-glucosidase increased the viability of insulin-resistant cells and the glucose consumption rate.	[174]
Hemp	Alkaline protease	TGLGR, SPVI, FY, and FR	Inhibited α-glucosidase by 70.02% (20 μL of the hydrolysate).	[175]
Lupin	Trypsin and pepsin hydrolysis	LTFPGSAED	Inhibited DPP-IV (IC_{50} of 207.5 μM)	[176]
Mulberry	N/A	WGVENAATYFWQTV	α-amylase and α-glucosidase inhibitory activity (IC_{50} of 16.25 and 12.56 μg/mL, respectively)	[177]
Oat	Trypsin hydrolysis	LQAFEPLR, EFLLAGNNK, LQAFEPLR, EFLLAGNNK.	Inhibited DPP-IV (IC_{50} 141.7 μM) and modulated its expression in Caco-2 cells.	[178]
Oat, buckwheat, and highland Barley	Alcalase and trypsin hydrolysis,	35 peptides	Inhibited DPP-IV (IC_{50} from 0.13 to 8.15 mg/mL)	[179]
Orange seed	Porcine pepsin, trypsin, and chymotrypsin hydrolysis	63 peptides	20 μL Inhibited α-amylase and α-glucosidase by 41.7 and 57%, respectively	[180]
Orange seed	Pepsin, trypsin, and chymotrypsin hydrolysis	978 peptides	0.1 and 1 mg/mL inhibited α-amylase and α-glucosidase by 42.35 and 45.39%, respectively	[181]
Pinto bean	Protamex®	PPHMLP, PLPTGGAGP, PPHMGGP, PLPLHMLP, LDDLGMGSLGALPVCM	100 μg/mL Inhibited α-amylase (IC_{50} of 0.31 mM)	[182]
Potato	Alcalase hydrolysis	DIKTNKPVIF	50 mg/kg/day lowered glucose and HbA1c levels in diabetic and non-diabetic mice.	[183]
Quinoa	α-chymotrypsin, Pronase E, and Bromelain hydrolysis	136 peptides	Inhibition of DPP-IV and α-glucosidase (IC_{50} of 1.12 and 1.86 mg/mL, respectively)	[184]
Quinoa	Pepsin and pancreatin hydrolysis	20 peptides	250 μM inhibited DPP-IV, α-amylase, and α-glucosidase by 17, 6.86, and 55.85%, respectively	[185]
Soy, lupin, and quinoa	Subtilisin, trypsin, and flavourzyme hydrolysis	20 peptides	Inhibited DPP-IV (IC_{50} 1.16 mg/mL)	[186]
Soybean	Pepsin and trypsin hydrolysis	140 peptides	2.5 mg/mL reduced the DPP-IV activity by 31.4%.	[149]

Table 5. *Cont.*

Plant Source	Obtaining	Peptides	Activity	Reference
Soybean	Pepsin and pancreatin, hydrolysis	12 peptides	Inhibited DPP-IV (IC_{50} of 0.91 mg/mL), α-amylase (IC_{50} of 1.70 mg/mL), and α-glucosidase (IC_{50} of 3.73 mg/mL).	[187]
Soybean	Naturally occurring	VSCNGVCSPFEMPPCGSSA CRCIPYGLVVGNCRHPSG	40 µg/[g·d] normalized fasting glucose levels in diabetic rats	[188]
Soybean	Trypsin hydrolysis	GSA, GAL	α-glucosidase inhibitory activity (IC_{50} 0.049 mg/mL) 3.52 mg/mL reduced fasting blood glucose in diabetic mouse by 42%	[189]
Soybean	Alkaline proteinase, papain, and trypsin hydrolysis	LLPLPVLK, SWLRL, WLRL *	α-glucosidase inhibitory effect (IC_{50} 162 µmol/L) 2.73 mg/mL inhibited the enzyme DPP-IV in 40.85%	[126]
Soybean	Trypsin and pepsin hydrolysis	IAVPTGVA	Inhibited DPP-IV (IC_{50} of 223.2 µM)	[176]
Tea	Fermentation	AADTDYRFS and AGDGTPYVR	Inhibited α-glucosidase with IC_{50} of 0.820 and 3.942, respectively. Increased expression levels of MDM2, IRS-1, Akt, PI3k and GLUT4.	[190]

N/A = Not available. * This symbol was placed when the reported activity is associated with a particular peptide sequence.

Notably, in the inhibition of α-glucosidase, the most potent plant peptides correspond to those containing three to six amino acids [174]. Similarly, another relevant factor is the properties of the amino acid residues at the N-terminal and C-terminal. In this regard, the analysis of several articles has led to the conclusion that α-glucosidase inhibitory activity is higher when arginine, tyrosine, lysine, threonine, or serine is located at the final N-terminal residue and when methionine or alanine is situated at the final C-terminus [175]. Another outstanding characteristic that has been observed is the presence of proline. Some investigations have documented that proline and a basic amino acid, such as lysine and arginine, can increase the bioactivity against α-glucosidase.

Regarding their net charge, peptides with 0 or +1 are the most effective in inhibiting α-glucosidase [191–193]. Docking simulation has shown that a glycine-serine-arginine sequence inhibits α-glucosidase by attaching to the pocket of the enzyme due to van der Waals forces [175,189,190]. The enzymes with which the hydrolysis is carried out play an important role in defining the biological activity of the peptides. Plant hydrolysates obtained with alkaline protease show the highest α-glucosidase inhibition rate [126]. It is worth mentioning that the most studied peptide inhibitors of α-glucosidase are between 5 and 6 amino acids; the tetrapeptides are the most potent inhibitors of this enzyme [194].

Plant peptides containing proline or alanine residues inhibited DPP-IV enzyme through competitive inhibition for its active site [195–197]. Regarding DPP-IV inhibition, the amino acids alanine and leucine have the most potential to interact with the catalytic site through hydrogen bonds and polar and non-polar interactions [171,184]. Also, peptides with high inhibition of DPP-IV contain isoleucine in the N1 position. This may be because isoleucine favors the formation of α-helix [172].

On the other hand, amino acids such as serine, threonine, and tyrosine interact with the α-amylase active site due to their hydroxyl group, inhibiting it. Furthermore, higher α-amylase inhibitory activities have been reported when these amino acids are located at the N-terminal position. On the other hand, proline or alanine near or at the C-terminal position promotes the formation of hydrogen bonds and electrostatic interactions of plant peptides with the catalytic site of α-amylase [187]. Our literature research found that the reports associate the amino acids glycine, alanine, valine, threonine, and proline with the most abundant amino acids in peptides with hypoglycemic activity.

2.6. Relation Between Plant Peptide Structure and Its Antimicrobial Activity

Medicinal plants used in traditional medicine are attractive sources of bioactive proteins and peptides that demonstrate a broad spectrum of activities, including antimicrobial [198]. Usually, antimicrobial peptides (AMPs) are broad-spectrum agents that act on bacteria, fungi, metazoans, and other parasites [199]. Most AMPs are short protein fragments composed of 10 to 50 amino acids with a net positive charge from +2 to +11. It has been observed that 50% of AMPs contain hydrophobic residues, which allows them to have a strong affinity to net negatively charged microbial membranes [200].

The enormous variety of plant AMPs causes difficulty in their classification. Considering the secondary structures, AMPs are classified into "α" family (with α-helical structures), "β" family (containing β-sheet structures stabilized by disulfide bonds), α-hairpinin (with a motif formed by antiparallel α-helices that are stabilized by two disulfide bridges), and "αβ" family (having both "α" and "β" structures) [201]. On the other hand, peptides with extended/combined conformation [202].

AMPs are classified considering their similarity to protein sequence, cysteine motifs, and distinctive patterns of disulfide bonds, determining the peptide folding. Therefore, they are commonly grouped as thionins, defensins, heveins, knottins, lipid transfer proteins, and cyclotides. Thionins are a family of antimicrobial peptides with low molecular weight (about 5 kDa), rich in arginine, lysine, and cysteine residues. Their structure includes two antiparallel α-helices and an antiparallel double-stranded β-sheet with three or four conserved disulfide linkages. They are positively charged with a neutral pH. Their toxic effect was postulated to arise from the lysis of the membranes of attaching cells. However, the precise mechanism underlying toxicity remains unknown [203]. Plant defensins consist of three antiparallel β-sheets and an α-helix parallel to them. They possess a variety of biological functions, such as inhibiting microbial growth and inhibiting enzyme activity [204]. The hevein family consists of positively charged peptides of 29 to 45 amino acids, with abundant glycine (6) and cysteine (8–10) and aromatic residues. They have a coil-β1-β2-coil-β3 structure that occurs by variations with the secondary structural motif in the presence of turns in two long coils in the β3 chain. Hevein domains bind to chitin, which is their primary target; usually, their action modes include degradation and disruption of the cell wall and plasma membrane due to their hydrolytic action, causing extravasation of plasma particles, so heveins have good antifungal activity [201]. Plant knottins contain approximately 30 amino acids. Their antimicrobial activity has been attributed to alterations in functional components of the plasma membrane. The typical structure of knottins involves conserved disulfide bonds between multiple cysteine pairs, forming a cystine knot [204]. The lipid transfer proteins (LTPs) family comprises cationic proteins of approximately 70 and 90 amino acids with eight cysteine residues. They share a defining structural feature, a conserved inner hydrophobic cavity surrounded by α-helices. They bind to a wide range of lipids, including fatty acids, phospholipids, prostaglandin B2, lyso-derivatives, and acyl-coenzyme A. Plant LTPs inhibit bacterial and fungal pathogens' growth by promoting pathogen membrane permeabilization [205]. Finally, cyclotides are ultra-stable peptides. They are around 30 amino acids and are disulfide-rich peptides from plants with a head-to-tail cyclic backbone and cystine knot arrangement of three conserved disulfide bonds [206].

Huan (2020) proposed that AMPs can also be classified based on amino acid-rich species, proline, tryptophan, arginine, histidine, and glycine-rich peptides [207,208]. Therefore, the data presented in Table 6 indicate that the most frequent residues were alanine, arginine, glycine, valine, and cysteine. Furthermore, an average isoelectric point of 8.5 and an average hydropathy of −0.15, showing the cationic and hydrophilic nature of most AMPs. Beyond these classical families of plant AMPs, recent studies have also pointed out that enzyme-derived cryptic peptides (cryptides) may represent an additional reservoir of antimicrobial molecules. For instance, ribosome-inactivating proteins (PD-L1/2) from *Phytolacca dioica* have been reported to release bioactive cryptides with antimicrobial and anti-biofilm activity [209]. Remarkably, these peptides exhibit structural plasticity, shifting between α-helix and β-sheet conformations depending on their environment (e.g., TFE, SDS, alginate, or LPS), directly influencing their bioactivity. More broadly, enzymes have been proposed as reservoirs of host defense peptides, underscoring the importance of considering cryptides as a relevant but underexplored source of plant-derived AMPs [210].

Table 6. Summarization of plant peptides with antimicrobial activity.

Plant Source (Common Name)	Obtaining	Peptides Sequences	Target Pathogen	Reference
Alfalfa	Pepsin	MDN, TMW, CVQ, AFR, ELAAAC, ILAAF, GNAPGAVA, LRDDF, ALRMSG, RDRFL, EYLIRKGWI, IEKDRSRGIF	*Escherichia coli*, *Micrococcus luteus*, *Bacillus subtilis*, and *Listeria innocua*.	[211]
Angel's trumpets	Naturally occurring	RHCESQSQRFKGTCLSEK NCASVCETEGFSGG DCRGLRRRCFCTRPC	*Staphylococcus epidermidis*.	[212]
Barrel medic	Naturally occurring	RNGCIVDPRCP YQQCRRPLYCRRR	*Sinorhizobium meliloti*	[213]
Bitter tomato	Naturally occurring	MKTIQGQSATTALT MEVARVQA	*Rhizoctonia solani* and *Colletotrichum gloeosporioides*.	[214]
Chickweed	Naturally occurring	VDPDVRAYCKH QCMSTRGDQAR KICESVCMRQD	*Alternaria alternata*, *Aspergillus niger*, *Bipolaris sorokiniana*, *Botrytis cinérea*, *Fusarium oxysporum*, *Fusarium solani*, *Phytophthora infestans* and *Pythium ultimum*.	[215]
Chili pepper	Naturally occurring	N-terminal: AVTXGQVDANLAPXV	*Colletotrichum lindemunthianum* and *Candida tropicalis*.	[216]
Cowpea	Naturally occurring	N-terminal sequence: KTCMT- Nested in Cp-thionin II	*Fusarium culmorum*.	[217]
Cowpea	Alcalase	CW, SC, WS, CR, WC, LA, GP, NV and RG	*Listeria monocytognes*, *Listeria innocua*, *Staphylococcus aureus*, *Sreptococcus pyogenes*, *Klebsiella pnemoniae*, *Pseudomonas aeruginosa*, *Escherichia coli* and *Salmonella typhimurium*.	[208]
Ground Bean	Naturally occurring	N-terminal sequence: KTCENLADTY	*Botrytis cinerea*, *Fusarium oxysporum*, *Mycosphaerella arachidicola*, *Escherichia coli*, *Proteus vulgaris*, *Mycobacterium phlei* and *Bacillus megaterium*.	[218]
Phytolocca dioica	Ribosome inactivating protein PD-L1/2 (precursor)	2 peptides	*Planktonic bacterial* cells	[209]
Potato	Acid digestion	AVCENDLNCC	*Candida albicans*.	[219]

Table 6. *Cont.*

Plant Source (Common Name)	Obtaining	Peptides Sequences	Target Pathogen	Referen
Rice	Pepsin	LRRHASEGGHGPHW FSKGVQRAAF EKLLGKQDKGVIIRA SSFSKGVQRAAF SSFSKGVQRAAF	*Porphyromonas gingivalis* and *Candida albicans*,	[220]
Rice	Naturally occurring	RRLMAAKAESRK	*Escherichia coli* and *Aggregatibacter actinomycetemcomitans*.	[221]
Rice bran	Bromelain	KVDHFPL	*Listeria monocytogenes*.	[222]
Sea onion	Naturally occurring	QIPLTGAHSIIGRA IPLSGPNAVIGRA	*Staphylococcus aureus* and *Pseudomonas aeruginosa*.	[223]
Soy	Synthesized	PGTAVFK IKAFKEATKVDKVVVLWTA	*Listeria monocytogenes* and *Pseudomonas aeruginosa*.	[224]
Soybean	Naturally occurring	PRPIPFPRPQP	*Escherichia coli, Staphylococcus aureus, Staphylococcus saprophyticus, Aeromonas hydrophila, Vibrio parahaemolyticus*, and *Salmonella enterica*.	[225]

Antimicrobial mechanisms of these small amino acid fragments are as heterogeneous as their structure, and some are based on breaking the membrane to cause lysis of bacterial cells [226]. Cationic AMPs generally exhibit a balance between hydrophobic and positively charged amino acid residues, allowing them to adopt an amphipathic conformation, which allows greater interaction with negatively charged bacterial membranes, which helps promote their insertion [227]. There are four models of interaction between an AMPs and the cell membrane. In (I) Barrel-stave model, AMPs insert vertically into the plasma membrane to form transmembrane pores. Here, hydrophobic regions of AMPs align with lipid tails. In (II) Carpet model, peptides are adsorbed parallel to the lipid bilayer to cover the cell surface. Here, peptides disrupt the membrane in a detergent-like manner, breaking the lipid bilayer into separate micelles. (III) The toroidal pore model is an intermediate type between the carpet and the barrel. Finally, (IV) in the disordered toroidal pore model, the pore formation is more random and involves fewer peptides, but additional peptides must stabilize the opening (Figure 2) [203,204,207,228].

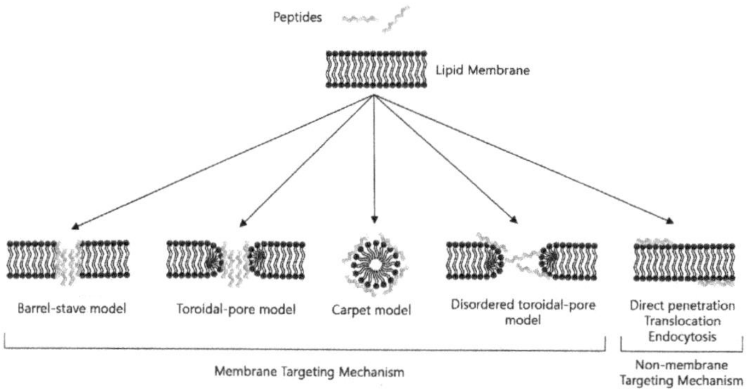

Figure 2. Schematic illustration of models of action by AMPs.

2.7. Relation Between Plant Peptides Structure and Its Antiviral Activity

Recent evidence highlights that some AMPs also present activity against various viruses, thus being called antiviral peptides (AVPs). Several properties may influence the antiviral activity of peptides, such as the topology, amino acid composition, charge, and many other chemical and structural characteristics. The overall biochemical features of AVPs are cationic and amphipathic characteristics and positive net charges. It has been observed that AVPs act primarily by directly inhibiting the viral particle, competing for the receptor on the target cell, and blocking its interaction/adsorption. However, they may also act at other levels of the viral cycle [229]. Miscellaneous AVPs target various steps in the viral cycle from receptor binding to replication and may be virucidal. In addition, some peptides can also be translocated into the cell cytoplasm and interact with intracellular targets, interfering with physiological and chemical functions, such as nucleic acid or protein synthesis [199]. We performed a bioinformatic analysis with the AVPs, as shown in Table 7. We observe that the most frequent residues in AVPs were cysteine, proline, glycine, isoleucine, and aspartic acid. In contrast, asparagine, proline, aspartic acid, isoleucine, and valine are found in more peptides. Likewise, we observed an average isoelectric point of 5.18 and an average hydropathy of −0.23.

Table 7. Summarization of plant peptides with antiviral activity.

Plant Source (Common Name) *	Peptides Sequence	Target Pathogen	Reference
Catechu	DHVTPDIAYNPRTYM DHVTPDIAYNPWAYF	Dengue virus	[230]
Chili pepper	Cb-1: GFPFLLNGPDQDQGDFIMFG Cb-1′: GFKGEQGVPQEMQNEQATIP	Pepper yellow mosaic virus	[231]
Devil's tree	CRPYGYRCDGVINQ CCDPYHCTPPLIGICL CRPYGYRCDGVINQ CCDPYRCTPPLIGICL CVPRFGRCDGIINQC CDPYLCTPPLVGICT CVPQYGVCDGIINQC CDPYYCSPPIYGHCI	Infectious bronchitis virus Dengue virus	[232]
Ground Bean	N-terminal sequence: KTCENLADTY	Human immunodeficiency virus	[218]
Oldenlandia affinis	CGETCVGGTCNTP GCTCSWPVCTRNGLPV	Human immunodeficiency virus	[233,234]
Sorghum	-	Herpes simplex virus type 1	[235]

* All peptides were naturally occurring.

To synthesize the information discussed in Sections 2.1–2.7, Table 8 summarizes the main structural and physicochemical features of plant-derived peptides associated with seven major biological activities. Table 8 integrates evidence on peptide length, net charge, secondary-structure tendencies, hydrophobicity patterns, and recurrent amino acids, highlighting general trends and mechanisms reported in the literature. This integrative overview provides a practical framework to connect peptide structural features with their observed bioactivities.

Table 8. Summary of structural and physicochemical features of plant-derived bioactive peptides by biological activity.

Biological Activity	Peptide Length (Amino Acids)	Net Charge	Secondary Structure	Key Hydrophobicity Traits	Key Amino Acids	Mechanism of Action	Reference
Antioxidant	2–21 (66% are 2–10)	0	Low content of random-coil and α-helix	Hydrophobic residues enhance membrane interaction	E, D, G, A, L, F, Y, H, M, C	Radical scavenging (DPPH, ABTS, ORAC); metal chelation; lipid peroxidation inhibition; activation of endogenous antioxidant enzymes	[41,51,62,78]
Antiproliferative	3–158 (63% are 3–10)	≥+1	β-pleated sheet peptides	>30% hydrophobic content enhances membrane interactions	E, L, S, F, A, K, R, P, W, G	Membrane pore formation; apoptosis (mitochondrial); autophagy-like death; interaction with HDAC1, MDM2; NK cell activation	[79,87,98]
Ace-inhibitory	2–10	0	β-turn/random-coil common; small, compact motifs	Hydrophobic pockets	L, I, P, Y, F, W, D, E	Competitive, noncompetitive, and uncompetitive inhibition	[109,112,118]
Hypolipidemic	2–20		Flexible; no strict fold requirement	≥4 hydrophobic residues promote lipid interactions	K, T, V, E, I, P, L	HMG-CoA reductase inhibition, LDLR upregulation, PCSK9 suppression, and inhibition of cholesterol esterase and pancreatic lipase	[132–134]
Hypoglycemic	3–6	0 or +1	Mostly flexible/random-coil; short motifs favored	Hydrophobic residues located near Proline enhance DPP-IV inhibition	P, A, L, I, R, K, T, S, Y, M	Inhibition of α-amylase, α-glucosidase, DPP-IV; GLUT4 modulation; insulin-like activity	[162,163,187]
Antimicrobial	10–50	+2 to +11	α-helix, β-sheet (disulfide-stabilized), αβ, hairpinin	~50% hydrophobic residues	R, K, A, G, V, C	Disrupt microbial membranes	[204,226,228]
Antiviral	8–2′		Flexible/α-helical motifs	Amphipathic balance enhances viral interaction	C, P, G, I, D, N, V	Adsorption/receptor blocking, virucidal effects, and intracellular disruption of nucleic acid/protein synthesis	[199,229]

3. New Plant Sources for Bioactive Peptides

For a long time, legumes and cereals were the primary sources of bioactive peptides, particularly from legumes. They represent 27% of the world's primary agricultural production, supplying around 15% of the protein worldwide [236]. However, obtaining peptides from legumes has become a concern, as it has been observed that these peptides may retain the allergenic sequence of various legume proteins [237]. As for cereal proteins, the trend of consuming gluten-free products has diminished their demand, even for gluten-free cereals such as maize, buckwheat, rice, millet, quinoa, etc., since it is known that they can be contaminated with gluten during processing, transportation, and handling [238]. For this reason, in recent years, peptides have been obtained from plant-derived sources such as leaves (e.g., *Moringa oleifera*, *Camelia sinensis*, *Spinaca oleracea*) or fruits (e.g., *Cucumis melo*, *Citrus* species, *Stenocereus pruinosus*), which have demonstrated bioactivities including antioxidant, antihypertensive, and anticancer effects, thus representing a safer alternative for consumers [239–241].

4. Current Scenario and Future Perspectives of Peptide-Based Drugs

Currently, the bioactive properties of peptides are often evaluated at non-physiological concentrations, which can lead to overestimating their antioxidant, anti-inflammatory, hypotensive, hypolipidemic, antimicrobial, and antiviral activities. For example, many *in vitro* studies report antioxidant activity at concentrations ranging from 0.1 to 1 mM. In contrast, physiological plasma concentrations after oral intake are typically in the low micromolar or even nanomolar range ($\leq$10 μM). Similarly, antihypertensive peptides derived from food proteins are commonly tested at doses of 0.5–5 mg/mL, which far exceed the levels likely to be achieved systemically in vivo [242]. This discrepancy underscores the importance of validating bioactivities under conditions that mimic physiological environments. In this sense, future research should prioritize the use of physiologically relevant concentrations and incorporate pharmacokinetic studies addressing absorption, distribution, metabolism, and excretion (ADME) to assess these molecules' bioavailability accurately. The application of in vivo models is also essential to determine therapeutic doses, safe administration periods, and potential toxicity profiles. Additionally, computational tools such as molecular docking and Density Functional Theory (DFT) simulations have become invaluable for screening thousands of peptide sequences, allowing researchers to identify candidates with higher chances of in vivo efficacy and pharmacological stability [242,243].

Beyond these aspects, other translational challenges also require attention. Purification remains a critical step, as many studies rely on different isolation and fractionation strategies—such as ultrafiltration and various chromatographic techniques—which makes results difficult to compare. The lack of standardized protocols for peptide production, purification, and characterization further complicates reproducibility across laboratories [244]. Moreover, peptide stability under physiological conditions is often compromised due to susceptibility to proteolytic degradation [245]. This highlights the need for innovative formulation strategies, such as microencapsulation, liposomes, or nanocarriers, which can improve stability, protect peptides from degradation, and enhance controlled release [246]. At the same time, the ability to target relevant tissues remains limited. Most bioactive peptides are studied in general systemic contexts, but specific delivery to organs such as the heart, liver, or brain will be essential for their therapeutic translation [247]. Recent work on peptide-based nanoparticles also emphasizes challenges related to stability, scalability, and tissue selectivity [248].

While bioactive peptides are generally considered safe, more systematic assessments of their safety profiles are required to confirm the absence of long-term or context-dependent adverse effects. Notably, evidence suggests that peptide activity can vary with factors such as dose, cellular context, or extracellular conditions, highlighting the need for deeper mechanistic studies under physiologically relevant scenarios [249].

These considerations highlight that while plant-derived peptides hold substantial promise as future therapeutic agents, further efforts in standardization, formulation, mechanistic elucidation, and safety assessment are necessary to bridge the gap between laboratory findings and clinical application. Together with the increasing consumer demand for plant-derived biopeptides, these strategies point toward a promising future for peptide-based drugs, provided that translational challenges are addressed through realistic experimental approaches [233,234].

5. Conclusions

Numerous diseases worldwide significantly impact the quality of life of large population groups. Consequently, searching for natural-origin molecules with therapeutic and preventive potential has gained considerable attention. In this context, peptides have emerged as promising agents against communicable and non-communicable diseases. Among these, plant-derived peptides stand out as an attractive option due to their vast botanical diversity, sustainability, and alignment with circular economy strategies. Compared to animal-derived peptides, they often present distinct structural and functional characteristics, such as lower allergenic potential and specific sequence–activity relationships. Recent scientific advances have established a clear relationship between the bioactivity of these molecules and their structural characteristics. This has underscored the importance of peptide purification and structural identification, which are now essential in studies evaluating their biological potential. These developments represent a significant step forward in peptide-based therapeutics. Nevertheless, critical challenges remain: the stability of peptides under physiological conditions, their bioavailability and tissue targeting, and the lack of standardized protocols for production and characterization continue to limit translation into real-world applications. Furthermore, depending on concentration, cell type, or extracellular environment, peptides may exert different or opposite effects, highlighting the need for deeper mechanistic studies under physiologically relevant conditions.

Integrating advanced purification methods, computational screening, and in vivo validation will be crucial to fully exploit the therapeutic promise of plant peptides. If these barriers are addressed, plant-derived bioactive peptides could evolve from experimental findings into reliable components of functional foods and peptide-based drugs, ultimately contributing to both human health and sustainable innovation.

Author Contributions: Conceptualization, S.A.-G.; formal analysis, S.A.-G., J.B.H. and E.P.G.-G.; investigation, S.A.-G., I.G.-A., L.A.J.-O., E.P.G.-G. and J.B.H.; writing—original draft preparation, S.A.-G., I.G.-A., L.A.J.-O., E.P.G.-G. and J.B.H.; editing, S.A.-G., J.B.H. and E.P.G.-G.; funding acquisitions, S.A.-G. and J.B.H. All authors have read and agreed to the published version of the manuscript.

Funding: This research was funded by CONACYT scholarship 597872 and CONFIE Sinaloa (PAIPC 2025).

Institutional Review Board Statement: Not applicable.

Informed Consent Statement: Not applicable.

Data Availability Statement: The data presented in this study are available upon reasonable request from the author (S.A.-G.).

Acknowledgments: We thank the Office of Graduate Studies at the Centro de Investigación en Alimentación y Desarrollo (CIAD).

Conflicts of Interest: The authors declare no conflicts of interest.

References

1. González, S. Dietary Bioactive Compounds and Human Health and Disease. *Nutrients* **2020**, *12*, 348. [CrossRef] [PubMed]
2. Cooper, B.M.; Iegre, J.; O'Donovan, D.H.; Halvarsson, M.Ö.; Spring, D.R. Peptides as a platform for targeted therapeutics fo cancer: Peptide-drug conjugates (PDCs). *Chem. Soc. Rev.* **2021**, *50*, 1480–1494. [CrossRef] [PubMed]
3. Singh, B.P.; Vij, S.; Hati, S. Functional significance of bioactive peptides derived from soybean. *Peptides* **2014**, *54*, 171–17 [CrossRef] [PubMed]
4. Pimenta, D.C.; Lebrun, I. Cryptides: Buried secrets in proteins. *Peptides* **2007**, *28*, 2403–2410. [CrossRef]
5. Marqus, S.; Pirogova, E.; Piva, T.J. Evaluation of the use of therapeutic peptides for cancer treatment. *J. Biomed. Sci.* **2017**, *24*, 2 [CrossRef]

Minkiewicz, P.; Iwaniak, A.; Darewicz, M. BIOPEP-UWM database of bioactive peptides: Current opportunities. *Int. J. Mol. Sci.* **2019**, *20*, 5978. [CrossRef]

Escamilla Rosales, M.F.; Olvera Rosales, L.; Jara Gutiérrez, C.E.; Jaimez Ordaz, J.; Santana Sepúlveda, P.A.; González Olivares, L.G. Proteins of milk, egg, and fish as a source of antioxidant peptides: Production, mechanism of action, and health benefits. *Food Rev. Int.* **2023**, *40*, 1600–1620. [CrossRef]

Sánchez, A.; Vázquez, A. Bioactive peptides: A review. *Food Qual. Saf.* **2017**, *1*, 29–46. [CrossRef]

Hernández-Jiménez, M.; Loa, J.; Rojas-Avelizapa, N. Agroindustrial plant wastes: Novel source of antimicrobial peptides. *Circ. Econ. Sustain.* **2025**, *5*, 2431–2465. [CrossRef]

Sun, L.; Liu, J.; He, Z.; Du, R. Plant-derived as alternatives to animal-derived bioactive peptides: A review of the preparation, bioactivities, structure–activity relationships, and applications in chronic diseases. *Nutrients* **2024**, *16*, 3277. [CrossRef]

Zhu, Z.; Xu, Z.; Li, Y.; Fan, Y.; Zhou, Y.; Song, K.; Meng, L. Antioxidant function and application of plant-derived peptides. *Antioxidants* **2024**, *13*, 1203. [CrossRef]

Nirmal, N.; Khanashyam, A.C.; Shah, K.; Awasti, N.; Babu, K.S.; Ucak, I.; Afreen, M.; Hassoun, A.; Tuanthong, A. Plant protein-derived peptides: Frontiers in sustainable food systems and applications. *Front. Sustain. Food Syst.* **2024**, *8*, 1292297. [CrossRef]

Cruz-Casas, D.E.; Aguilar, C.N.; Ascacio-Valdés, J.A.; Rodríguez-Herrera, R.; Chávez-González, M.L.; Flores-Gallegos, A.C. Enzymatic hydrolysis and microbial fermentation: The most favorable biotechnological methods for the release of bioactive peptides. *Food Chem. Mol. Sci.* **2021**, *3*, 100047. [CrossRef]

Heng, X.; Chen, H.; Lu, C.; Fen, T.; Li, K.; Gao, E. Study on synergistic fermentation of bean dregs and soybean meal by multiple strains and proteases. *LWT* **2022**, *154*, 112626. [CrossRef]

Farid, M.S.; Anjum, R.; Yang, Y.; Tu, M.; Zhang, T.; Pan, D.; Sun, Y.; Wu, Z. Recent trends in fermented plant-based analogues and products, bioactive peptides, and novel technologies-assisted fermentation. *Trends Food Sci. Technol.* **2024**, *149*, 104529. [CrossRef]

de Souza, T.S.P.; de Andrade, C.J.; Koblitz, M.G.B.; Fai, A.E.C. Microbial peptidase in food processing: Current state of the art and future trends. *Catal. Lett.* **2023**, *153*, 114–137. [CrossRef]

Kullman, W. *Enzymatic Peptide Synthesis*; CRC Press: Boca Raton, FL, USA, 2018.

Li, W.; Huang, J.; Zheng, L.; Liu, W.; Fan, L.; Sun, B.; Su, G.; Xu, J.; Zhao, M. A fast stop-flow two-dimensional liquid chromatography tandem mass spectrometry and its application in food-derived protein hydrolysates. *Food Chem.* **2023**, *406*, 135000. [CrossRef]

Léonil, J.; Gagnaire, V.; Mollé, D.; Pezennec, S.; Bouhallab, S. Application of chromatography and mass spectrometry to the characterization of food proteins and derived peptides. *J. Chromatogr. A* **2000**, *881*, 1–21. [CrossRef]

Noor, Z.; Ahn, S.B.; Baker, M.S.; Ranganathan, S.; Mohamedali, A. Mass spectrometry–based protein identification in proteomics—A review. *Brief. Bioinform.* **2021**, *22*, 1620–1638. [CrossRef]

Sies, H. Oxidative stress: Concept and some practical aspects. *Antioxidants* **2020**, *9*, 852. [CrossRef]

Thanan, R.; Oikawa, S.; Hiraku, Y.; Ohnishi, S.; Ma, N.; Pinlaor, S.; Yongvanit, P.; Kawanishi, S.; Murata, M. Oxidative stress and its significant roles in neurodegenerative diseases and cancer. *Int. J. Mol. Sci.* **2015**, *16*, 193–217. [CrossRef]

Akbari, B.; Baghaei-Yazdi, N.; Bahmaie, M.; Mahdavi Abhari, F. The role of plant-derived natural antioxidants in reduction of oxidative stress. *BioFactors* **2022**, *48*, 611–633. [CrossRef]

Pisoschi, A.M.; Pop, A.; Cimpeanu, C.; Predoi, G. Antioxidant capacity determination in plants and plant-derived products: A review. *Oxidative Med. Cell. Longev.* **2016**, *2016*, 9130976. [CrossRef] [PubMed]

He, R.; Ju, X.; Yuan, J.; Wang, L.; Girgih, A.T.; Aluko, R.E. Antioxidant activities of rapeseed peptides produced by solid state fermentation. *Food Res. Int.* **2012**, *49*, 432–438. [CrossRef]

Torres-Fuentes, C.; del Mar Contreras, M.; Recio, I.; Alaiz, M.; Vioque, J. Identification and characterization of antioxidant peptides from chickpea protein hydrolysates. *Food Chem.* **2015**, *180*, 194–202. [CrossRef] [PubMed]

Yu, Y.-P.; Lai, S.-J.; Chang, C.-R.; Yu, R.C.; Hsieh, J.-F.; Chen, H.-M. Peptidomic analysis of low molecular weight antioxidative peptides prepared by lotus (*Nelumbo nucifera* Gaertn.) seed protein hydrolysates. *LWT* **2021**, *144*, 111138. [CrossRef]

Huang, C.; Tang, X.; Liu, Z.; Huang, W.; Ye, Y. Enzymes-dependent antioxidant activity of sweet apricot kernel protein hydrolysates. *LWT* **2022**, *154*, 112825. [CrossRef]

Montone, C.M.; Zenezini Chiozzi, R.; Marchetti, N.; Capriotti, A.L.; Cerrato, A.; Laganà, A.; Fai, P.B.S.; Cavaliere, C. Peptidomic approach for the identification of peptides with potential antioxidant and anti-hyperthensive effects derived from asparagus by-products. *Molecules* **2019**, *24*, 3627. [CrossRef]

Intiquilla, A.; Jiménez-Aliaga, K.; Guzmán, F.; Pérez-Rea, D.; Cárdenas, M.; Zavaleta, A.I. Novel antioxidant peptides obtained by alcalase hydrolysis of *Erythrina edulis* (pajuro) protein. *J. Sci. Food Agric.* **2019**, *99*, 2420–2427. [CrossRef]

Siow, H.-L.; Gan, C.-Y. Extraction of antioxidative and antihypertensive bioactive peptides from *Parkia speciosa* seeds. *Food Chem.* **2013**, *141*, 3435–3442. [CrossRef]

32. Zhu, S.; Du, C.; Yu, T.; Wu, T.; Zhu, H.; Li, S. Antioxidant activity of selenium-enriched peptides from the protein hydrolysate c *Cardamine violifolia*. *J. Food Sci.* **2019**, *84*, 3504–3511. [CrossRef]
33. García, M.C.; Endermann, J.; Gonzalez-Garcia, E.; Marina, M.L. HPLC-Q-TOF-MS identification of antioxidant and antihyperten sive peptides recovered from cherry (*Prunus cerasus* L.) subproducts. *J. Agric. Food Chem.* **2015**, *63*, 1514–1520. [CrossRef]
34. Wali, A.; Mijiti, Y.; Yanhua, G.; Maimaitituersun, W.; Abudureheman, B. Isolation and Identification of a Novel Antioxidan Peptide from Chickpea (*Cicer arietinum* L.) Sprout Protein Hydrolysates. *Int. J. Pept. Res. Ther.* **2020**, *27*, 219–227. [CrossRef]
35. Zhang, T.; Li, Y.; Miao, M.; Jiang, B. Purification and characterisation of a new antioxidant peptide from chickpea (*Cicer arietiun* L.) protein hydrolysates. *Food Chem.* **2011**, *128*, 28–33. [CrossRef]
36. Ghribi, A.M.; Sila, A.; Przybylski, R.; Ghorbel, R.; Bougatef, A.; Cheba, B.A.; Nedjar-Arroume, N. Purification and identificatio of novel antioxidant peptides from enzymatic hydrolysate of chickpea (*Cicer arietinum* L.) protein concentrate. *J. Funct. Food* **2015**, *12*, 516–525. [CrossRef]
37. Kou, X.; Gao, J.; Xue, Z.; Zhang, H.; Wang, H. Purification and identification of antioxidant peptides from chickpea (*Cicer arietinur* L.) albumin hydrolysates. *LWT-Food Sci. Technol.* **2013**, *50*, 591–598. [CrossRef]
38. Jin, D.-X.; Liu, X.-L.; Zheng, X.-Q.; Wang, X.-J.; He, J.-F. Preparation of antioxidative corn protein hydrolysates, purification an evaluation of three novel corn antioxidant peptides. *Food Chem.* **2016**, *204*, 427–436. [CrossRef] [PubMed]
39. Zhang, S.; Zhang, M.; Yang, R.; Zhang, S.; Lin, S. Preparation, identification, and activity evaluation of antioxidant peptides fror protein hydrolysate of corn germ meal. *J. Food Process Preserv.* **2019**, *43*, e14160. [CrossRef]
40. Wang, Y.; Chen, H.; Wang, X.; Wang, J.; Liu, H.; Li, H.; Zhang, Z. Isolation and identification of a novel peptide from zein wit antioxidant and antihypertensive activities. *Food Funct.* **2015**, *6*, 3799–3806. [CrossRef] [PubMed]
41. Jiang, Y.; Zhang, M.; Lin, S.; Cheng, S. Contribution of specific amino acid and secondary structure to the antioxidant property c corn gluten proteins. *Food Res. Int.* **2018**, *105*, 836–844. [CrossRef]
42. Zhuang, H.; Tang, N.; Yuan, Y. Purification and identification of antioxidant peptides from corn gluten meal. *J. Funct. Foods* **201** *5*, 1810–1821. [CrossRef]
43. Tang, X.; He, Z.; Dai, Y.; Zheng, F.; Tan, M.; Li, L. Peptide fractionation and free radical scavenging activity of zein hydrolysate. *Agric. Food Chem.* **2010**, *58*, 587–593. [CrossRef] [PubMed]
44. Wang, L.; Ma, M.; Yu, Z.; Du, S.-K. Preparation and identification of antioxidant peptides from cottonseed proteins. *Food Chen* **2021**, *352*, 129399. [CrossRef] [PubMed]
45. Girgih, A.T.; He, R.; Malomo, S.A.; Offengenden, M.; Wu, J.; Aluko, R.E. Structural and functional characterization of hemp see (*Cannabis sativa* L.) protein-derived antioxidant and antihypertensive peptides. *J. Funct. Foods* **2014**, *6*, 384–394. [CrossRef]
46. Lu, R.-R.; Qian, P.; Sun, Z.; Zheng, X.Q.; Qi, G.J. Hempseed protein derived antioxidative peptides: Purification, identificatio and protection from hydrogen peroxide-induced apoptosis in PC12 cells. *Food Chem.* **2010**, *123*, 1210–1218. [CrossRef]
47. Gao, J.; Li, T.; Chen, D.; Gu, H.; Mao, X. Identification and molecular docking of antioxidant peptides from hemp seed protei hydrolysates. *LWT* **2021**, *147*, 111453. [CrossRef]
48. Samaei, S.P.; Martini, S.; Tagliazucchi, D.; Gianotti, A.; Babini, E. Antioxidant and angiotensin I-converting enzyme (ACE inhibitory peptides obtained from alcalase protein hydrolysate fractions of hemp (*Cannabis sativa* L.) bran. *J. Agric. Food Chen* **2021**, *69*, 9220–9228. [CrossRef]
49. Peng, L.; Kong, X.; Wang, Z.; Ai-lati, A.; Ji, Z.; Mao, J. Baijiu vinasse as a new source of bioactive peptides with antioxidant an anti-inflammatory activity. *Food Chem.* **2021**, *339*, 128159. [CrossRef]
50. Memarpoor-Yazdi, M.; Mahaki, H.; Zare-Zardini, H. Antioxidant activity of protein hydrolysates and purified peptides fror *Zizyphus jujuba* fruits. *J. Funct. Foods* **2013**, *5*, 62–70. [CrossRef]
51. Yang, R.; Li, X.; Lin, S.; Zhang, Z.; Chen, F. Identification of novel peptides from 3 to 10 kDa pine nut (*Pinus koraiensis*) mea protein, with an exploration of the relationship between their antioxidant activities and secondary structure. *Food Chem.* **2017** *219*, 311–320. [CrossRef]
52. García-Mora, P.; Martín-Martínez, M.; Bonache, M.A.; González-Miret, M.L.; Peñas, E.; Frias, J.; Martínez-Villaluenga, C Identification, functional gastrointestinal stability and molecular docking studies of lentil peptides with dual antioxidant an angiotensin I converting enzyme inhibitory activities. *Food Chem.* **2017**, *221*, 464–472. [CrossRef]
53. Avilés-Gaxiola, S.; León-Félix, J.; Jiménez-Nevárez, Y.B.; Gutiérrez-Grijalva, E.P.; Araiza-Macias, M.J.; Heredia, J.B. Antioxidan and anti-inflammatory properties of novel peptides from Moringa oleifera Lam. leaves. *S. Afr. J. Bot.* **2021**, *141*, 466–47 [CrossRef]
54. Sun, C.; Tang, X.; Ren, Y.; Yang, D.; Lu, W.; Xiong, W. Novel Antioxidant Peptides Purified from Mulberry (*Morus atropurpure* Roxb.) Leaf Protein Hydrolysates with Hemolysis Inhibition Ability and Cellular Antioxidant Activity. *J. Agric. Food Chem.* **201** *67*, 7650–7659. [CrossRef]
55. Kusumah, J.; Real Hernandez, L.M.; de Mejia, E.G. Antioxidant potential of mung bean (*Vigna radiata*) albumin peptides produce by enzymatic hydrolysis analyzed by biochemical and in silico methods. *Foods* **2020**, *9*, 1241. [CrossRef]

Chunkao, S.; Youravong, W.; Yupanqui, C.T.; Alashi, A.M.; Aluko, R.E. Structure and Function of Mung Bean Protein-Derived Iron-Binding Antioxidant Peptides. *Foods* **2020**, *9*, 1406. [CrossRef] [PubMed]

Sonklin, C.; Alashi, A.M.; Laohakunjit, N.; Aluko, R.E. Functional characterization of mung bean meal protein-derived antioxidant peptides. *Molecules* **2021**, *26*, 1515. [CrossRef] [PubMed]

Xia, J.; Song, H.; Huang, K.; Li, S.; Guan, X. Purification and characterization of antioxidant peptides from enzymatic hydrolysate of mungbean protein. *J. Food Sci.* **2020**, *85*, 1735–1741. [CrossRef] [PubMed]

Zarei, M.; Ebrahimpour, A.; Abdul-Hamid, A.; Anwar, F.; Saari, N. Identification and characterization of papain-generated antioxidant peptides from palm kernel cake proteins. *Food Res. Int.* **2014**, *62*, 726–734. [CrossRef]

Agrawal, H.; Joshi, R.; Gupta, M. Isolation, purification and characterization of antioxidative peptide of pearl millet (*Pennisetum glaucum*) protein hydrolysate. *Food Chem.* **2016**, *204*, 365–372. [CrossRef]

Wang, M.; Li, C.; Li, H.; Li, J.; Zhao, F.; Jiang, L. *In vitro* and in silico antioxidant activity of novel peptides prepared from *Paeonia ostii* 'Feng Dan' hydrolysate. *Antioxidants* **2019**, *8*, 433. [CrossRef]

Yang, J.; Hu, L.; Cai, T.; Li, P.; Zheng, Z.; Cheng, X.; Liu, J.; Zhou, H.; Li, H.; Fu, H. Purification and identification of two novel antioxidant peptides from perilla (*Perilla frutescens* L. Britton) seed protein hydrolysates. *PLoS ONE* **2018**, *13*, e0200021. [CrossRef] [PubMed]

Kim, J.M.; Liceaga, A.M.; Yoon, K.Y. Purification and identification of an antioxidant peptide from perilla seed (*Perilla frutescens*) meal protein hydrolysate. *Food Sci. Nutr.* **2019**, *7*, 1645–1655. [CrossRef] [PubMed]

Yan, Q.-J.; Huang, L.-H.; Sun, Q.; Jiang, Z.-Q.; Wu, X. Isolation, identification and synthesis of four novel antioxidant peptides from rice residue protein hydrolyzed by multiple proteases. *Food Chem.* **2015**, *179*, 290–295. [CrossRef]

Wang, X.; Chen, H.; Fu, X.; Li, S.; Wei, J. A novel antioxidant and ACE inhibitory peptide from rice bran protein: Biochemical characterization and molecular docking study. *LWT* **2017**, *75*, 93–99. [CrossRef]

Wattanasiritham, L.; Theerakulkait, C.; Wickramasekara, S.; Maier, C.S.; Stevens, J.F. Isolation and identification of antioxidant peptides from enzymatically hydrolyzed rice bran protein. *Food Chem.* **2016**, *192*, 156–162. [CrossRef]

Ren, L.-K.; Yang, Y.; Ma, C.-M.; Liu, D.-Y.; Wu, K.; Wang, T.-W.; Feng, W.-H.; Li, Y. Identification and in silico analysis of novel antioxidant peptides in broken rice protein hydrolysate and its cytoprotective effect against H_2O_2-induced 2BS cell model. *Food Res. Int.* **2022**, *162*, 112108. [CrossRef]

Garzón, A.G.; Veras, F.F.; Brandelli, A.; Drago, S.R. Purification, identification and in silico studies of antioxidant, antidiabetogenic and antibacterial peptides obtained from sorghum spent grain hydrolysate. *LWT* **2022**, *153*, 112414. [CrossRef]

Ma, H.; Liu, R.; Zhao, Z.; Song, D.; Yang, J.; Ni, Y. A novel peptide from soybean protein isolate significantly enhances resistance of the organism under oxidative stress. *PLoS ONE* **2016**, *11*, e0159938. [CrossRef]

Tonolo, F.; Moretto, L.; Grinzato, A.; Fiorese, F.; Folda, A.; Scalcon, V.; Marin, O.; Bindoli, A.; Feller, E.; Bellamio, M.; et al. Fermented Soy-Derived Bioactive Peptides Selected by a Molecular Docking Approach Show Antioxidant Properties Involving the Keap1/Nrf2 Pathway. *Antioxidants* **2020**, *9*, 1306. [CrossRef]

Zhang, M.; Mu, T.-H.; Sun, M.-J. Purification and identification of antioxidant peptides from sweet potato protein hydrolysates by Alcalase. *J. Funct. Foods* **2014**, *7*, 191–200. [CrossRef]

Meshginfar, N.; Mahoonak, A.S.; Hosseinian, F.; Ghorbani, M.; Tsopmo, A. Production of antioxidant peptide fractions from a by-product of tomato processing: Mass spectrometry identification of peptides and stability to gastrointestinal digestion. *J. Food Sci. Technol.* **2018**, *55*, 3498–3507. [CrossRef]

Ce, N.; Tianhong, C.; Min, C.; Jianing, D.; Xueli, Z.; Juanjuan, L. Enzymatic hydrolysis preparation and activity analysis of antioxidant peptides derived from walnut dregs. *Food Mach.* **2024**, *40*, 51–61. [CrossRef]

Zhao, Y.; Zhao, Q.; Lu, Q. Purification, structural analysis, and stability of antioxidant peptides from purple wheat bran. *BMC Chem.* **2020**, *14*, 58. [CrossRef] [PubMed]

Li, Q.; Chao, W.; Qiu, L. Therapeutic peptides: Chemical strategies fortify peptides for enhanced disease treatment efficacy. *Amino Acids* **2025**, *57*, 25. [CrossRef] [PubMed]

Fan, H.; Liu, H.; Zhang, Y.; Liao, M.; Wang, A.; Wang, J. Review on plant-derived bioactive peptides: Biological activities, mechanism of action and utilizations in food development. *J. Future Foods* **2022**, *2*, 143–159. [CrossRef]

Liu, X.; Hu, Q.; Shen, Y.; Xiao, H.; Li, C.; Wang, S.; Chen, J.; Liu, G. Research Progress on Antioxidant Peptides from Fish By-Products: Purification, Identification, and Structure-Activity Relationship. *Metabolites* **2024**, *14*, 561. [CrossRef]

Selamassakul, O.; Laohakunjit, N.; Kerdchoechuen, O.; Yang, L.; Maier, C.S. Isolation and characterisation of antioxidative peptides from bromelain-hydrolysed brown rice protein by proteomic technique. *Process Biochem.* **2018**, *70*, 179–187. [CrossRef]

Chatupheeraphat, C.; Roytrakul, S.; Phaonakrop, N.; Rattanabunyong, S.; Viyanant, V.; Rinaldi, G.; Brindley, P.J.; Berriman, M.; Hongeng, S.; Loukas, A.; et al. A novel peptide derived from ginger induces apoptosis through the modulation of p53, BAX, and BCL_2 expression in leukemic cell lines. *Planta Medica* **2021**, *87*, 560–569. [CrossRef]

Fosgerau, K.; Hoffmann, T. Peptide therapeutics: Current status and future directions. *Drug Discov. Today* **2015**, *20*, 122–128. [CrossRef]

81. Luna-Vital, D.A.; de Mejía, E.G.; Loarca-Piña, G. Selective mechanism of action of dietary peptides from common bean on HCT11 human colorectal cancer cells through loss of mitochondrial membrane potential and DNA damage. *J. Funct. Foods* **2016**, *2* 24–39. [CrossRef]
82. Clemente, A.; Marín-Manzano, M.C.; Jiménez, E.; Arques, M.C.; Domoney, C. The anti-proliferative effect of TI1B, a majc Bowman–Birk isoinhibitor from pea (*Pisum sativum* L.), on HT29 colon cancer cells is mediated through protease inhibition. *Br. Nutr.* **2012**, *108*, S135–S144. [CrossRef] [PubMed]
83. Chiangjong, W.; Chutipongtanate, S.; Hongeng, S. Anticancer peptide: Physicochemical property, functional aspect and trend i clinical application. *Int. J. Oncol.* **2020**, *57*, 678–696. [CrossRef] [PubMed]
84. He, Z.; Lin, J.; Peng, D.; Lin, L.; Yang, B.; Lin, Z. Peptide fractions from Sacha inchi induced apoptosis in HepG2 cells via P5 activation and a mitochondria-mediated pathway. *J. Sci. Food Agric.* **2023**, *103*, 7621–7630. [CrossRef] [PubMed]
85. Schaduangrat, N.; Nantasenamat, C.; Prachayasittikul, V.; Shoombuatong, W. ACPred: A computational tool for the predictio and analysis of anticancer peptides. *Molecules* **2019**, *24*, 1973. [CrossRef] [PubMed]
86. Ma, S.; Huang, D.; Zhai, M.; Li, D.; Wang, Z.; Duan, H.; Ji, H. Isolation of a novel bio-peptide from walnut residual protei inducing apoptosis and autophagy on cancer cells. *BMC Complement. Altern. Med.* **2015**, *15*, 413. [CrossRef]
87. Gupta, N.; Bhagyawant, S.S. Bioactive peptide of Cicer arietinum L. induces apoptosis in human endometrial cancer via DN fragmentation and cell cycle arrest. *3 Biotech* **2021**, *11*, 63. [CrossRef]
88. Vital, D.A.L.; Loarca-Piña, G.; Dia, V.P.; de Mejía, E.G. Peptides extracted from common bean (*Phaseolus vulgaris* L.) non-digestib fraction caused differential gene expression of HCT116 and RKO human colorectal cancer cells. *Food Res. Int.* **2014**, *62*, 193–20 [CrossRef]
89. Wang, H.; Ng, T. Isolation and characterization of an antifungal peptide with antiproliferative activity from seeds of Phaseolu vulgaris cv. 'Spotted Bean'. *Appl. Microbiol. Biotechnol.* **2007**, *74*, 125–130. [CrossRef]
90. Kuerban, A.; Al-Malki, A.L.; Kumosani, T.A.; Yagoub, A.E.A.; Abuzinadah, M.F.; Al-Jaouni, S.K.; Kamal, M.A.; Al-Ghamd S.N.; Hassan, H.M. Identification, protein antiglycation, antioxidant, antiproliferative, and molecular docking of novel bioactiv peptides produced from hydrolysis of *Lens culinaris*. *J. Food Biochem.* **2020**, *44*, e13494. [CrossRef]
91. Caccialupi, P.; Ceci, L.R.; Siciliano, R.A.; Duranti, M.; Morale, C.U.; Zacheo, G. Bowman-Birk inhibitors in lentil: Heterologou expression, functional characterisation and anti-proliferative properties in human colon cancer cells. *Food Chem.* **2010**, *12* 1058–1066. [CrossRef]
92. Ye, N.; Hu, P.; Xu, S.; Xu, S.; Liu, L.; Wang, K.; Li, J.; Wang, J. Preparation and characterization of antioxidant peptides from carrc seed protein. *J. Food Qual.* **2018**, *2018*, 8579094. [CrossRef]
93. Ortiz-Martinez, M.; de Mejia, E.G.; García-Lara, S.; Oseguera-Toledo, M.E. Antiproliferative effect of peptide fractions isolate from a quality protein maize, a white hybrid maize, and their derived peptides on hepatocarcinoma human HepG2 cells. *J. Func Foods* **2017**, *34*, 36–48. [CrossRef]
94. Chu, K.T.; Ng, T.B. First report of a glutamine-rich antifungal peptide with immunomodulatory and antiproliferative activitie from family Amaryllidaceae. *Biochem. Biophys. Res. Commun.* **2004**, *325*, 167–173. [CrossRef] [PubMed]
95. Pinto, M.E.F.; Najas, J.Z.; Magalhães, L.G.; Cilli, E.M.; de Souza, J.; Raddi, M.S.G.; de Azevedo, R.A.; de Camargo, M.S.; Said, S de Oliveira, L.S.; et al. Inhibition of breast cancer cell migration by cyclotides isolated from Pombalia calceolaria. *J. Nat. Proc* **2018**, *81*, 1203–1208. [CrossRef] [PubMed]
96. Fan, X.; Guo, H.; Teng, C.; Yang, X.; Wang, Z.; Li, Y.; Zhang, Z. Anti-colon cancer activity of novel peptides isolated from *in vitr* digestion of quinoa protein in Caco-2 cells. *Foods* **2022**, *11*, 194. [CrossRef]
97. Wang, L.; Zhang, J.; Yuan, Q.; Sun, Q.; Yao, W.; Guo, H. Separation and purification of an anti-tumor peptide from rapesee (*Brassica campestris* L.) and the effect on cell apoptosis. *Food Funct.* **2016**, *7*, 2239–2248. [CrossRef]
98. Ma, K.; Wang, Z.; Ju, X.; Huang, J.; He, R. Rapeseed peptide inhibits HepG2 cell proliferation by regulating the mitochondrial an P53 signaling pathways. *J. Sci. Food Agric.* **2023**, *103*, 1474–1483. [CrossRef]
99. Kannan, A.; Hettiarachchy, N.S.; Lay, J.O.; Liyanage, R. Human cancer cell proliferation inhibition by a pentapeptide isolated an characterized from rice bran. *Peptides* **2010**, *31*, 1629–1634. [CrossRef]
100. Chen, Z.; Li, W.; Santhanam, R.K.; Lim, S.H.E.; Kim, H. Bioactive peptide with antioxidant and anticancer activities from blac soybean [*Glycine max* (L.) Merr.] byproduct: Isolation, identification and molecular docking study. *Eur. Food Res. Technol.* **201** *245*, 677–689. [CrossRef]
101. Gao, C.; Sun, R.; Xie, Y.-R.; Fan, S.; Ma, G.; Ma, F.; Hou, B. The soy-derived peptide Vglycin inhibits the growth of colon cance cells *in vitro* and in vivo. *Exp. Biol. Med.* **2017**, *242*, 1034–1043. [CrossRef]
102. González-Montoya, M.; Hernández-Ledesma, B.; Silván, J.M.; Mora-Escobedo, R.; Martínez-Villaluenga, C. Peptides derived fron *in vitro* gastrointestinal digestion of germinated soybean proteins inhibit human colon cancer cells proliferation and inflammatior *Food Chem.* **2018**, *242*, 75–82. [CrossRef]
103. Rayaprolu, S.J.; Hettiarachchy, N.S.; Horax, R.; Satchithanandam, E.; Chen, P.; Mauromoustakos, A. Purification and characteriza tion of a peptide from soybean with cancer cell proliferation inhibition. *J. Food Biochem.* **2017**, *41*, e12374. [CrossRef]

4. Cruz-Huerta, E.; Fernández-Tomé, S.; Arques, M.C.; Clemente, A.; Amigo, L.; Recio, I.; Hernández-Ledesma, B. The protective role of the Bowman-Birk protease inhibitor in soybean lunasin digestion: The effect of released peptides on colon cancer growth. *Food Funct.* **2015**, *6*, 2626–2635. [CrossRef] [PubMed]
5. Chang, V.H.-S ; Yang, D.H.-A.; Lin, H.-H.; Hsu, J.-C.; Liu, Y.-C.; Hsu, C.-L. IbACP, a sixteen-amino-acid peptide isolated from *Ipomoea batatas* leaves, induces carcinoma cell apoptosis. *Peptides* **2013**, *47*, 148–156. [CrossRef] [PubMed]
6. Xie, M.; Liu, D.; Yang, Y. Anti-cancer peptides: Classification, mechanism of action, reconstruction and modification. *Open Biol.* **2020**, *10*, 200004. [CrossRef]
7. Akbarmehr, A.; Peighambardoust, S.H.; Ghanbarzadeh, B.; Sarabandi, K. Physicochemical, antioxidant, antimicrobial, and *in vitro* cytotoxic activities of corn pollen protein hydrolysates obtained by different peptidases. *Food Sci. Nutr.* **2023**, *11*, 2403–2417. [CrossRef]
8. Kaur, A.; Kehinde, B.A.; Sharma, P.; Sharma, D.; Kaur, S. Recently isolated food-derived antihypertensive hydrolysates and peptides: A review. *Food Chem.* **2021**, *346*, 128719. [CrossRef]
9. Udenigwe, C.C.; Aluko, R.E. Food Protein-Derived Bioactive Peptides: Production, Processing, and Potential Health Benefits. *J. Food Sci.* **2012**, *77*, R11–R24. [CrossRef]
0. Aluko, R.E. Structure and function of plant protein-derived antihypertensive peptides. *Curr. Opin. Food Sci.* **2015**, *4*, 44–50. [CrossRef]
1. Chen, L.; Cheng, F.; Chen, H.; Shu, G. Preparation and identification of novel angiotensin-I-converting enzyme inhibitory peptides from *Moringa oleifera* leaf. *LWT* **2024**, *205*, 116472. [CrossRef]
2. Wu, J.; Aluko, R.E.; Nakai, S. Structural Requirements of Angiotensin I-Converting Enzyme Inhibitory Peptides: Quantitative Structure−Activity Relationship Study of Di- and Tripeptides. *J. Agric. Food Chem.* **2006**, *54*, 732–738. [CrossRef] [PubMed]
3. Wu, S.; Wang, X.; Qi, W.; Guo, Q. Bioactive protein/peptides of flaxseed: A review. *Trends Food Sci. Technol.* **2019**, *92*, 184–193. [CrossRef]
4. Gallego, M.; Mora, L.; Toldrá, F. Health relevance of antihypertensive peptides in foods. *Curr. Opin. Food Sci.* **2018**, *19*, 8–14. [CrossRef]
5. Tang, S.; Chen, D.; Shen, H.; Guo, X.; Yuan, Y.; Jiang, X.; Zheng, B.; Zeng, S.; Huang, M. Discovery of two novel ACE inhibitory peptides from soybeans: Stability, molecular interactions, and in vivo antihypertensive effects. *Int. J. Biol. Macromol.* **2025**, *308*, 142247. [CrossRef]
6. Wu, J.-S.; Li, J.-M.; Lo, H.-Y.; Hsiang, C.-Y.; Ho, T.-Y. Anti-hypertensive and angiotensin-converting enzyme inhibitory effects of Radix Astragali and its bioactive peptide AM-1. *J. Ethnopharmacol.* **2020**, *254*, 112724. [CrossRef]
7. Wei, D.; Fan, W.-l.; Xu, Y. Identification of water-soluble peptides in distilled spent grain and its angiotensin converting enzyme (ACE) inhibitory activity based on UPLC-Q-TOF-MS and proteomics analysis. *Food Chem.* **2021**, *353*, 129521. [CrossRef]
8. Sangiorgio, S.; Vidović, N.; Boschin, G.; Arnoldi, A.; Lammi, C. Preparation, characterization and *in vitro* stability of a novel ACE-inhibitory peptide from soybean protein. *Foods* **2022**, *11*, 2667. [CrossRef]
9. Aiello, G.; Lammi, C.; Boschin, G.; Zanoni, C.; Arnoldi, A. Exploration of Potentially Bioactive Peptides Generated from the Enzymatic Hydrolysis of Hempseed Proteins. *J. Agric. Food Chem.* **2017**, *65*, 10174–10184. [CrossRef]
20. Nuchprapha, A.; Paisansak, S.; Sangtanoo, P.; Srimongkol, P.; Saisavoey, T.; Reamtong, O.; Karnchanatat, A. Two novel ACE inhibitory peptides isolated from longan seeds: Purification, inhibitory kinetics and mechanisms. *RSC Adv.* **2020**, *10*, 12711–12720. [CrossRef]
21. Ma, K.; Wang, Y.; Wang, M.; Wang, J.; Guo, C.; Yang, Y. Antihypertensive activity of the ACE–renin inhibitory peptide derived from *Moringa oleifera* protein. *Food Funct.* **2021**, *12*, 8994–9006. [CrossRef]
22. Esteve, C.; Marina, M.L.; Garcia, M.C. Novel strategy for the revalorization of olive (*Olea europaea*) residues based on the extraction of bioactive peptides. *Food Chem.* **2015**, *167*, 272–280. [CrossRef]
23. Siow, H.-L.; Gan, C.-Y. Extraction, identification, and structure-activity relationship of antioxidative and α-amylase inhibitory peptides from cumin seeds (*Cuminum cyminum*). *J. Funct. Foods* **2016**, *22*, 1–12. [CrossRef]
24. Gonzalez-Garcia, E.; Marina, M.L.; Garcia, M. Plum (*Prunus domestica* L.) by-product as a new and cheap source of bioactive peptides: Extraction method and peptides characterization. *J. Funct. Foods* **2014**, *11*, 428–437. [CrossRef]
25. Zhang, L.; Miao, J.; Guo, J.; Liu, S.; Wang, R.; Ding, Y. Two novel angiotensin I-converting enzyme (ACE) inhibitory peptides from rice (*Oryza sativa* L.) bran protein. *J. Agric. Food Chem.* **2023**, *71*, 4153–4162. [CrossRef] [PubMed]
26. Wang, R.; Zhao, H.; Pan, X.; Orfila, C.; Lu, W.; Ma, Y. Preparation of bioactive peptides with antidiabetic, antihypertensive, and antioxidant activities and identification of alpha-glucosidase inhibitory peptides from soy protein. *Food Sci. Nutr.* **2019**, *7*, 1848–1856. [CrossRef]
27. Xu, Z.; Wu, C.; Sun-Waterhouse, D.; Li, D.; Zhao, W.; Su, G.; Ma, H. Identification of post-digestion angiotensin-I converting enzyme (ACE) inhibitory peptides from soybean protein Isolate: Their production conditions and in silico molecular docking with ACE. *Food Chem.* **2021**, *345*, 128855. [CrossRef]

128. Foltz, M.; van der Pijl, P.C.; Duchateau, G. Current *In Vitro* Testing of Bioactive Peptides Is Not Valuable. *J. Nutr.* **2010**, *14*
117–118. [CrossRef]
129. Shen, W.; Matsui, T. Current knowledge of intestinal absorption of bioactive peptides. *Food Funct.* **2017**, *8*, 4306–4314. [CrossRe
130. Lammi, C.; Aiello, G.; Boschin, G.; Arnoldi, A. Multifunctional peptides for the prevention of cardiovascular disease: A ne concept in the area of bioactive food-derived peptides. *J. Funct. Foods* **2019**, *55*, 135–145. [CrossRef]
131. Ye, H.; Xu, Y.; Sun, Y.; Ye, S.; Tang, H.; Wei, X. Purification, identification and hypolipidemic activities of three novel hypolipidemi peptides from tea protein. *Food Res. Int.* **2023**, *165*, 112450. [CrossRef]
132. Soares, R.A.M.; Mendonça, S.; de Castro, L.Í.A.; Menezes, A.C.C.C.C.; Arêas, J.A.G. Major peptides from Amaranth (*Amaranthu cruentus*) protein inhibit HMG-CoA reductase activity. *Int. J. Mol. Sci.* **2015**, *16*, 4150–4160. [CrossRef]
133. Gomes, M.J.C.; Lima, S.L.S.; Alves, N.E.G.; Silva, A.G.M.; Morais, H.A.; Pimenta, D.C.; Vidigal, P.V.T.; Arêas, J.A.G. Commo bean protein hydrolysate modulates lipid metabolism and prevents endothelial dysfunction in BALB/c mice fed an atherogeni diet. *Nutr. Metab. Cardiovasc. Dis.* **2020**, *30*, 141–150. [CrossRef]
134. Garzón, A.G.; Cian, R.E.; Aquino, M.E.; Drago, S.R. Isolation and identification of cholesterol esterase and pancreatic lipas inhibitory peptides from brewer's spent grain by consecutive chromatography and mass spectrometry. *Food Funct.* **2020**, *1* 4994–5003. [CrossRef] [PubMed]
135. Pérez-Míguez, R.; Marina, M.L.; Castro-Puyana, M. High resolution liquid chromatography tandem mass spectrometry for th separation and identification of peptides in coffee silverskin protein hydrolysates. *Microchem. J.* **2019**, *149*, 103951. [CrossRef]
136. Marques, M.R.; Fontanari, G.G.; Pimenta, D.C.; Soares-Freitas, R.M.; Arêas, J.A.G. Proteolytic hydrolysis of cowpea proteins i able to release peptides with hypocholesterolemic activity. *Food Res. Int.* **2015**, *77*, 43–48. [CrossRef]
137. Zanoni, C.; Aiello, G.; Arnoldi, A.; Lammi, C. Hempseed Peptides Exert Hypocholesterolemic Effects with a Statin-Lik Mechanism. *J. Agric. Food Chem.* **2017**, *65*, 8829–8838. [CrossRef] [PubMed]
138. Liu, M.; Li, Y.; Li, W.; Yang, B.; Wang, X.; Tian, Z.; Zhang, X. Hemp seed protein exerts its hypoglycemic and hypolipidemic effect through degradation into short peptides. *Food Chem.* **2025**, *484*, 144406. [CrossRef]
139. Lammi, C.; Zanoni, C.; Arnoldi, A.; Aiello, G. YDFYPSSTKDQQS (P3), a peptide from lupin protein, absorbed by Caco-2 cell modulates cholesterol metabolism in HepG2 cells via SREBP-1 activation. *J. Food Biochem.* **2018**, *42*, e12757. [CrossRef]
140. Lammi, C.; Bollati, C.; Lecca, D.; Pia Abbracchio, M.; Arnoldi, A. Lupin peptide T9 (GQEQSHQDEGVIVR) modulates the mutan PCSK9^{D374Y} pathway: *In vitro* characterization of its dual hypocholesterolemic behavior. *Nutrients* **2019**, *11*, 1665. [CrossRef]
141. Lammi, C.; Aiello, G.; Dellafiora, L.; Galaverna, G.; Vistoli, G.; Arnoldi, A. Assessment of the multifunctional behavior of lupi peptide P7 and its metabolite using an integrated strategy. *J. Agric. Food Chem.* **2020**, *68*, 13179–13188. [CrossRef]
142. Lammi, C.; Aiello, G.; Bollati, C.; Vistoli, G.; Abbracchio, M.P.; Arnoldi, A. Trans-epithelial transport, metabolism and biologica activity assessment of the multi-target lupin peptide lilpkhsdad (P5) and its metabolite lpkhsdad (p5-met). *Nutrients* **2021**, *13*, 86
[CrossRef]
143. Prados, I.M.; Plaza, M.; Marina, M.L.; García, M.C. Evaluation of the relationship between the peptide profiles and the lipid lowering properties of olive seed hydrolysates as a tool for tuning hypocholesterolemic functionality. *Food Funct.* **2020**, *1* 4973–4981. [CrossRef] [PubMed]
144. Ngoh, Y.-Y.; Choi, S.B.; Gan, C.-Y. The potential roles of Pinto bean (*Phaseolus vulgaris* cv. Pinto) bioactive peptides in regulatin physiological functions: Protease activating, lipase inhibiting and bile acid binding activities. *J. Funct. Foods* **2017**, *33*, 67–7
[CrossRef]
145. Yang, F.; Huang, J.; He, H.; Li, Y.; Zhang, Z.; Chen, Z. Study on the hypolipidemic activity of rapeseed protein-derived peptide *Food Chem.* **2023**, *423*, 136315. [CrossRef] [PubMed]
146. Tong, L.T.; Ju, Z.; Qiu, J.; Zhang, H.L.; Wang, Y.Q.; Wang, F. Peptide GEQQQQPGM derived from rice α-globulin reduces the ris of atherosclerosis in hamsters by improving vascular endothelial cells injury. *RSC Adv.* **2017**, *7*, 49194–49203. [CrossRef]
147. Wakasa, Y.; Tamakoshi, C.; Ohno, T.; Hayashi, S.; Ozawa, K.; Takaiwa, F. The hypocholesterolemic activity of transgenic rice see accumulating lactostatin, a bioactive peptide derived from bovine milk β-lactoglobulin. *J. Agric. Food Chem.* **2011**, *59*, 3845–385
[CrossRef]
148. Aiello, G.; Ferruzza, S.; Ranaldi, G.; Sambuy, Y.; Vistoli, G.; Arnoldi, A.; Lammi, C. Behavior of three hypocholesterolemi peptides from soy protein in an intestinal model based on differentiated Caco-2 cell. *J. Funct. Foods* **2018**, *45*, 363–370. [CrossRef
149. Lammi, C.; Arnoldi, A.; Aiello, G. Soybean peptides exert multifunctional bioactivity modulating 3-hydroxy-3-methylglutaryl CoA reductase and dipeptidyl peptidase-IV targets *in vitro*. *J. Agric. Food Chem.* **2019**, *67*, 4824–4830. [CrossRef]
150. Lammi, C.; Zanoni, C.; Arnoldi, A.; Vistoli, G. Two peptides from soy β-conglycinin induce a hypocholesterolemic effect i HepG2 cells by a statin-like mechanism: Comparative *in vitro* and in silico modeling studies. *J. Agric. Food Chem.* **2015**, *63* 7945–7951. [CrossRef]
151. Yao, S.; Agyei, D.; Udenigwe, C.C. Structural basis of bioactivity of food peptides in promoting metabolic health. In *Advances i Food and Nutrition Research*; Elsevier: Amsterdam, The Netherlands, 2018; Volume 84, pp. 145–181.

2. Karami, Z.; Akbari-adergani, B. Bioactive food derived peptides: A review on correlation between the structure of bioactive peptides and their functional properties. *J. Food Sci. Technol.* **2019**, *56*, 535–547. [CrossRef]
3. Nagaoka, S. Structure-function properties of hypolipidemic peptides. *J. Food Biochem.* **2019**, *43*, e12539. [CrossRef] [PubMed]
4. Orona-Tamayo, D.; Valverde, M.E.; Paredes-López, O. Bioactive peptides from selected Latin American food crops–A nutraceutical and molecular approach. *Crit. Rev. Food Sci. Nutr.* **2019**, *59*, 1949–1975. [CrossRef] [PubMed]
5. Chatterjee, C.; Gleddie, S.; Xiao, C.W. Soybean bioactive peptides and their functional properties. *Nutrients* **2018**, *10*, 1211. [CrossRef] [PubMed]
6. Maestri, E.; Marmiroli, M.; Marmiroli, N. Bioactive peptides in plant-derived foodstuffs. *J. Proteom.* **2016**, *147*, 140–155. [CrossRef]
7. Peighambardoust, S.H.; Karami, Z.; Pateiro, M.; Lorenzo, J.M. A review on health-promoting, biological, and functional aspects of bioactive peptides in food applications. *Biomolecules* **2021**, *11*, 631. [CrossRef]
8. Görgüç, A.; Gençdağ, E.; Yılmaz, F.M. Bioactive peptides derived from plant origin by-products: Biological activities and techno-functional utilizations in food developments—A review. *Food Res. Int.* **2020**, *136*, 109504. [CrossRef]
9. García, M.C.; Puchalska, P.; Esteve, C.; Marina, M.L. Vegetable foods: A cheap source of proteins and peptides with antihypertensive, antioxidant, and other less occurrence bioactivities. *Talanta* **2013**, *106*, 328–349. [CrossRef]
0. Meena, S.; Kanthaliya, B.; Joshi, A.; Khan, F.; Arora, J. Biologia futura: Medicinal plants-derived bioactive peptides in functional perspective—A review. *Biol. Futur.* **2020**, *71*, 195–208. [CrossRef]
1. Moreno-Valdespino, C.A.; Luna-Vital, D.; Camacho-Ruiz, R.M.; Mojica, L. Bioactive proteins and phytochemicals from legumes: Mechanisms of action preventing obesity and type-2 diabetes. *Food Res. Int.* **2020**, *130*, 108905. [CrossRef]
2. Patil, S.P.; Goswami, A.; Kalia, K.; Kate, A.S. Plant-Derived Bioactive Peptides: A Treatment to Cure Diabetes. *Int. J. Pept. Res. Ther.* **2020**, *26*, 955–968. [CrossRef]
3. Marya; Khan, H.; Nabavi, S.M.; Habtemariam, S. Anti-diabetic potential of peptides: Future prospects as therapeutic agents. *Life Sci.* **2018**, *193*, 153–158. [CrossRef]
4. Jakubczyk, A.; Karaś, M.; Złotek, U.; Szymanowska, U. Identification of potential inhibitory peptides of enzymes involved in the metabolic syndrome obtained by simulated gastrointestinal digestion of fermented bean (*Phaseolus vulgaris* L.) seeds. *Food Res. Int.* **2017**, *100*, 489–496. [CrossRef] [PubMed]
5. Mojica, L.; Luna-Vital, D.A.; González de Mejía, E. Characterization of peptides from common bean protein isolates and their potential to inhibit markers of type-2 diabetes, hypertension, and oxidative stress. *J. Sci. Food Agric.* **2017**, *97*, 2401–2410. [CrossRef] [PubMed]
6. Oseguera-Toledo, M.E.; Gonzalez de Mejia, E.; Amaya-Llano, S.L. Hard-to-cook bean (*Phaseolus vulgaris* L.) proteins hydrolyzed by alcalase and bromelain produced bioactive peptide fractions that inhibit targets of type-2 diabetes and oxidative stress. *Food Res. Int.* **2015**, *76*, 839–851. [CrossRef] [PubMed]
7. Hsu, P.K.; Pan, F.F.C.; Hsieh, C.S. McIRBP-19 of bitter melon peptide effectively regulates diabetes mellitus (DM) patients' blood sugar levels. *Nutrients* **2020**, *12*, 1252. [CrossRef]
8. Jin, J.; Huang, W.-L.; Rao, G.-L.; Zhang, X.-Q.; Xu, D.-Q.; Ye, H.-L. Total Synthesis of MC2, a Peptide from Momordica Charantia and Its Hypoglycemic Activity. *Chin. J. Nat. Med.* **2011**, *9*, 58–60. [CrossRef]
9. Yang, B.; Li, X.; Zhang, C.; Deng, Y.; Zhang, Z.; Rao, Z.; Ji, A.; Xu, L. Design, synthesis and biological evaluation of novel peptide MC2 analogues from *Momordica charantia* as potential anti-diabetic agents. *Org. Biomol. Chem.* **2015**, *13*, 4551–4561. [CrossRef]
0. Lo, H.Y.; Li, C.C.; Ho, T.Y.; Hsiang, C.Y. Identification of the bioactive and consensus peptide motif from *Momordica charantia* insulin receptor-binding protein. *Food Chem.* **2016**, *204*, 298–305. [CrossRef]
1. Mojica, L.; De Mejía, E.G. Optimization of enzymatic production of anti-diabetic peptides from black bean (*Phaseolus vulgaris* L.) proteins, their characterization and biological potential. *Food Funct.* **2016**, *7*, 713–727. [CrossRef]
2. Cermeño, M.; Connolly, A.; O'Keeffe, M.B.; Flynn, C.; Alashi, A.M.; Aluko, R.E.; FitzGerald, R.J. Identification of bioactive peptides from brewers' spent grain and contribution of Leu/Ile to bioactive potency. *J. Funct. Foods* **2019**, *60*, 103455. [CrossRef]
3. Chandrasekaran, S.; Luna-Vital, D.; de Mejia, E.G. Identification and comparison of peptides from chickpea protein hydrolysates using either bromelain or gastrointestinal enzymes and their relationship with markers of type 2 diabetes and bitterness. *Nutrients* **2020**, *12*, 3843. [CrossRef] [PubMed]
4. Li, P.; Chen, G.; Liang, R.; Zhu, Y.; Li, Y.; Liu, B. Identification and function analysis of Novel hypoglycemic and antioxidant peptides from Chickpea. *Plant Foods Hum. Nutr.* **2024**, *79*, 834–842. [CrossRef] [PubMed]
5. Cai, L.; Wu, S. Jia, C.; Cui, C.; Sun-Waterhouse, D. Active peptides with hypoglycemic effect obtained from hemp (*Cannabis sativa* L.) protein through identification, molecular docking, and virtual screening. *Food Chem.* **2023**, *429*, 136912. [CrossRef] [PubMed]
6. Lammi, C.; Bollati, C.; Ferruzza, S.; Ranaldi, G.; Sambuy, Y.; Canali, R.; Virgili, F.; Arnoldi, A. Soybean-and lupin-derived peptides inhibit DPP-IV activity on in situ human intestinal Caco-2 cells and ex vivo human serum. *Nutrients* **2018**, *10*, 1082. [CrossRef]
7. Jha, S.; Gupta, S.K.; Bhattacharyya, P.; Ghosh, A.; Mandal, P. *In vitro* Antioxidant and Antidiabetic Activity of Oligopeptides Derived from Different Mulberry (*Morus alba* L.) Cultivars. *Pharmacogn. Res.* **2018**, *10*, 361–367. [CrossRef]

178. Wang, F.; Zhang, Y.; Yu, T.; Zhang, B.; Tian, S.; Fan, J.; Mao, F. Oat globulin peptides regulate antidiabetic drug targets and gluco transporters in Caco-2 cells. *J. Funct. Foods* **2018**, *42*, 12–20. [CrossRef]
179. Wang, F.; Yu, G.; Zhang, Y.; Zhang, B.; Fan, J. Dipeptidyl peptidase IV inhibitory peptides derived from oat (*Avena sativa* L buckwheat (*Fagopyrum esculentum*), and highland barley (*Hordeum vulgare trifurcatum* (L.) Trofim) proteins. *J. Agric. Food Che* **2015**, *63*, 9543–9549. [CrossRef]
180. Mazloomi, S.N.; Mora, L.; Aristoy, M.-C.; Toldrá, F.; Mahoonak, A.S.; Ghorbani, M. Impact of simulated gastrointestinal digestic on the biological activity of an alcalase hydrolysate of orange seed (*Siavaraze, Citrus sinensis*) by-products. *Foods* **2020**, *9*, 121 [CrossRef]
181. Mazloomi, S.N.; Mahoonak, A.S.; Mora, L.; Aristoy, M.-C.; Ghorbani, M.; Toldrá, F. Pepsin hydrolysis of orange by-products fc the production of bioactive peptides with gastrointestinal resistant properties. *Foods* **2021**, *10*, 679. [CrossRef]
182. Ngoh, Y.-Y.; Tye, G.J.; Gan, C.-Y. The investigation of α-amylase inhibitory activity of selected Pinto bean peptides via preclinic study using AR42J cell. *J. Funct. Foods* **2017**, *35*, 641–647. [CrossRef]
183. Asokan, S.M.; Wang, T.; Su, W.T.; Lin, W.T. Antidiabetic effects of a short peptide of potato protein hydrolysate in STZ-induce diabetic mice. *Nutrients* **2019**, *11*, 779. [CrossRef]
184. Mudgil, P.; Kilari, B.P.; Kamal, H.; Ghuman, B.S.; Maqsood, S. Multifunctional bioactive peptides derived from quinoa protei hydrolysates: Inhibition of α-glucosidase, dipeptidyl peptidase-IV and angiotensin I converting enzymes. *J. Cereal Sci.* **202** *96*, 103130. [CrossRef]
185. Vilcacundo, R.; Martínez-Villaluenga, C.; Hernández-Ledesma, B. Release of dipeptidyl peptidase IV, α-amylase and α-glucosida inhibitory peptides from quinoa (*Chenopodium quinoa* Willd.) during *in vitro* simulated gastrointestinal digestion. *J. Funct. Foo* **2017**, *35*, 531–539. [CrossRef]
186. Rivero-Pino, F.; Espejo-Carpio, F.J.; Guadix, E.M. Identification of dipeptidyl peptidase IV (DPP-IV) inhibitory peptides fro vegetable protein sources. *Food Chem.* **2021**, *354*, 129473. [CrossRef] [PubMed]
187. González-Montoya, M.; Hernández-Ledesma, B.; Mora-Escobedo, R.; Martínez-Villaluenga, C. Bioactive peptides from germ nated soybean with anti-diabetic potential by inhibition of dipeptidyl peptidase-IV, α-amylase, and α-glucosidase enzymes. *Int. Mol. Sci.* **2018**, *19*, 2883. [CrossRef]
188. Jiang, H.; Feng, J.; Du, Z.; He, X.; Zhang, C.; Gao, C.; Yang, M.; Sun, R.; Ma, G.; Xie, Y. Oral administration of soybean peptid Vglycin normalizes fasting glucose and restores impaired pancreatic function in Type 2 diabetic Wistar rats. *J. Nutr. Biochem.* **201** *25*, 954–963. [CrossRef] [PubMed]
189. Jiang, M.Z.; Yan, H.; He, R.H.; Ma, Y.K. Purification and a molecular docking study of alpha-glucosidase-inhibitory peptides fro a soybean protein hydrolysate with ultrasonic pretreatment. *Eur. Food Res. Technol.* **2018**, *244*, 1995–2005. [CrossRef]
190. Wang, T.; Bo, N.; Sha, G.; Li, Y.; Sun, Q.; Huang, W.; Wang, Z.; Li, Z.; Ma, C. Identification and molecular mechanism of nov hypoglycemic peptide in ripened pu-erh tea: Molecular docking, dynamic simulation, and cell experiments. *Food Res. Int.* **202** *194*, 114930. [CrossRef]
191. Ibrahim, M.A.; Bester, M.J.; Neitz, A.W.H.; Gaspar, A.R.M. Structural properties of bioactive peptides with α-glucosidas inhibitory activity. *Chem. Biol. Drug Des.* **2018**, *91*, 370–379. [CrossRef]
192. Yan, J.; Zhao, J.; Yang, R.; Zhao, W. Bioactive peptides with antidiabetic properties: A review. *Int. J. Food Sci. Technol.* **2019**, *5* 1909–1919. [CrossRef]
193. Rivero-Pino, F.; Espejo-Carpio, F.J.; Guadix, E.M. Antidiabetic food-derived peptides for functional feeding: Production, functio ality and in vivo evidences. *Foods* **2020**, *9*, 983. [CrossRef]
194. Acquah, C.; Dzuvor, C.K.O.; Tosh, S.; Agyei, D. Anti-diabetic effects of bioactive peptides: Recent advances and clinica implications. *Crit. Rev. Food Sci. Nutr.* **2020**, *62*, 2158–2171. [CrossRef]
195. Castañeda-Pérez, E.; Jiménez-Morales, K.; Quintal-Novelo, C.; Segura-Campos, M.; Chel-Guerrero, L.; Betancur-Ancon D. Enzymatic protein hydrolysates and ultrafiltered peptide fractions from Cowpea Vigna unguiculata L bean with *in vitr* antidiabetic potential. *J. Iran. Chem. Soc.* **2019**, *16*, 1773–1781. [CrossRef]
196. Liu, R.; Cheng, J.; Wu, H. Discovery of food-derived dipeptidyl peptidase IV inhibitory peptides: A review. *Int. J. Mol. Sci.* **201** *20*, 463. [CrossRef] [PubMed]
197. Nongonierma, A.B.; FitzGerald, R.J. Features of dipeptidyl peptidase IV (DPP-IV) inhibitory peptides from dietary proteins. *Food Biochem.* **2019**, *43*, e12451. [CrossRef] [PubMed]
198. Panya, A.; Sawasdee, N.; Songprakhon, P.; Tragoolpua, Y.; Rotarayanont, S.; Choowongkomon, K.; Yenchitsomanus, P.T. Synthetic Bioactive Peptide Derived from the Asian Medicinal Plant Acacia catechu Binds to Dengue Virus and Inhibits Ce Entry. *Viruses* **2020**, *12*, 1267. [CrossRef]
199. Galdiero, S.; Falanga, A.; Tarallo, R.; Vitiello, M.; Raiola, J.; Galdiero, M.; Vitiello, G. Peptide inhibitors against herpes simple virus infections. *J. Pept. Sci.* **2013**, *19*, 148–158. [CrossRef]
200. Divyashree, M.; Mani, M.K.; Reddy, D.; Kumavath, R.; Ghosh, P.; Azevedo, V.; Barh, D. Clinical Applications of Antimicrobia Peptides (AMPs): Where do we Stand Now? *Protein Pept. Lett.* **2020**, *27*, 120–134. [CrossRef]

1. dos Santos-Silva, C.A.; Zupin, L.; Oliveira-Lima, M.; Ricci-Azevedo, R.; Guimarães, A.G.; da Cunha, M.; da Silva, A.A. Plant Antimicrobial Peptides: State of the Art, In Silico Prediction and Perspectives in the Omics Era. *Bioinform. Biol. Insights* **2020**, *14*, 1177932220952739. [CrossRef]
2. Scocchi, M.; Mardirossian, M.; Runti, G.; Benincasa, M. Non-Membrane Permeabilizing Modes of Action of Antimicrobial Peptides on Bacteria. *Curr. Top. Med. Chem.* **2016**, *16*, 76–88. [CrossRef]
3. Nawrot, R.; Barylski, J.; Nowicki, G.; Broniarczyk, J.; Buchwald, W.; Goździcka-Józefiak, A. Plant antimicrobial peptides. *Folia Microbiol.* **2014**, *59*, 181–196. [CrossRef]
4. Li, J.; Hu, S.; Jian, W.; Xie, C.; Yang, X. Plant antimicrobial peptides: Structures, functions, and applications. *Bot. Stud.* **2021**, *62*, 5. [CrossRef]
5. Tam, J.P.; Wang, S.; Wong, K.H.; Tan, W.L. Antimicrobial Peptides from Plants. *Pharmaceuticals* **2015**, *8*, 711–757. [CrossRef]
6. Craik, D.J.; Du, J. Cyclotides as drug design scaffolds. *Curr. Opin. Chem. Biol.* **2017**, *38*, 8–16. [CrossRef]
7. Huan, Y.; Kong, Q.; Mou, H.; Yi, H. Antimicrobial Peptides: Classification, Design, Application and Research Progress in Multiple Fields. *Front. Microbiol.* **2020**, *11*, 2559. [CrossRef] [PubMed]
8. Osman, A.; Enan, G.; Al-Mohammadi, A.-R.; Badr, A.; El-Didamony, G.; El-Gazzar, N.; El-Badry, M. Antibacterial peptides produced by Alcalase from cowpea seed proteins. *Antibiotics* **2021**, *10*, 870. [CrossRef] [PubMed]
9. Pizzo, E.; Pane, K.; Bosso, A.; Gaglione, R.; Di Maro, A.; Chambery, A.; Parente, A.; Arciello, A. Novel bioactive peptides from PD-L1/2, a type 1 ribosome inactivating protein from Phytolacca dioica L. Evaluation of their antimicrobial properties and anti-biofilm activities. *Biochim. Biophys. Acta (BBA)—Biomembr.* **2018**, *1860*, 1425–1435. [CrossRef] [PubMed]
10. Bosso, A.; Di Maro, A.; Cafaro, V.; Gaglione, R.; Pizzo, E.; Varcamonti, M.; Parente, A.; Arciello, A. Enzymes as a reservoir of host defence peptides. *Curr. Top. Med. Chem.* **2020**, *20*, 1310–1323. [CrossRef]
11. Kobbi, S.; Nedjar, N.; Chihib, N.; Guesmi, A.; Bachère, E.; Arroume, N. Synthesis and antibacterial activity of new peptides from Alfalfa RuBisCO protein hydrolysates and mode of action via a membrane damage mechanism against *Listeria innocua*. *Microb. Pathog.* **2018**, *115*, 41–49. [CrossRef]
12. Kaewklom, S.; Wongchai, M.; Petvises, S.; Hanpithakphong, W.; Aunpad, R. Structural and biological features of a novel plant defensin from *Brugmansia* × *candida*. *PLoS ONE* **2018**, *13*, e0201668. [CrossRef]
13. Farkas, A.; Maróti, G.; Dürgő, H. Medicago truncatula symbiotic peptide NCR247 contributes to bacteroid differentiation through multiple mechanisms. *Proc. Natl. Acad. Sci. USA* **2014**, *111*, 5183. [CrossRef] [PubMed]
14. Chen, C.-S.; Chen, C.-Y.; Ravinath, D.M.; Lin, C.-H.; Chuang, C.-C.; Chen, J.-Y.; Tzen, T.C. Functional characterization of chitin-binding lectin from *Solanum integrifolium* containing anti-fungal and insecticidal activities. *BMC Plant Biol.* **2018**, *18*, 3. [CrossRef] [PubMed]
15. Slavokhotova, A.A.; Rogozhin, E.A.; Musolyamov, A.K.; Andreev, Y.A.; Oparin, P.B.; Berkut, A.A.; Egorov, T.A.; Grishin, E.V.; Odintsova, T.I. Novel antifungal α-hairpinin peptide from *Stellaria media* seeds: Structure, biosynthesis, gene structure and evolution. *Plant Mol. Biol.* **2014**, *84*, 189–202. [CrossRef] [PubMed]
16. Diz, M.S.; Carvalho, A.O.; Ribeiro, S.F.F.; Da Cunha, M.; Gadelha, C.A.; Gomes, V.M. Characterisation, immunolocalisation and antifungal activity of a lipid transfer protein from chili pepper (*Capsicum annuum*) seeds with novel α-amylase inhibitory properties. *Physiol. Plant.* **2011**, *142*, 233–246. [CrossRef]
17. Schmidt, M.; Arendt, E.K.; Thery, T.L.C. Isolation and characterisation of the antifungal activity of the cowpea defensin Cp-thionin II. *Food Microbiol.* **2019**, *82*, 504–514. [CrossRef]
18. Wong, J.H.; Ng, T.B. Sesquin, a potent defensin-like antimicrobial peptide from ground beans with inhibitory activities toward tumor cells and HIV-1 reverse transcriptase. *Peptides* **2005**, *26*, 1120–1126. [CrossRef]
19. Lee, J.-K.; Gopal, R.; Seo, C.H.; Cheong, H.; Park, Y. Isolation and Purification of a Novel Deca-Antifungal Peptide from Potato (*Solanum tuberosum* L. cv. Jopung) Against *Candida albicans*. *Int. J. Mol. Sci.* **2012**, *13*, 4021–4032. [CrossRef]
20. Taniguchi, M.; Kameda, M.; Namae, T.; Ochiai, A.; Saitoh, E.; Tanaka, T. Identification and characterization of multifunctional cationic peptides derived from peptic hydrolysates of rice bran protein. *J. Funct. Foods* **2017**, *34*, 287–296. [CrossRef]
21. Takayama, S.; Hashimoto, K.; Kokubu, E.; Ishihara, K.; Saito, K. Inhibitory effects of a novel cationic dodecapeptide [CL(14–25)] derived from cyanate lyase of rice on endotoxic activities of LPSs from *Escherichia coli* and periodontopathic *Aggregatibacter actinomycetemcomitans*. *Microb. Pathog.* **2016**, *94*, 2–11. [CrossRef]
22. Pu, C.; Tang, W. The antibacterial and antibiofilm efficacies of a liposomal peptide originating from rice bran protein against *Listeria monocytogenes*. *Food Funct.* **2017**, *8*, 4159–4169. [CrossRef]
23. Cunsolo, V.; Schicchi, R.; Chiaramonte, M.; Salmeri, M.; Vella, F.M.; Foti, S.; Geraci, C. Identification of New Antimicrobial Peptides from Mediterranean Medical Plant *Charybdis pancration* (Steinh.) Speta. *Antibiotics* **2020**, *9*, 747. [CrossRef] [PubMed]
24. Dhayakaran, R.; Neethirajan, S.; Weng, X. Investigation of the antimicrobial activity of soy peptides by developing a high throughput drug screening assay. *Biochem. Biophys. Rep.* **2016**, *6*, 149–157. [CrossRef] [PubMed]

225. Freitas, C.S.; Vericimo, M.A.; da Silva, M.L.; Motta, M.C.; da Silva, A.F.; Leite, S.G.; Rozental, S.; de Alencar, A.C.; Uchoa, A.F.; de Freitas, C.M.; et al. Encrypted antimicrobial and antitumoral peptides recovered from a protein-rich soybean (*Glycine ma* by-product. *J. Funct. Foods* **2019**, *54*, 187–198. [CrossRef]
226. Rashid, R.; Veleba, M.; Kline, K.A. Focal Targeting of the Bacterial Envelope by Antimicrobial Peptides. *Front. Cell Dev. Biol.* **201** *4*, 55. [CrossRef]
227. Bechinger, B.; Gorr, S.U. Antimicrobial Peptides: Mechanisms of Action and Resistance. *J. Dent. Res.* **2017**, *96*, 254–260. [CrossRef
228. Shwaiki, L.N.; Lynch, K.M.; Arendt, E.K. Future of antimicrobial peptides derived from plants in food application—A focus o synthetic peptides. *Trends Food Sci. Technol.* **2021**, *112*, 312–324. [CrossRef]
229. Vilas Boas, L.C.P.; Campos, M.L.; Berlanda, R.L.A.; de Carvalho Neves, N.; Franco, O.L. Antiviral peptides as promisin therapeutic drugs. *Cell. Mol. Life Sci.* **2019**, *76*, 3525–3542. [CrossRef]
230. Panya, A.; Yongpitakwattana, P.; Budchart, P.; Sawasdee, N.; Songprakhon, P.; Choowongkomon, K.; Yenchitsomanus, P.T. Nove bioactive peptides demonstrating anti-dengue virus activity isolated from the Asian medicinal plant *Acacia Catechu*. *Chem. Bio Drug Des.* **2019**, *93*, 100–109. [CrossRef]
231. Moulin, M.M.; Rodrigues, R.; Ribeiro, S.F.F.; Nogueira, F.C.; de Freitas, R.G.; Soares, M.R.; da Cunha, M.; Aragão, F.J.; Gome V.M. Trypsin inhibitors from *Capsicum baccatum* var. pendulum leaves involved in Pepper yellow mosaic virus resistance. *Gene Mol. Res.* **2014**, *13*, 9229–9243. [CrossRef]
232. Nguyen, P.Q.T.; Ooi, J.S.G.; Nguyen, N.T.K.; He, W.S.; Tan, T.Y.T.; Fang, C.M.; Chen, W.N.; Tam, J.P.; Ooi, L.S.M. Antiviral Cystin Knot α-Amylase Inhibitors from *Alstonia scholaris*. *J. Biol. Chem.* **2015**, *290*, 31138–31150. [CrossRef]
233. Henriques, S.T.; Huang, Y.-H.; Rosengren, K.J.; Franquelim, H.G.; Carvalho, F.A.; Johnson, A.; Sonza, S.; Tachedjian, G.; Castanh M.A.; Daly, N.L.; et al. Decoding the membrane activity of the cyclotide kalata B1: The importance of phosphatidylethanolamin phospholipids and lipid organization on hemolytic and anti-HIV activities. *J. Biol. Chem.* **2011**, *286*, 24231–24241. [CrossRef] [PubMed]
234. Nawae, W.; Hannongbua, S.; Ruengjitchatchawalya, M. Molecular dynamics exploration of poration and leaking caused b Kalata B1 in HIV-infected cell membrane compared to host and HIV membranes. *Sci. Rep.* **2017**, *7*, 3638. [CrossRef] [PubMed]
235. Camargo Filho, I.; Cortez, D.A.G.; Ueda-Nakamura, T.; Nakamura, C.V.; Dias Filho, B.P. Antiviral activity and mode of action of peptide isolated from *Sorghum bicolor*. *Phytomedicine* **2008**, *15*, 202–208. [CrossRef] [PubMed]
236. Carbonaro, M.; Nucara, A. Legume proteins and peptides as compounds in nutraceuticals: A structural basis for dietary health effects. *Nutrients* **2022**, *14*, 1188. [CrossRef]
237. Belsito, D.V.; Hill, R.A.; Klaassen, C.D.; Liebler, D.C.; Marks, J.G., Jr.; Shank, R.C.; Slaga, T.J.; Snyder, P.W.; Andersen, F.A. *Safety Assessment of Plant-Derived Proteins and Peptides as Used in Cosmetics*; Cosmetic Ingredient Review: Washington, DC, USA, 2017.
238. Bustamante, M.Á.; Fernández-Gil, M.P.; Churruca, I.; Pinto, Á.; Sarria, B.; Rodriguez, M.A.; Landa, B.B. Evolution of gluten content in cereal-based gluten-free products: An overview from 1998 to 2016. *Nutrients* **2017**, *9*, 21. [CrossRef]
239. Yuan, H.; Luo, Z.; Ban, Z.; Liu, D.; Wu, S.; Cheng, Z.; Li, Y.; Chen, M. Bioactive peptides of plant origin: Distribution, functionality and evidence of benefits in food and health. *Food Funct.* **2022**, *13*, 3133–3158. [CrossRef]
240. Avilés-Gaxiola, S.; Contreras-Angulo, L.A.; García-Aguiar, I.; Heredia, J.B. *Moringa oleifera* Lam. leaf peptides: Antioxidant and antiproliferative activity in human colon cancer Caco-2 cell line. *Antioxidants* **2024**, *13*, 1367. [CrossRef]
241. Rasouli, H.; Parvaneh, S.; Mahnam, A.; Kiamari, A.M.; Barzegar, A. Anti-angiogenic potential of trypsin inhibitor purified from *Cucumis melo* seeds: Homology modeling and molecular docking perspective. *Int. J. Biol. Macromol.* **2017**, *96*, 118–128. [CrossRef
242. Nardo, A.E.; Suárez, S.; Quiroga, A.V.; Añón, M.C. Amaranth as a source of antihypertensive peptides. *Front. Plant Sci.* **2020** *11*, 578631. [CrossRef]
243. Gao, X.; Bu, F.; Yi, D.; Shi, H.; Yang, Y.; Jiang, X.; Zheng, S. Molecular docking and antihypertensive effects of a novel angiotensin-converting enzyme inhibitory peptide from yak bone. *Front. Nutr.* **2022**, *9*, 993744. [CrossRef]
244. Jiang, Y.; Sun, J.; Chandrapala, J.; Liu, Z.; Wang, S.; Chen, J.; Liu, G. Recent progress of food-derived bioactive peptides: Extraction purification, function, and encapsulation. *Food Front.* **2024**, *5*, 1240–1264. [CrossRef]
245. Pérez-Pérez, V.; Jiménez-Martínez, C.; González-Escobar, J.L.; Corzo-Ríos, L.J. Exploring the impact of encapsulation on the stability and bioactivity of peptides extracted from botanical sources: Trends and opportunities. *Front. Chem.* **2024**, *12*, 142350 [CrossRef]
246. Atma, Y.; Murray, B.S.; Sadeghpour, A.; Goycoolea, F.M. Encapsulation of short-chain bioactive peptides (BAPs) for gastrointestinal delivery: A review. *Food Funct.* **2024**, *15*, 3959–3979. [CrossRef] [PubMed]
247. Xiao, W.; Jiang, W.; Chen, Z.; Hang, Z.; Liu, C.; Han, K.; Li, J.; Zhang, S.; Wei, D.; Wang, T.; et al. Advance in peptide-based drug development: Delivery platforms, therapeutics and vaccines. *Signal Transduct. Target. Ther.* **2025**, *10*, 74. [CrossRef] [PubMed]

3. Thongpon, P.; Tang, M.; Cong, Z. Peptide-based nanoparticle for tumor therapy. *Biomedicines* **2025**, *13*, 1415. [CrossRef]
9. Mirzapour-Kouhdasht, A.; McClements, D.J.; Taghizadeh, M.S.; Niazi, A.; Garcia-Vaquero, M. Strategies for oral delivery of bioactive peptides with focus on debittering and masking. *NPJ Sci. Food* **2023**, *7*, 22. [CrossRef]

molecules

MDPI

Review

Natural Products as Novel Therapeutic Agents for Triple-Negative Breast Cancer: Current Evidence, Mechanisms, Challenges, and Opportunities

Qingzhou Li [1,†], Zhen Ye [2,†], Guilin Wang [2], Yuhui Chen [2], Jinghong Deng [2], Dong Wang [1,2,*] and Yumei Wang [1,2,*]

1 State Key Laboratory of Southwestern Chinese Medicine Resources, School of Pharmacy, Chengdu University of Traditional Chinese Medicine, Chengdu 611137, China; liqingzhou@stu.cdutcm.edu.cn

2 School of Basic Medical Sciences, Chengdu University of Traditional Chinese Medicine, Chengdu 611137, China; yezhen@stu.cdutcm.edu.cn (Z.Y.); guilinwang@stu.cdutcm.edu.cn (G.W.); yhchen@cdutcm.edu.cn (Y.C.); wddengjinghong@163.com (J.D.)

* Correspondence: dwang@cdutcm.edu.cn (D.W.); yumeiwang@cdutcm.edu.cn (Y.W.)

† These authors contributed equally to this work.

Abstract: Breast cancer (BC) tops the list of causes for female fatalities globally, with the elusive triple-negative breast cancer (TNBC) constituting 10–20% of all cases. Current clinical strategies for combating TNBC encompass a multifaceted approach, including surgical intervention, radiation therapy, chemotherapy, and advanced targeted drugs and immunotherapies. While these modalities have catalyzed significant advancements in TNBC management, lingering limitations continue to pose formidable challenges. There is an acute need for novel therapeutics in the realm of TNBC treatment. Natural products (NPs) have emerged as a rich reservoir for pharmaceutical innovation, owing to their extraordinary range of structures and physicochemical properties. Scholars have reported diverse evidence of NPs' efficacy against TNBC. This review aims to comprehensively explore the bioactive constituents, specifics and commonalities of chemical structure, and pharmacological mechanisms of NPs, specifically examining their multifaceted roles in impeding TNBC. NPs, which have recently garnered significant interest, are intriguing in terms of their capacity to combat TNBC through multifaceted mechanisms, including the suppression of tumor cell proliferation, the induction of apoptosis, and the inhibition of tumor metastasis. These natural agents primarily encompass a range of compounds, including terpenoids, glycosides, phenolic compounds, and alkaloids. An in-depth exploration has unveiled their involvement in key signaling pathways, including the transforming growth factor-beta (TGF-β), vascular endothelial growth factor A (VEGFA), phosphoinositide 3-kinase/protein kinase B (PI3K/AKT), Wingless/Int-1 (Wnt) /β-catenin, and mitogen-activated protein kinase (MAPK) pathways. Meanwhile, this review also looks at the challenges and opportunities that arise from harnessing natural compounds to influence TNBC, while outlining the prospective trajectory for future research in the field of NPs.

Keywords: triple-negative breast cancer; herbal medicine; natural products; therapeutic strategies; mechanism

1. Introduction

Breast cancer (BC), which accounted for 670,000 deaths globally in 2022, has been defined as a highly life-threatening disease among the female population by the World Health Organization (WHO) [1]. The complexity of BC is strongly related to its heterogeneity, which results in diverse clinical and therapeutic outcomes. Among the various

 https://doi.org/10.3390/molecules3006120

subtypes, TNBC is particularly aggressive. Unlike luminal A/B BC, human epidermal growth receptor 2 (HER2)-positive BC, and basal-like BC, TNBC lacks expression of estrogen receptors (ERs), progesterone receptors (PRs), and HER2, making it unresponsive to hormone therapy, endocrine therapy, and HER2-targeted therapy [2]. TNBC treatment currently relies primarily on chemotherapy as a therapeutic modality [3]. For instance, paclitaxel- and anthracycline-based chemotherapy drugs are currently the principal choices for treating TNBC, with platinum-based compounds demonstrating promising effects in certain treatment contexts [4–7]. However, the induction of endogenous or acquired resistance in early-stage treatment, significant adverse effects, and potential non-specific off-target effects severely limit the utility of these chemotherapy drugs [8]. Consequently, novel therapeutic strategies, such as targeted therapy and immunotherapy, are being continuously developed to improve their treatment efficacy and survival rates [9,10]. For example, poly adenosinediphosphate-ribose polymerase (PARP) inhibitors have shown therapeutic potential in patients with breast cancer gene (BRCA) mutations. Moreover, other synthetic small-molecule drugs, such as tyrosine kinase inhibitors, including Lapatinib and Afatinib, have also demonstrated commendable efficacy by targeting multiple tyrosine kinase receptors, thereby disrupting the signal transduction pathways of cancer cells [11]. In addition, immune checkpoint inhibitors, such as pembrolizumab and atezolizumab, have displayed breakthrough activity in certain TNBC patients.

Traditional Chinese medicines (TCMs) are regarded as valuable complementary and alternative therapies with significant anticancer potential [12,13]. Herbal medicines (HMs), which have served as the primary remedies in many ancient civilizations, have garnered significant interest in cancer drug discovery. Notably, numerous NPs derived from HMs have demonstrated antitumor effects, a discovery that gained considerable attention following the identification of paclitaxel's remarkable anticancer activity. As a result, HMs-associated NPs have become a focal point in drug discovery, with research in this area experiencing substantial growth. Recently, it was found that approximately 30% of drugs approved by the U.S. Food and Drug Administration (FDA) are HMs extracts and HM-associated NPs [14]. Paclitaxel, for instance, originally extracted from the herbal plant Taxus chinensis, remains a first-line treatment for various carcinomas. However, many NPs with potential therapeutic benefits are yet to be fully explored. HMs are recognized as a rich source of medicinal compounds, providing advantages such as cost-effectiveness and a wide array of structural and functional variations. The therapeutic efficacy of HMs and associated NPs in treating TNBC has been demonstrated through their ability to target multiple signaling pathways and exhibit various bioactivities.

In this review, we present a comprehensive analysis of the chemical characteristics, bioactivities, and botanical origins of NPs identified as potential adjuvant treatment for TNBC. The primary compounds within these NPs include terpenoids, glycosides, phenolics, and alkaloids. These NPs have been shown to inhibit TNBC through multiple mechanisms, such as the suppression of angiogenesis, prevention of metastasis, induction of apoptosis, interruption of the cell cycle, and regulation of autophagy. The key pathways involved include TGF-β, VEGFA, PI3K/AKT, Wnt/β-catenin, and MAPK. Consequently, NPs offer significant potential for the development of innovative and effective therapies for TNBC.

2. Activity of Natural Products in Treatment of Triple-Negative Breast Cancer

2.1. Terpenoids

Terpenoids, which represent the largest class of compounds produced by plants, are derived from mevalonic acid, and are characterized by a molecular framework of isoprene units [15]. The compounds are connected in various ways to form polymers: monoterpenes consist of two isoprene units, sesquiterpenes consist of three, diterpenes consist of four, etc. Terpenoids hold significant importance in the fields of biology, pharmacology, and chemistry. Among their numerous biological properties, their antitumor effects are particularly noteworthy, encompassing diverse activities such as anti-proliferative, apoptotic, anti-angiogenic, and anti-metastatic actions [1,16].

Terpenoids undergo enzymatic elongation and cyclization reactions, resulting in compounds with high hydrophobicity and structural complexity. Compared to non-terpenoid NPs, terpenoids exhibit more intricate ring structures and chiral stereochemistry. In proteins, the number of hydrophobic amino acids correlates with protein folding; thus, hydrophobic residues are prevalent in protein active pockets and interfaces that are crucial for ligand binding. The hydrophobic effect is considered a significant driving force for molecular recognition in protein–ligand interactions. Consequently, hydrophobic terpenoids may demonstrate favorable properties in protein–ligand binding. Some terpenoid compounds have been found to possess inhibitory activity against protein–protein interactions, targeting various drug sites, highlighting their broad applicability in drug discovery endeavors [17,18]. Terpenoids are widely distributed in nature and exhibit diverse biological activities, including anti-TNBC effects (Figure 1, Table 1 and Table S1).

2.1.1. Anti-Metastasis

Terpenoids have been reported to induce apoptosis in TNBC cells, which is a form of programmed cell death crucial for the development, homeostasis, and defense mechanisms of multicellular organisms. Appropriate apoptosis function can prevent tumorigenesis [19]. Terpenoids such as atractylenolide-1 (AT-1), demethylzeylasteral (T-96), isotoosendanin (ITSN), toosendanin (TSN), Eupalinolide O (EO), and Jatamanvaltrate P have been documented to modulate the expression of apoptosis-related proteins, including B-cell lymphoma 2 (Bcl2), Bcl-2-associated X protein (Bax), and Caspase-3, thereby inducing death of TNBC cells [20–24]. Apoptosis is regulated by multiple signaling pathways. T-96 significantly reduces lysine-specific demethylase 1 (LSD1) protein levels, increases its target protein phosphatase and tensin homolog (PTEN), and enhances histone methylation. This action ultimately downregulates the PI3K/AKT signaling pathway, effectively inhibiting the growth of TNBC cells and promoting apoptosis [25]. Maackiain (MA) regulates mitochondrial apoptosis signaling in TNBC by promoting GADD45A through the inhibition of microRNA(miR)-374a [26]. EO may induce apoptosis in TNBC cells by modulating reactive oxygen species (ROS) production and the AKT/p38 MAPK pathway [23]. Terpene compounds such as Cycloart-23E-ene-3β, 25-diol, germacrone, and β-Bisabolene have also been reported to induce apoptosis in TNBC cells [27–29].

2.1.2. Regulation of Cell Cycle

Disruptions in the cell cycle, which is essential for regulated cell proliferation, are pivotal in tumor development and progression, and research into targeting these disruptions offers promising strategies for cancer treatment. Terpenoids, such as AT-1, Jatamanvaltrate P, Cucurbitacin B, Leariifolianoid E, and germacrone, have been shown to inhibit the cell cycle, effectively suppressing the growth and proliferation of TNBC cells [21,24,27,29–32]. They regulate the expression of cell cycle proteins, such as cyclin-dependent kinase (CDK1) and Cyclin D1 (CCND1), leading to cell cycle arrest and inhibition of TNBC cell growth [28,32]. The progression of the cell cycle also depends on the regulation of intracellular signaling pathways. Cucurbitacin B may inhibit the proliferation of human BC cells by disrupting the microtubule network, downregulating Cellular myelocytomatosis oncogene (c-Myc) and nucleophosmin/B23, and interfering with the trafficking of nucleophosmin/B23 from the nucleolus to the nucleoplasm, leading to G2/M arrest [33,34].

2.1.3. Modulation of Autophagy

Autophagy is a self-degradative process that is crucial for regulating metabolism, maintaining energy balance, and controlling cell development and death in response to stress or starvation [35]. In cancer cells, autophagy has a dual role, both promoting tumor cell survival and growth, and potentially inhibiting tumor initiation and progression [36]. As research advances, a better understanding of the complex mechanisms of autophagy may offer new directions for cancer treatment. Terpenoid compounds such as Anomanolide C (AC), ITSN, TSN, and Jatamanvaltrate P have been reported to modulate autophagy. Treatment with these compounds increases the LC3B-II/LC3B-I ratio in TNBC cells, leading to autophagy activation and subsequent inhibition of TNBC [20,24,37–39].

Terpenoid compounds show significant potential in inhibiting TNBC by targeting multiple mechanisms, including metastasis suppression and apoptosis induction, demonstrating their multifaceted abilities to hinder the progression of TNBC. For instance, AC has been demonstrated to concurrently impede cell metastasis and modulate autophagy. AT-1 exhibits the potential to inhibit metastasis, arrest the cell cycle, and provoke apoptosis. Similarly, Jatamanvaltrate P has been noted for its ability to halt the cell cycle, influence autophagy, and induce apoptosis. Collectively, these active terpenoid compounds present substantial potential as therapeutic agents for TNBC. Nonetheless, a more profound understanding of their underlying mechanisms is essential, necessitating further research. Additionally, extensive preclinical studies will be pivotal in advancing their translation into clinical practice.

Table 1. Profiles of terpenoids in NPs with therapeutic effects on TNBC.

Natural Products	Structure	Sources	Concentration In Vitro	Dosage In Vivo	Biological Activities/Mechanisms	Potential Targets	Reference
Anomanolide C	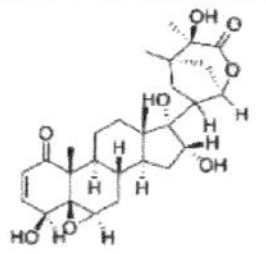	*Tubocapsicum anomalum (S. Vidal)* M. J. Pires	0.25, 1, 2 μM	25, 50 mg/kg	Induces glutathione peroxidase 4 (GPX4) ubiquitination to trigger autophagy-dependent ferroptosis in TNBC.	GPX4	[38]
Atractylenolide-1	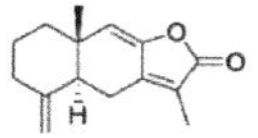	*Atractylodes lancea*	25, 50 μM	20, 50 mg/kg	Inhibits phosphorylation of the Extracellular signal-regulated kinase (ERK) family; suppresses the expression of Triosephosphate Isomerase 1 (TPI1) and Glucose-6-phosphate isomerase GPI (GPI), thereby affecting the glycolytic/gluconeogenic pathway.	TPI1 and GPI	[21]
Britannin	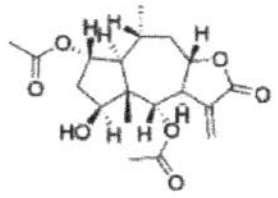	*Inula japonica* Thunb	5, 10, 20 μM	5, 10 mg/kg	Induces Zinc finger e-box binding homeobox 1 (ZEB1) degradation, inhibiting invasion and metastasis of TNBC cells.	ZEB1	[40]
Demethylzeylasteral	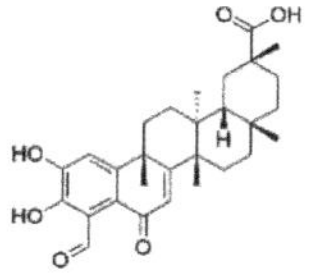	*Tripterygium wilfordii*	5, 10 μM	4 mg/kg	Reduces the expression of LSD1 protein, enhances the expression of its target protein PTEN, and suppresses the PI3K/AKT signaling pathway.	LSD1	[25]
Isotoosendanin	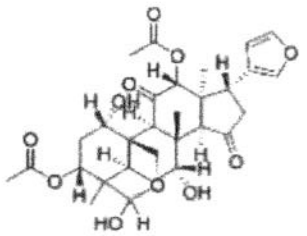	*Melia azedarach* L.	0.1, 0.3, 1 μM	0.1, 1 mg/kg	Directly engages with the kinase domain of TGF-β receptor type-1 (TGFβR1), thereby abolishing TGF-β-induced epithelial–mesenchymal transition (EMT) in the TNBC microenvironment.	TGFβR1	[20,37]
Toosendanin	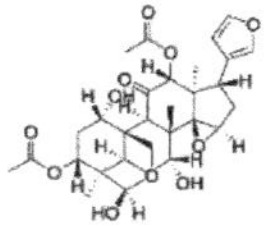	*Melia azedarach* L.	0.1, 1, 5 μM	/	Inhibits autophagy by alkalizing lysosomal pH.	/	[20,39]
Jatamanvaltrate P	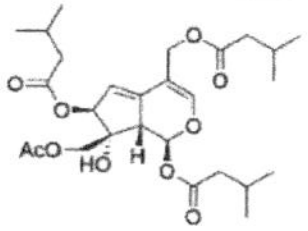	*Valeriana jatamansi* A. DC.	2.5, 5 μM	15 mg/kg	Enhances PARP and caspase cleavage, indicating apoptosis induction, and reducing Cyclin B1, CCND1, and Cell division cycle 2 (Cdc-2) expression, which are crucial for cell cycle progression.	/	[24]

Table 1. *Cont.*

Natural Products	Structure	Sources	Concentration In Vitro	Dosage In Vivo	Biological Activities/Mechanisms	Potential Targets	Reference
Maackiain	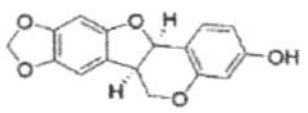	*Spatholobus suberectus* H. S. Lo	1, 2, 2.5, 5 μM	25, 50 mg/kg	Enhances the induction of GADD45α protein and mRNA through the inhibition of miR-374a.	/	[26]
Triptonide	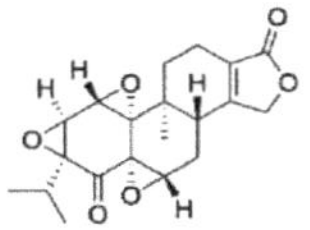	*Tripterygium wilfordii* Hook. f.	0.2 μM	5 mg/kg	Suppresses the tumorigenesis and metastasis of TNBC via triggering the degradation of Twist1 and Notch1 oncoproteins.	/	[41,42]
Eupalinolide O	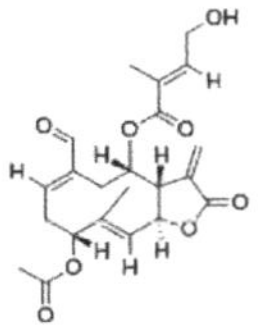	*Eupatorium lindleyanum* DC.	5, 10 μM	15, 30 mg/kg	Induces apoptosis, which is orchestrated through ROS generation and the AKT/p38 MAPK signaling pathway.	/	[23]

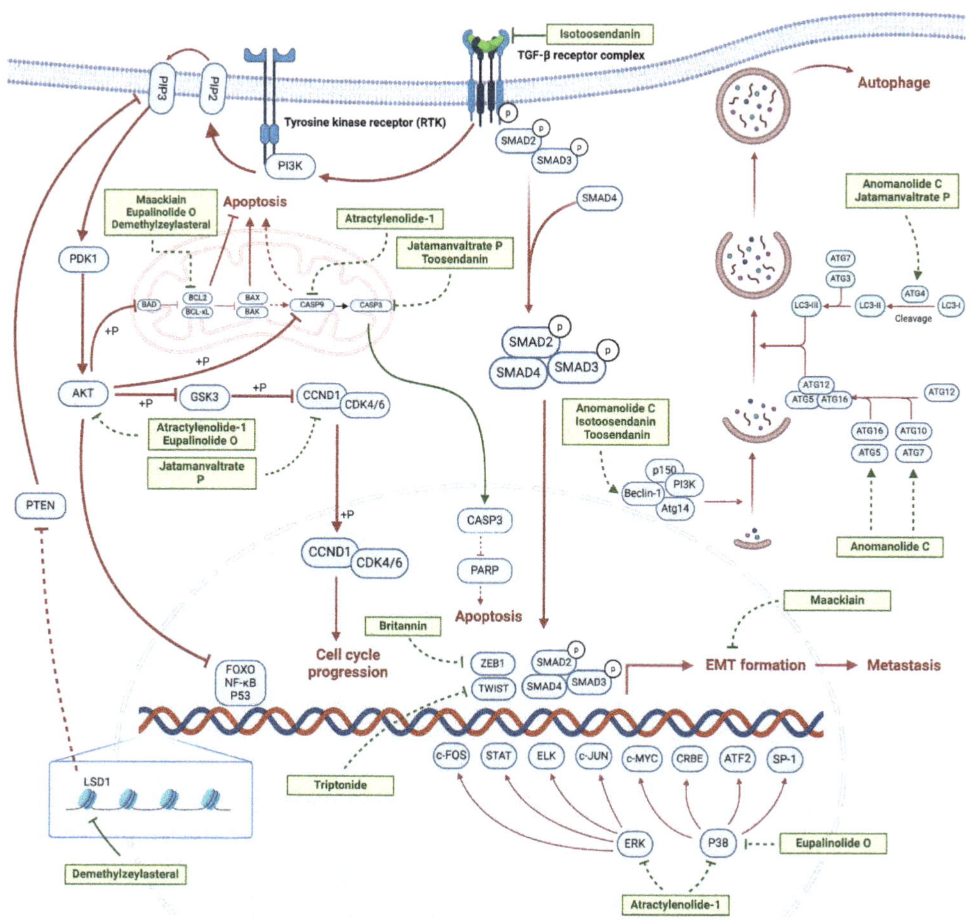

Figure 1. The regulation of terpenoids in TNBC.

2.2. Glycosides

Glycosides are characterized by the presence of a sugar moiety linked to a non-sugar portion known as the aglycone. These compounds display distinct biological activities due to their varied aglycone structures, while their shared sugar moieties contribute to similar functionalities.

NPs contain numerous glycoside components, and research has documented their inhibitory effects on TNBC (Figure 2, Table 2 and Table S1).

2.2.1. Anti-Metastasis

Recent studies have highlighted that glycoside compounds, like icariin, Vanicoside B, asiaticoside (AS), Astragaloside IV (AS-IV), scutellarin (SC), Platycodin D (PD), and rutin, have the potential to inhibit the metastasis of TNBC cells through various mechanisms [43–53]. Icariin, for instance, has been reported to reduce TNBC cell invasion by targeting and suppressing the c-Jun N-terminal kinase (JNK)/c-Jun signaling pathway [47]. Vanicoside B has demonstrated significant anti-metastatic properties [43]. One proposed mechanism is its ability to counteract Cyclin-Dependent Kinase 8 (CDK8)-mediated signaling pathways, which are involved in tumor progression [43]. AS inhibits the P2RX7-mediated TGF-β/SMAD signaling pathway by enhancing PPARG expression, which in turn reduces EMT in TNBC [44]. AS-IV has been shown to inhibit both the proliferation and invasion of TNBC cells and to suppress tumor growth in animal models. This is achieved through mechanisms involving the downregulation of Vav Family Member 3 (Vav3) and the inactivation of the Rac1/MAPK signaling pathway [54]. SC reduces TNBC metastasis by targeting the tumor necrosis factor alpha (TNFα)/Tumor necrosis factor receptor 2 (TNFR2) pathway, which leads to the activation of runt-related transcription factor 1 (RUNX1) and increased production of granulocyte-colony stimulating factor (G-CSF) in TNBC-associated endothelial cells [52,53]. PD inhibits the migration, invasion, and adhesion of highly metastatic TNBC cells by downregulating matrix metalloproteinase-9 (MMP-9) and inhibiting epidermal growth factor receptor (EGFR), as well as the MAPK and PI3K/AKT pathways [49]. Ophiopogonin D has the potential to suppress TGF-β1-mediated metastatic behavior in TNBC cells by modulating the ITGB1/FAK/Src/AKT/β-catenin/MMP-9 signaling axis [55]. Rutin inhibits the migration and invasion of TNBC cells by reducing c-Met phosphorylation and blocking its downstream signaling pathways [51,56].

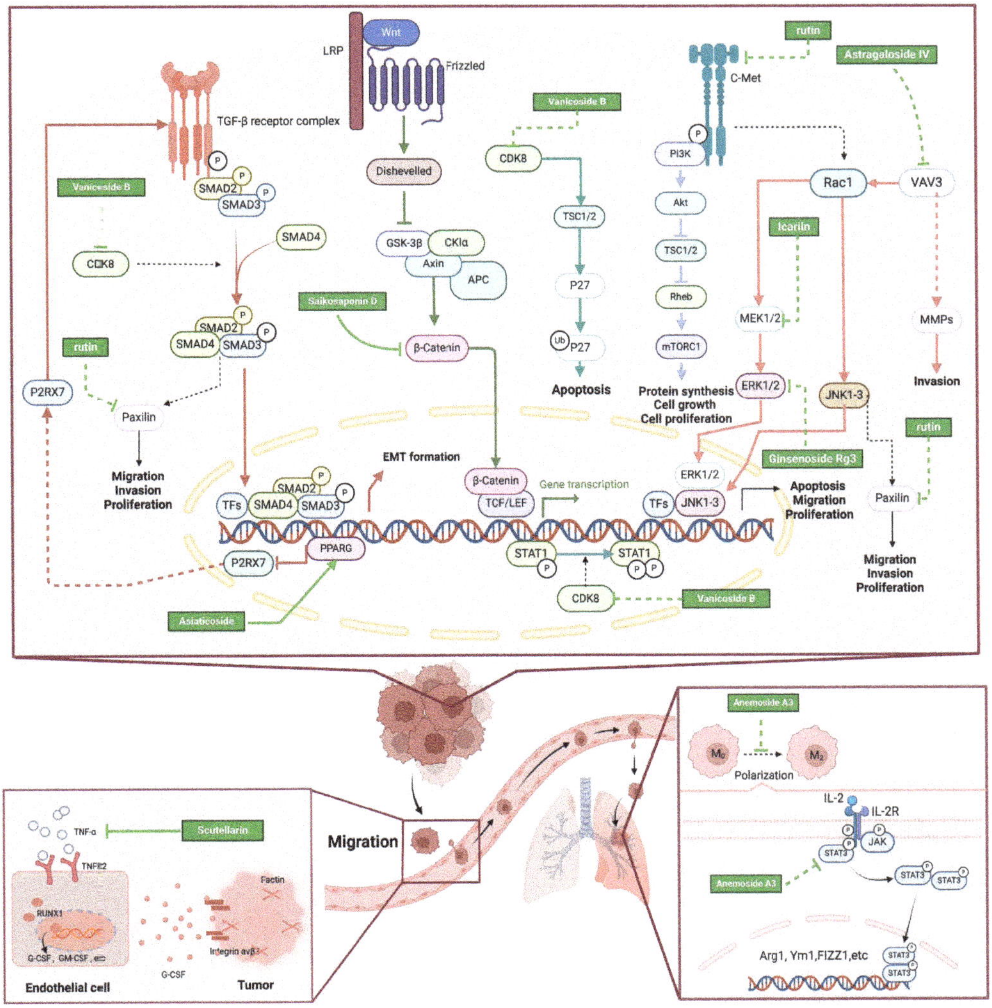

Figure 2. The regulation of glycosides in TNBC.

2.2.2. Pro-Apoptosis

Studies have shown that glycoside compounds, such as icariin, Vanicoside B, Ginsenoside Rg3, Saikosaponin D, DLBS1425, Saxifragifolin A, and neohesperidin, can also induce apoptosis in TNBC cells through various mechanisms [43,47,57–63]. Icariin has been reported to induce apoptosis by causing an excessive accumulation of ROS in cells [47]. Vanicoside B induces apoptosis in TNBC cells through activation of the Skp2-p27 signaling pathway, achieved by inhibiting CDK8 activity [43]. Rg3 induces apoptosis in TNBC cells through the classical mitochondria-dependent caspase activation pathway. Additionally, Rg3 has been reported to induce apoptosis in MDA-MB-231 cells by blocking Nuclear Factor kappa B (NF-κB) signaling, inactivating ERK and AKT, and destabilizing mutant p53 [64,65]. SSD exhibits a strong anti-proliferative effect on TNBC cells by inducing apoptosis. Its potential molecular mechanism involves inhibiting the Wnt/β-catenin signaling pathway, rather than targeting EGFR and Neurotensin receptor 1 (NTSR1) [59]. Glycoside compounds have demonstrated potential as therapeutic agents for TNBC by inducing apoptosis through various mechanisms.

2.2.3. Other Anti-TNBC Effects

In addition to inhibiting metastasis and inducing apoptosis in TNBC cells, glycoside compounds also exhibit other anti-TNBC activities, such as synergistic effects with other drugs, enhancing drug efficacy, and modulating the tumor microenvironment. Ginsenoside Rg1 enhances the cytotoxic effects of doxorubicin through a synergistic mechanism [66]. Additionally, Ginsenoside Rg1 has been shown to protect effectively against cisplatin-induced liver damage, reducing associated histological damage. This protective effect is mediated by inhibiting the Kelch-like epichlorohydrin-associated protein 1 (Keap1)-Nrf2 interaction, partly due to p62 accumulation, and primarily through increasing the production of Nrf2-related antioxidant proteins [67]. The combination of oxymatrine and AS-IV can enhance the immune system to combat TNBC by improving the tumor microenvironment and boosting the antitumor activity of T cells [45]. Anemoside A3 prevents the polarization of macrophages into the M2 type by decreasing the phosphorylation of Signal transducer and activator of transcription 3 (STAT3) protein, thereby suppressing M2 macrophage-driven tumor metastasis [68]. Glycoside compounds have shown potential in inhibiting TNBC. We can further explore additional candidate compounds with anti-TNBC activity.

Glycoside compounds inhibit the migration, invasion, and growth of TNBC cells by modulating various signaling pathways, such as MAPK, PI3K/AKT, CDK8, and ITGB1/FAK/Src/AKT. These compounds induce apoptosis in TNBC cells through multiple mechanisms: promoting excessive ROS accumulation, inhibiting CDK8 to activate the Skp2-p27 signaling pathway, triggering the mitochondria-dependent caspase activation pathway, blocking NF-κB signaling, inhibiting ERK and AKT, destabilizing mutant p53, and suppressing the Wnt/β-catenin signaling pathway. Furthermore, glycoside compounds may improve chemotherapy efficacy and reduce drug toxicity, offering better treatment options for immune-suppressed TNBC patients. However, the precise mechanisms are not yet fully understood, and need further investigation in order to enhance the use of glycoside compounds in treating TNBC.

Table 2. Profiles of glycosides in TCM with therapeutic effects on TNBC.

Natural Products	Structure	Sources	Concentration In Vitro	Dosage In Vivo	Biological Activities/Mechanisms	Potential Targets	Reference
Rutin	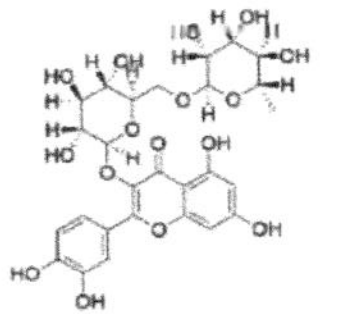	*Acacia erioloba* E. Mey.	150, 300 μM	30 mg/kg	Rutin acts as an inhibitor of c-Met Kinase and effectively inhibits the proliferation of TNBC cells both in vitro and in vivo.	/	[56]
Icariin	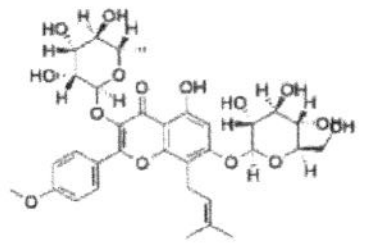	*Epimedium wushanense* Maxim.	5, 10, 20 μM	20, 40 mg/kg	Induces apoptosis and suppresses the migration of TNBC cells through modulation of the Sirtuin (Sirt) 6 /NF-κB signaling pathway.	/	[47]
Vanicoside B	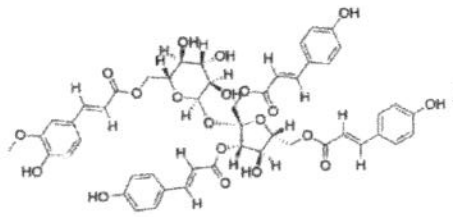	*Persicaria dissitiflora* H. Hara	5, 10, 20 μM	5, 10 mg/kg	The inhibition of CDK8-mediated signaling pathways leads to downregulation of EMT proteins and instigates cell cycle arrest and apoptosis in TNBC cells.	/	[43]
Anemoside A3	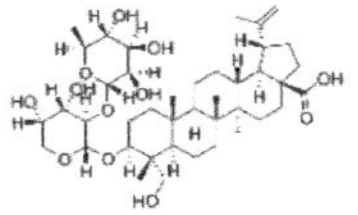	*Pulsatilla chinensis* (Maxim.) Klob.	50, 100 μg/mL	10, 20 mg/kg	Promotes differentiation of M0 macrophages into the M1 phenotype, which hinges upon the Toll-like receptor 4 (TLR4) /NF-κB/MAPK signaling cascade.	TLR4	[68]
Asiaticoside	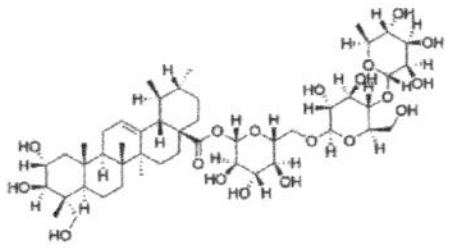	*Centella asiatica* (L.) Urban	0.25 μM	1, 2, 3 mg/kg	Inhibits P2RX7-mediated TGF-β /SMAD signaling through upregulation of PPARγ expression, thus attenuating EMT in TNBC.	PPARγ	[44]
Astragaloside IV	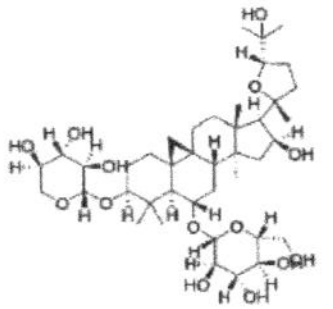	*Astragalus membranaceus* Bunge	20, 40 μg/mL	20, 40 mg/kg	Reduces Vav3 expression.	/	[54]

Table 2. *Cont.*

Natural Products	Structure	Sources	Concentration In Vitro	Dosage In Vivo	Biological Activities/Mechanisms	Potential Targets	Reference
Ginsenoside Rg1	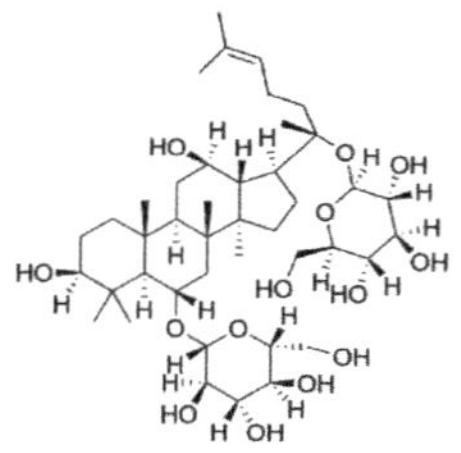	*Panax ginseng* C. A. Mey.	10 μM	/	Activates the DNA damage-response elements [Ataxia telangiectasia mutated protein kinase (ATM), H2A histone family member X (H2AX), Radiation-sensitive 51 (RAD51), X-ray repair cross-complementing protein 1(XRCC1), apoptosis-related genes (P21, TP53, Apoptotic peptidase activating factor 1 (APAF1), BAX, Caspase 3(CASP3), and Caspase 9(CASP9))] in MDA-MB-231 cells.	/	[66,69]
Ginsenosides Rg3	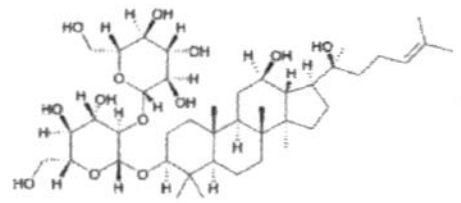	*Panax ginseng* C. A. Mey.	300 mg/mL	/	Increases the Bax/Bcl-2 ratio, causes mitochondrial membrane potential depolarization, leads to cytochrome c release, and induces caspase-3 and PARP cleavage.	/	[64,65]
Saikosaponin D	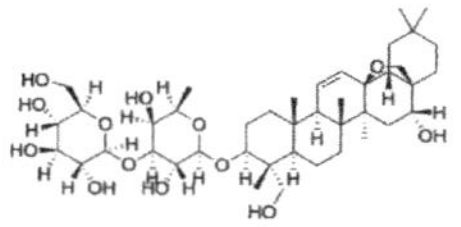	*Bupleurum rotundifolium* L.	10, 15, 30 μM	/	Suppresses the proliferation of TNBC cells through modulation of the β-catenin signaling cascade.	β-catenin	[70]
Scutellarin	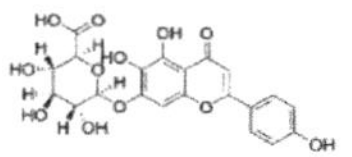	*Erigeron breviscapus* (Vaniot) Hand. -Mazz.	20 μM	10 mg/kg	Reduces TNBC metastasis through the targeting of TNFα/TNFR2-initiated activation of RUNX1 and subsequent production of G-CSF in TNBC-associated endothelial cells.	TNFα/TNFR2	[71]

2.3. Phenolics

Phenolic compounds are composed of aromatic rings and hydroxyl groups. These compounds demonstrate a variety of biological activities, such as antioxidant, anti-inflammatory, antitumor, and antimicrobial effects [72–75]. Their antioxidant properties arise from the hydroxyl groups on the aromatic rings, which can donate electrons [75]. The natural world is rich in phenolic compounds, and recent studies have shown that some of these compounds can effectively combat TNBC (Figure 3, Table 3 and Table S1).

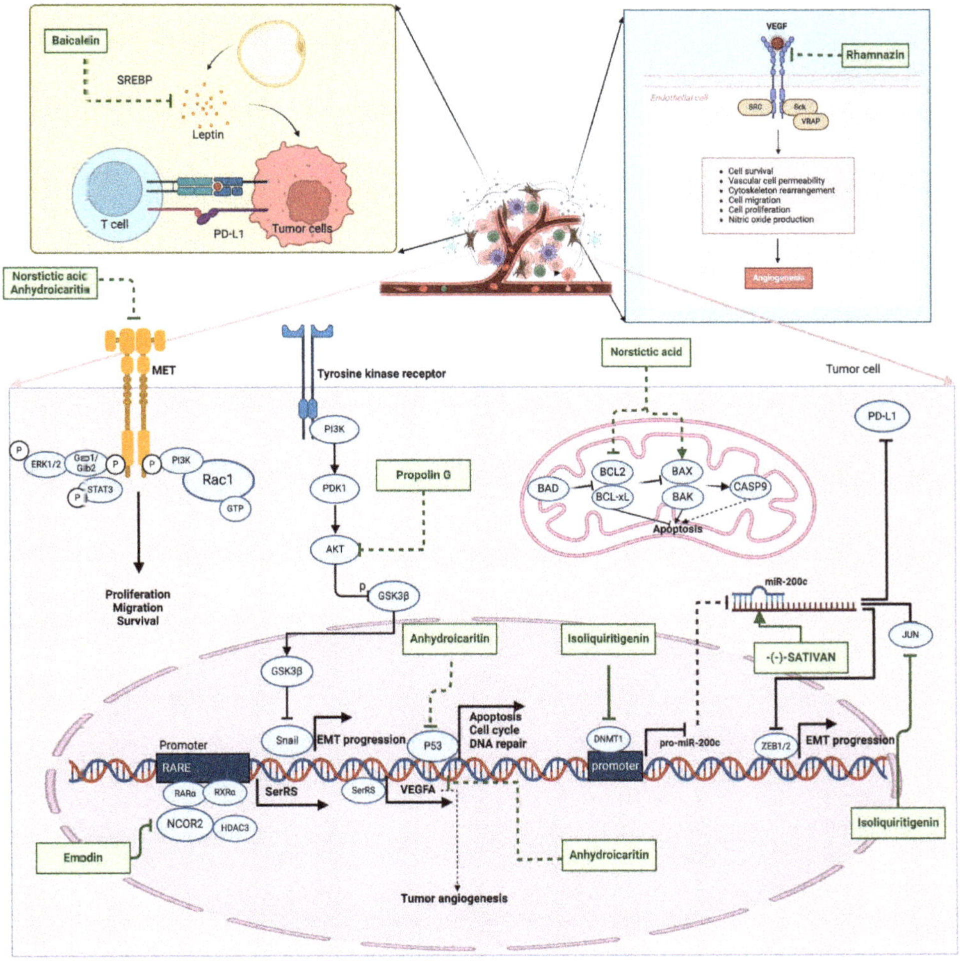

Figure 3. The regulation of phenolics in TNBC.

2.3.1. Anti-Metastasis

Recent studies indicate that phenolic compounds can inhibit the metastasis of TNBC cells. For instance, (-)-sativan may regulate miR-200c, thereby affecting the EMT process and Programmed death-ligand 1 (PD-L1) expression in TNBC cells [76]. Similarly, isoliquiritigenin can also increase miR-200c levels, thereby suppressing the EMT process [77]. Propolin G inhibits the EMT in TNBC cells by degrading the cytoskeletal protein vimentin. This process is mediated by Glycogen synthase kinase 3β (GSK-3β) and regulated by Snail and Histone deacetylase 6 (HDAC6) [78]. Baicalein effectively reverses the expression of

Interferon-induced protein with tetratricopeptide repeats 2 (IFIT2), which is associated with metastasis, recurrence, and poor prognosis in TNBC patients [79]. Norstictic acid significantly inhibits the proliferation, migration, and invasion of TNBC cells, while it shows minimal toxicity to non-tumorigenic MCF-10A mammary epithelial cells. Its potential mechanism involves the inhibition of c-Met phosphorylation [80]. Gigantol prevents TNBC cell migration and invasion by suppressing the Wnt/β-catenin signaling pathway [81].

2.3.2. Anti-Angiogenesis

Phenolic compounds can inhibit the progression of TNBC by suppressing angiogenesis, primarily through the inhibition of VEGFA. Emodin exhibits anti-angiogenic effects in TNBC by targeting the transcriptional regulators nuclear receptor corepressor 2 (NCOR2) and Seryl-tRNA synthetase (SerRS), thereby suppressing VEGFA transcription [82]. Research shows that the hydroxyl group at the C-3 position of Emodin is crucial for its anticancer activity [83]. Rhamnazin, a new vascular endothelial growth factor receptor 2 (VEGFR2) inhibitor, blocks tumor angiogenesis and growth. Rhamnazin significantly reduced the proliferation, migration, and tube formation of human umbilical vascular endothelial cells (HUVECs) in vitro, and also inhibited sprout formation in rat aorta ring assays [84]. Cirsimaritin reduced angiogenesis in MDA-MB-231 cells by lowering the levels of vascular endothelial growth factor (VEGF), p-AKT, and p-ERK [85].

2.3.3. Other Anti-TNBC Effects

Phenolic compounds not only inhibit angiogenesis and metastasis in TNBC, but also suppress its progression through other mechanisms. Anhydroicaritin induces apoptosis in TNBC cells [86]. Isoliquiritigenin inhibits the proliferation and migration of TNBC cells by reducing the expression of miR-374a, which negatively regulates PTEN and results in increased apoptosis and decreased invasiveness by inactivating the AKT pathway [77]. Kaempferol inhibits the expression of Sirt3 and Sirt6, which are potential therapeutic targets for TNBC due to their roles in cancer stem cell activity, chemotherapy resistance, and metastasis [87]. The combination of Morin and doxorubicin enhances the efficacy against TNBC [88]. Baicalein increases the sensitivity of TNBC cells to doxorubicin by promoting autophagy-mediated downregulation of CDK1 [89,90].

Recent studies have reported that phenolic compounds suppress metastasis and angiogenesis, induce apoptosis in TNBC cells, and enhance the sensitivity of TNBC to chemotherapy drugs. The inhibition of angiogenesis by phenolic compounds is closely linked to VEGFA, which plays a crucial role in this process by promoting the proliferation, migration, and lumen formation of endothelial cells, thereby directly influencing vascular growth and development. In the tumor microenvironment, VEGFA overexpression is strongly associated with tumor angiogenesis, creating conditions that support tumor growth [91,92]. Phenolic compounds affect TNBC cells through various mechanisms to prevent the onset and progression of the disease. For example, isoliquiritigenin has been shown to simultaneously suppress TNBC cell metastasis and induce apoptosis. Despite the promising results of phenolic compounds in anti-TNBC research, further studies are necessary to evaluate their potential for clinical application.

2.4. Alkaloids

Alkaloids are a group of naturally occurring compounds characterized by nitrogen-containing heterocyclic structures, and are frequently found in various plants. Current investigations indicate that alkaloids such as berberine, matrine, fangchinoline, and Dragmacidin D, sourced from diverse organisms, exhibit therapeutic promise against TNBC through the inhibition of proliferation [93–98] (Figure 4, Table 4 and Table S1).

Table 3. Profiles of phenolics in TCM with therapeutic effects on TNBC.

Natural Products	Structure	Sources	Concentration In Vitro	Dosage In Vivo	Biological Activities/Mechanisms	Potential Targets	Reference
(-)-Sativan(SA)	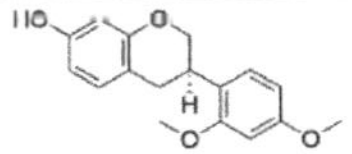	*Spatholobus suberectus* Dunn	5, 10, 20 μM	25, 50 mg/kg	Upregulates miR-200c to suppress the process of EMT.	/	[76]
Anhydroicaritin	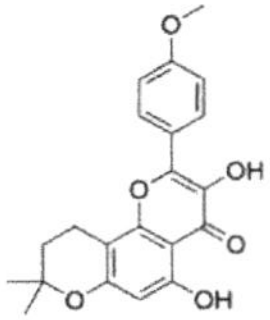	*Epimedium brevicornu* Maxim.	10, 20, 30 μM	20 mg/kg	Inhibits the HIF-1α/VEGFA signaling pathway.	/	[86]
Emodin	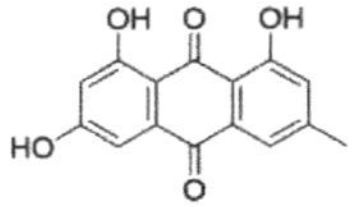	*Rheum palmatum* L.	0.25, 10, 25 μM	40 mg/kg	Regulates the activity of NCOR2; inhibits SIK3; suppresses TGF-β1.	NCOR2	[82]
Isoliquiritigenin	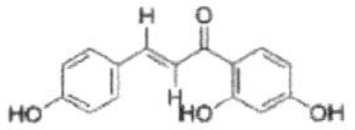	*Spatholobus suberectus* Dunn	10 μM	20, 40 mg/kg	Reduces miR-374a expression; elevates BAX protein and mRNA levels by inhibiting miR-374a; enhances miR-200c expression.	/	[99,100]
Kaempferol	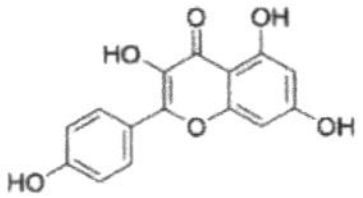	*Caragana frutex* (L.) C. Koch	12.5, 25 μg/mL,	/	Inhibits Sirt3 and Sirt6 proteins, inhibiting the PI3K/AKT pathway.	Sirts	[87]
Morin	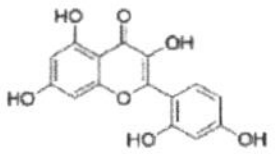	*Maclura pomifera* (Raf.) C. K. Schneid., *Petasites formosanus* Kitam.	10, 50, 100, 200 μM	10 mg/kg	Inhibits the growth and invasion of the highly metastatic BC cell line MDA-MB-231, which is partly achieved through the suppression of the AKT pathway.	/	[100,101]
Propolin G	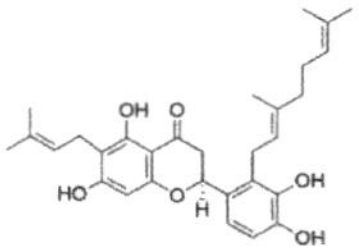	Taiwanese propolis	5, 10, 15, 20 μM	/	Inhibits the stabilization of vimentin protein, mediated by GSK-3β-Snail, and interferes with HDAC6, thereby impeding the suppression of cell migration and invasion in TNBC.	GSK-3β, HDAC6	[78]

Table 3. *Cont.*

Natural Products	Structure	Sources	Concentration In Vitro	Dosage In Vivo	Biological Activities/Mechanisms	Potential Targets	Reference
Rhamnazin	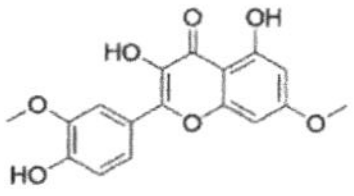	*Callicarpa kwangtungensis chun*, *Alocnemum strobilaceum*	10, 15, 20 μM	200 mg/kg	Suppresses VEGF-induced phosphorylation of VEGFR2 and its downstream signaling mediators in HUVECs.	VEGFR2	[84]
Baicalein	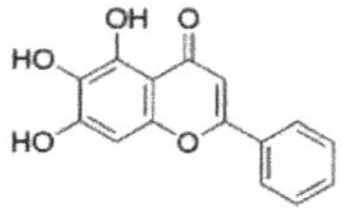	*Scutellaria baicalensis* Georgi	50, 100 ng/ml	50 mg/kg	Reverses the expression of IFIT2.	/	[89,102]
Norstictic acid	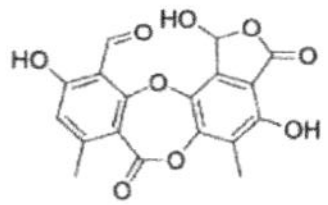	*Dimelaena oreina*, *Umbilicaria virginis*	15, 30 μM	15 mg/kg	Inhibits the proliferation, migration, and invasion of TNBC MDA-MB-231 cells.	/	[80]

Anti-Proliferation

Alkaloids have been reported to inhibit the proliferation of TNBC, with the PI3K/AKT pathway playing a crucial role. Fangchinoline inhibits the proliferation of MDA-MB-231 cells by targeting the AKT/GSK-3β/CCND1 signaling pathway. Additionally, fangchinoline has been shown to induce G1 phase cell cycle arrest and promote apoptosis in these cells [94]. Matrine disrupts the PI3K/AKT signaling pathway by reducing levels of PI3K, AKT, phosphorylated AKT, and phosphoglycerate kinase 1, leading to the inhibition of TNBC cell growth. Furthermore, matrine inhibits TNBC cell proliferation and induces apoptosis by suppressing HN1 expression [97]. Harmine inhibits the proliferation, invasion, and migration of MDA-MB-231 cells, while promoting apoptosis. This effect may result from the suppression of the EMT process and the PI3K/AKT signaling pathway [103]. Berberine inhibits TNBC cell growth by targeting the GPER1/NF-κB pathway, and also influences TNBC cell migration by reducing IL-6 secretion, and LH2 expression by modulating glycolysis [104,105]. Dragmacidin D effectively induces apoptosis in spheroids of TNBC [98]. Ungeremine exhibits cytotoxic effects on MDA-MB-231 cells by inducing ferroptosis, necroptotic apoptosis, and autophagy, alongside apoptosis, via caspase activation. It also affects mitochondrial membrane potential and increases ROS production [106].

In general, alkaloids inhibit cell migration, promote apoptosis, and block the cell cycle, thereby exerting antitumor effects in TNBC. Fangchinoline has been reported to inhibit TNBC cell growth, migration, and proliferation. Research indicates that substituting benzyl units in fangchinoline is crucial for significantly enhancing its antitumor activity [107]. This insight directs our efforts to synthesize fangchinoline derivatives, with the goal of finding more effective and less toxic candidates. Dragmacidin D effectively induces apoptosis in TNBC spheroids. However, it does not show cytotoxic effects on the same cell lines after prolonged incubation under traditional 2D growth conditions [98]. This suggests that establishing suitable in vitro models is essential. Most of these alkaloids have demonstrated anticancer activity in vitro, but lack data in vivo. Further in-depth research is needed.

While this review focuses on terpenoids, glycosides, phenolics, and alkaloids as the predominant anti-TNBC agents identified in our screening, a comprehensive list of additional compounds with reported activity (including the steroid, withanolide, unsaturated fatty acids, and furanocoumarins) is systematically cataloged in Supplementary Table S1, annotated with their chemical structures, source organisms, and experimental validation data.

Table 4. Profiles of alkaloids in TCM with therapeutic effects on TNBC.

Natural Products	Structure	Sources	Concentration In Vitro	Dosage In Vivo	Biological Activities/Mechanisms	Potential Targets	Reference
Fangchinoline	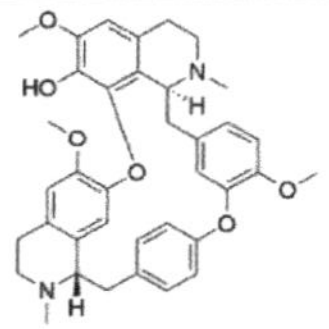	*Stephania tetrandra* (Radix)	6, 12, 24 μM5, 10, 30 μM	/	Inhibits the AKT/Gsk 3β/CCND1 signaling pathway; blocks cell cycle progression at the G1 phase; induces apoptosis.	/	[94,95]
Dragmacidin D	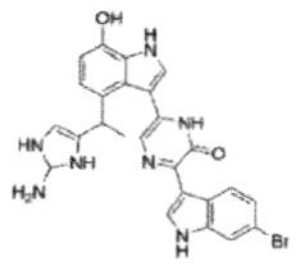	*Genus Spongosorites*	3.8, 7.6 μM	/	Induces apoptosis in TNBC spheroids.	/	[98]
Protopine	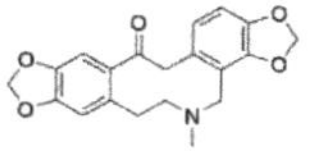	*Corydalis yanhusuo*	1, 10, 100 μM	/	Specifically inhibits the cell adhesion ability of MDA-MB-231.	/	[108]
Subereamolline A	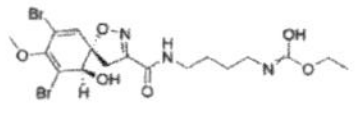	*Pseudoceratina arabica*	2 μM	/	Inhibits the migration and invasion of MDA-MB-231 cells.	/	[109]
Harmine	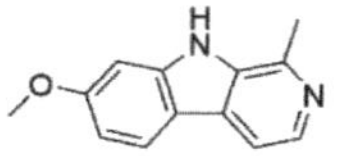	*Peganum harmala*	10 μM	/	Inhibits the process of EMT, associated with the PI3K/AKT signaling pathway.	/	[110]
Ungeremine	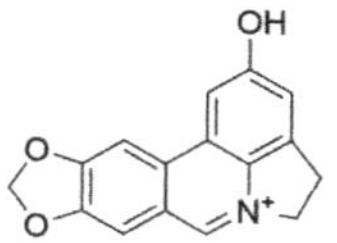	*Crinum zeylanicum*	3.67 μM	/	Induces ferroptosis, necroptosis, and autophagy, as well as apoptosis, mediated by caspase activation, MMP alteration, and increase ROS production.	/	[106]
Berberine	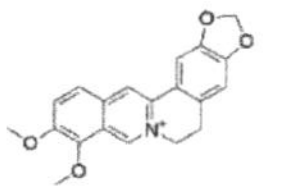	*Coptis chinensis*	1.25, 2.5, 5 μM	15, 30 mg/mL	Reduces IL-6 secretion and LH2 expression; modulates the GPER1/NF-κB pathway; inhibits the NLRP3 inflammasome pathway.	LH2	[107,111]
Matrine	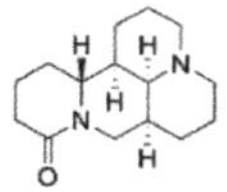	*Sophora flavescens*	2, 4 μM	/	Inhibits the PI3K/AKT signaling pathway; suppresses HN1 expression.	/	[93,97]

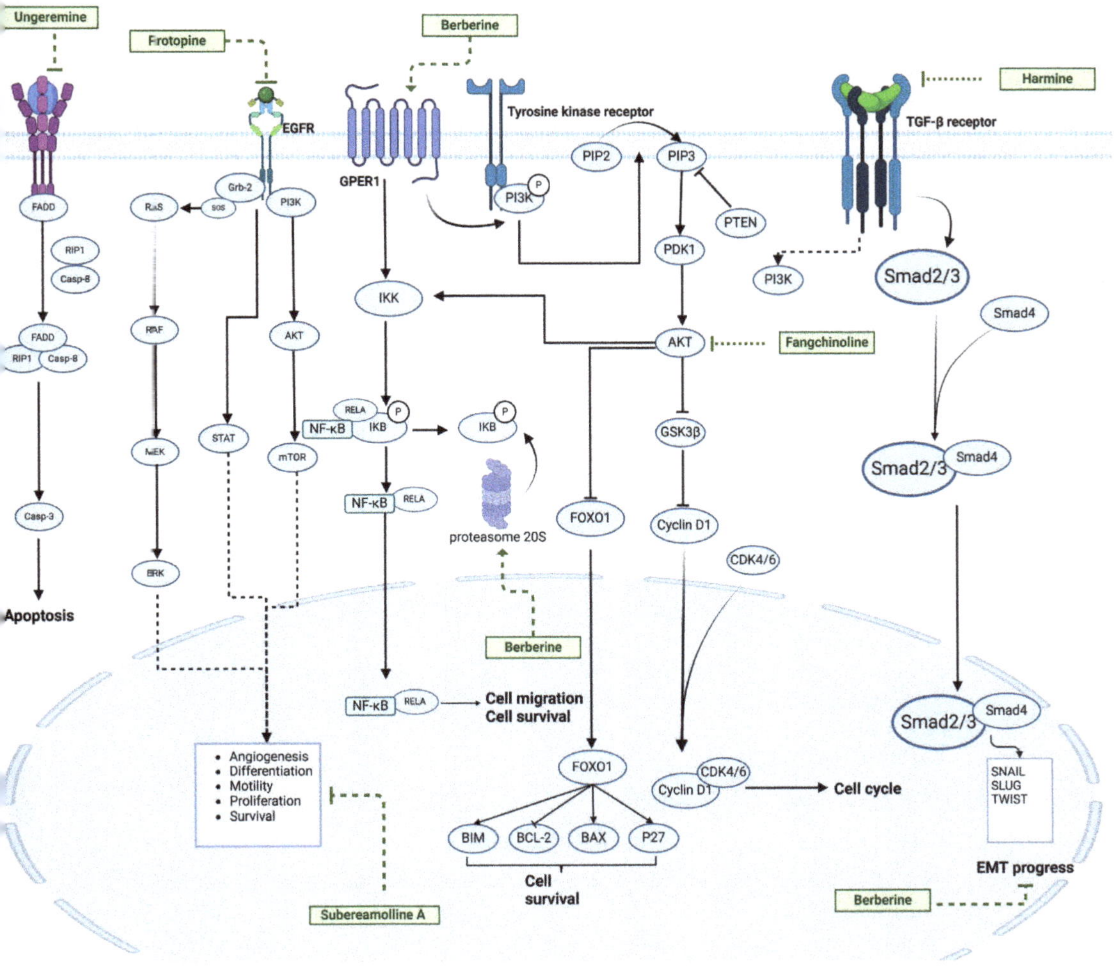

Figure 4. The regulation of alkaloids in TNBC.

3. Natural Products: Mechanisms of Inhibiting TNBC

Emerging evidence highlights the multi-target mechanisms of NPs—including terpenes, glycosides, phenolics, and alkaloids—against TNBC. These compounds suppress tumor progression through diverse pathways, such as metastasis inhibition, apoptosis induction, cell cycle arrest, and tumor microenvironment modulation. Their polypharmacological effects often arise from modulating crosstalk between key signaling nodes, particularly the TGF-β, VEGFA, PI3K/AKT, Wnt/β-catenin, and MAPK pathways. By elucidating the interplay of these critical pathways, the following section further illuminates how NPs exert their therapeutic potential in TNBC.

3.1. TGF-β Signaling Pathway

Transcriptional regulation of EMT induced by TGF-β involves a complex cascade of events. Initially, in response to TGF-β, Smad2 and Smad3 become activated and form complexes with Smad4. These complexes then regulate the transcription of target genes by interacting with other DNA-binding transcription factors. During the induction of EMT, the activated Smads play a pivotal role in mediating transcriptional regulation through three distinct families of transcription factors. This process ultimately leads to the repression of epithelial marker gene expression and the activation of mesenchymal gene expression, which are hallmarks of EMT.

Additionally, TGF-β directly activates non-Smad signaling pathways. This activation occurs through the interaction of signaling mediators either directly with the TβRII and/or TβRI receptors, or via adaptor proteins. Among these non-Smad signaling responses, the activation of Erk MAP kinases, Rho family GTPases (Rho GTPases), and the PI3K /AKT pathway in response to TGF-β has been specifically linked to TGF-β-induced EMT. These pathways regulate various cellular processes, including cytoskeleton organization and cell growth, survival, migration, and invasion, which are crucial for the EMT process. The discovery of ITSN, AS, and harmine demonstrates their capacity to inhibit the TGF-β pathway, thereby suppressing EMT. This inhibition, in turn, contributes to the suppression of TNBC. Therefore, screening NPs that modulate the TGF-β pathway represents a promising strategy for the development of anti-TNBC therapies.

3.2. VEGFA Signaling Pathway

VEGFA is a key regulator of angiogenesis, stimulating endothelial cell proliferation, migration, and new blood vessel formation by its interaction with its primary receptors, VEGFR1 and VEGFR2 [112,113]. Tumor progression is frequently linked to enhanced angiogenesis, with VEGFA playing a central role in this process. As the tumor grows, increased VEGFA expression stimulates the formation of new blood vessels that provide essential nutrients to the tumor and support the spread of cancer cells [114,115]. Research indicates that elevated VEGFA levels are associated with poor clinical outcomes [116]. Compounds such as anhydroicaritin, rhamnazin, and emodin inhibit angiogenesis and metastasis by targeting the VEGFA pathway. VEGFA serves as a key target protein for treating TNBC, and developing and screening NPs that inhibit the VEGFA pathway could provide new therapeutic approaches for TNBC.

3.3. PI3K Signaling Pathway

The PI3K/AKT pathway is one of the most critical intracellular signaling pathways, playing a key role in various biological processes, such as cell growth, proliferation, metabolism, and survival. PI3K is a lipid kinase that activates downstream signaling by phosphorylating phosphatidylinositol 4,5-bisphosphate to form phosphatidylinositol 3,4,5-trisphosphate. The generation of PIP3 promotes the recruitment and activation of AKT, which regulates various downstream effectors, like mTOR, involved in processes like cell proliferation, survival, and metabolism [117]. Overactivation of the PI3K/AKT pathway is often associated with various cancer types. Studies have shown that abnormal activation of this pathway can result from mechanisms such as PIK3CA gene mutations or PTEN deletion or inactivation [118]. These genetic alterations lead to persistent activation of the pathway, promoting tumor cell proliferation and survival. Compounds such as AT-1, rhamnazin, norstictic acid, harmine, T-96, and matrine inhibit the PI3K/AKT pathway by targeting either PI3K or the phosphorylated forms of AKT. This inhibition induces apoptosis, suppresses cell migration, and reduces cell proliferation in TNBC. AKT promotes tumor cell survival by phosphorylating pro-apoptotic proteins, such as Bcl-2-associated death promoter and Caspase-9, thereby enhancing the cells' ability to evade apoptosis, and it also plays a crucial role in tumor invasion and metastasis [119]. Overactivation of the PI3K/AKT pathway can enhance the motility and invasiveness of tumor cells, allowing them to breach the basement membrane and invade surrounding tissues [120]. Additionally, the PI3K/AKT pathway enhances the activation of tumor-associated fibroblasts, which supports tumor metastasis. Tumor growth and metastasis depend on an effective blood supply, and the PI3K/AKT pathway plays a crucial role in tumor angiogenesis. This pathway regulates the expression of VEGF, promoting the formation of new blood vessels [121]. Therefore, developing NP-based inhibitors of the PI3K/AKT pathway could provide new therapeutic strategies for TNBC.

3.4. *Wnt/β-Catenin Signaling Pathway*

The Wnt/β-catenin signaling pathway is essential for various physiological processes, including cell proliferation, differentiation, migration, and tissue homeostasis [122,123]. When Wnt proteins bind to Frizzled receptors on the cell surface, this pathway is activated, leading to the activation of β-catenin's transcriptional activity. In the absence of Wnt signals, β-catenin is degraded in the cytoplasm. However, when Wnt signals are present, β-catenin accumulates and translocates to the nucleus, where it binds to T-cell factor/lymphoid enhancer factor to initiate the transcription of target genes [122].

Inhibiting the Wnt/β-catenin signaling pathway can effectively suppress tumor stem cell renewal, reduce cell proliferation, enhance sensitivity to chemotherapy, and inhibit tumor metastasis [123]. Saikosaponin D, a compound shown to significantly repress β-catenin and its downstream target genes, induces apoptosis in TNBC cells. Thus, developing NPs that target and inhibit the Wnt/β-catenin signaling pathway represents a promising strategy for treating TNBC.

3.5. *MAPK Signaling Pathway*

The MAPK signaling cascade involves three key kinases: MAPKKK, MAPKK, and MAPK, collectively known as mitogen-activated protein kinases. These pathways play a crucial role in the onset and progression of cancer [124]. Abnormal activation of MAPK pathways, often due to mutations in Ras proteins and rapidly accelerated fibrosarcoma proteins, leads to the continuous activation of the ERK1/2 pathway [125]. Research has shown that compounds such as EO, AT-1, Icariin, and AS-IV can inhibit the MAPK pathway, thereby suppressing TNBC.

NPs offer potential for targeting TNBC through the inhibition of other various critical pathways. For instance, kaempferol has been shown to suppress the expression of Sirt3 and Sirt6, which are associated with cancer stem cell functions, chemotherapy resistance, and metastasis. Another important target in TNBC therapy is Methionine (MET); research indicates that norstictic acid and rutin can effectively inhibit MET activity.

AC enhances the ubiquitination of GPX4, while britannin promotes the ubiquitination of ZEB1, a key transcription factor involved in EMT, thereby reducing TNBC cell metastasis. EGFR, through its interaction with ligands like epidermal growth factor (EGF), activates downstream signaling pathways that regulate cell proliferation, migration, and survival. Protopine has been identified as an inhibitor of EGFR expression, contributing to the suppression of TNBC. In addition, compounds such as ungeremine, Jatamanvaltrate P, ITSN, and AC can modulate autophagy, and consequently inhibit TNBC. Therefore, future research should delve deeper into the mechanisms of these pathways to discover natural compounds that can target these processes, offering new therapeutic avenues for TNBC treatment.

4. Challenges and Opportunities

TNBC, characterized by the absence of ER, PR, and HER2 immunostaining, is refractory to both endocrine therapy and conventional molecularly targeted therapies, making chemotherapy the primary systemic treatment approach. NPs are hailed as a primary source for drug discovery, due to their immense structural and physicochemical diversity.

Each type of compound that inhibits TNBC exhibits unique characteristics and commonalities. The antitumor effects of terpenoids encompass a broad spectrum of activities, including anti-proliferative, apoptotic, anti-angiogenic, and anti-metastatic actions. Glycosides primarily exert their anticancer effects through inhibiting metastasis, inducing apoptosis, and synergizing with other medications to enhance therapeutic efficacy and modulate the tumor microenvironment. Phenolic compounds manifest their anticancer effects mainly by inhibiting angiogenesis and metastasis in TNBC, and by increasing the cancer cells' sensitivity to chemotherapeutic agents. Alkaloids predominantly achieve their anticancer effects through anti-proliferation. These effects are closely linked to their structural features. For example, the substitution of benzyl units in fangchinoline is a key pharmacophore that enhances its anti-proliferative effect on cancer cells. Meanwhile, research shows that the introduction of fluorine atoms or fluorinated groups significantly enhances the antitumor efficacy of the compound and prolongs its pharmacokinetic profile in vivo [107]. By systematically categorizing and organizing these NPs, we might infer the potential anticancer properties of compounds with similar structures, such as 3-aryl isocoumarins and 8-hydroxy-3-aryl isocoumarin, which are akin to kaempferol, morin, propolin G, rhamnazin, and baicalein, and may also exhibit certain anti-TNBC activities [126,127].

To comprehensively assess the therapeutic potential of NPs in TNBC, we compared their in vitro and in vivo efficacy with current standard treatments. NPs such as AC, AT-1, Vanicoside B, SA, and emodin demonstrated promising anticancer activity in cellular and animal models. For instance, AC exhibited potent effects against MDA-MB-231 cells (IC_{50} = 1.02 μM) and significantly inhibited tumor growth in murine TNBC models. Rutin reduced tumor volume by 53.85% and 65.72% in mouse TNBC models. Compounds like T-96, ITSN, and AS showed remarkable efficacy at low doses in vivo. While these findings are derived from preclinical studies, they provide critical scientific rationale for further development of NPs. Paclitaxel, a microtubule inhibitor, stabilizes microtubules to block cell cycle progression and induce apoptosis. Clinical trials have demonstrated its efficacy in TNBC, particularly when combined with carboplatin, improving overall and disease-free survival rates [128]. PARP inhibitors (e.g., Olaparib, Talazoparib) target BRCA-mutated TNBC by impairing DNA repair mechanisms. These agents are approved for metastatic TNBC patients with germline BRCA1/2 mutations [129]. NPs such as AC and T-96 have shown minimal toxicity in preclinical models, with no significant hepatorenal damage or adverse effects observed during prolonged administration. In contrast, paclitaxel is associated with dose-limiting neurotoxicity, myelosuppression, and hypersensitivity reactions, while PARP inhibitors may cause hematologic and gastrointestinal toxicities [130]. Although current research on NPs remains preclinical, advances in translational medicine are bridging the gap between laboratory discoveries and clinical applications. Further studies should prioritize evaluating their therapeutic potential in specific TNBC subtypes (e.g., BRCA-mutated tumors), for which NPs may offer distinct advantages in efficacy and tolerability.

NPs possess unparalleled structural diversity, a hallmark advantage derived from the vast biochemical repertoire of plants, animals, and microorganisms. These evolutionarily refined compounds exhibit intricate and unique architectures that far surpass the structural complexity of most synthetic counterparts. Such chemical distinctiveness enables NPs to engage with diverse biological targets through multifaceted interactions, positioning them as invaluable resources for drug discovery. In the context of TNBC—a malignancy characterized by dysregulation of multiple signaling pathways (e.g., PI3K/AKT/mTOR, VEGFA) and aberrant crosstalk within oncogenic networks—monotherapies targeting single pathways often yield suboptimal outcomes, due to compensatory resistance mechanisms. The polypharmacological properties of NPs enable them to simultaneously modulate multiple therapeutic targets, offering a promising strategy to address the heterogeneity and adaptive resistance of TNBC. For instance, AT-1 has been demonstrated to inhibit phosphorylation of the ERK family and suppress the expression of TPI1 and GPI, demonstrating significant anti-TNBC activity in preclinical models [21]. This multi-targeted approach enhances therapeutic efficacy and mitigates compensatory resistance mechanisms that are commonly observed in TNBC.

However, the clinical translation of NPs for TNBC faces multiple challenges, requiring prioritized research. First, there is an intricate relationship between the structural complexity of NPs and their toxicity; for example, the pentacyclic ring structure of TSN, which contains a carbonyl group, has been associated with marked hepatotoxicity, whereas ITSN, featuring an oxygen bridge, exhibits significantly reduced toxicity [37,131,132]. Second, conventional screening paradigms overemphasizing direct cytotoxicity (e.g., oxoflavidin's prioritization based solely on MDA-MB-231 cell inhibition) risk overlooking multimodal therapeutic mechanisms, despite murine models remaining indispensable for preclinical validation [133]. Furthermore, critical gaps persist in understanding NP-associated safety profiles and comparative advantages over existing therapies. Future investigations must integrate structural optimization, mechanism elucidation, comprehensive safety assessments, and comparative efficacy studies to advance clinically viable NP-based TNBC therapeutics.

5. Conclusions

NPs exhibit unparalleled chemical diversity, underpinning their remarkable potential as reservoirs of bioactive compounds with multifaceted pharmacological properties. Emerging evidence highlights their growing significance in addressing TNBC, with numerous NPs demonstrating antitumor efficacy through diverse mechanisms.

In this review, we have compiled information on NPs with anti-TNBC properties. These natural agents predominantly comprise a variety of compounds, such as terpenoids, glycosides, phenolic substances, and alkaloids. A profound exploration has revealed their participation in crucial signaling pathways, specifically the TGF-β, VEGFA, PI3K/AKT, Wnt/β-catenin, and MAPK signaling pathways. Although these NPs exhibit potential anticancer activities, the studies on them remain at the preclinical basic research stage, and further exploration is still required.

Supplementary Materials: The following supporting information can be downloaded at https://www.mdpi.com/article/10.3390/molecules30061201/s1. A comprehensive list of additional compounds with reported activity (including the steroid, withanolide, unsaturated fatty acids, and furanocoumarins) [134–183].

Author Contributions: Conceptualization, Q.L. and Z.Y.; methodology, Q.L.; investigation, G.W.; resources, Y.C.; data curation, J.D.; writing—original draft preparation, Q.L.; writing—review and editing, Z.Y., D.W. and Y.W.; visualization, Q.L. and Z.Y.; supervision, D.W.; project administration, Y.W.; funding acquisition, D.W. All authors have read and agreed to the published version of the manuscript.

Funding: The National Natural Science Foundation of China, grant number 82172723; the Science and Technology Department of Sichuan Province, grant number 2024YFFK0156; the Key Projects of Science and Technology Plan of Inner Mongolia Autonomous Region, grant number 201802115; and the Innovation Team and Talents Cultivation Program of the National Administration of Traditional Chinese Medicine. The APC was funded by the National Key R&D Program of China, grant number 2023YFF0720300.

Institutional Review Board Statement: Not applicable.

Informed Consent Statement: Not applicable.

Data Availability Statement: No new data were created or analyzed in this study. Data sharing is not applicable to this article.

Acknowledgments: The figures were created with BioRender.com (accessed on 4 March 2025).

Conflicts of Interest: The authors declare no conflicts of interest.

Abbreviations

The following abbreviations are used in this manuscript:

TNBC	Triple-negative breast cancer
NPs	Natural products
BC	Breast cancer
WHO	World Health Organization
HER2	Human epidermal growth receptor 2
ER	Estrogen receptor
PR	Progesterone receptor
TCMs	Traditional Chinese medicines
HMs	Herbal medicines
FDA	Food and Drug Administration
AT-1	Atractylenolide-1
EMT	Epithelial–mesenchymal transition
AS	Asiaticoside
ITSN	Isotoosendanin
GPX4	Glutathione peroxidase 4
T-96	Demethylzeylasteral
AC	Anomanolide C
TPI1	Triosephosphate Isomerase 1
GPI	Glucose-6-phosphate isomerase
EO	Eupalinolide O
LSD1	Lysine-specific demethylase 1
PTEN	Phosphatase and tensin homolog
MA	Maackiain
ROS	Reactive oxygen species
CDK1	Cyclin-dependent kinase
CCND1	Cyclin D1
PARP	Poly adenosinediphosphate-ribose polymerase
BRCA	Breast cancer gene
TGF-β	Transforming growth factor-beta
VEGF	Vascular endothelial growth factor
PI3K	Phosphoinositide3kinase
AKT	Protein kinase B
Wnt	Wingless/Int-1
AS- IV	Astragaloside IV
SC	Scutellarin
PD	Platycodin D
TNFα	Tumor necrosis factor alpha
TNFR2	Tumor necrosis factor receptor 2
TGFβR1	TGF-β receptor type-1
ERK	Extracellular signal-regulated kinase
JNK	c-Jun N-terminal kinase
Rho GTPase	Rho family GTPases
NF-κB	Nuclear Factor kappa B
Bcl2	B-cell lymphoma 2
Bax	Bcl-2-associated X protein

c-Myc	Cellular myelocytomatosis oncogene
ZEB1	Zinc finger e-box binding homeobox 1
Cdc-2	Cell division cycle 2
MAPK	Mitogen-activated protein kinase
PTP1B	Protein tyrosine phosphatase 1B
CDK8	Cyclin-Dependent Kinase 8
Vav3	Vav Family Member 3
RUNX1	Runt-related transcription factor 1
G-CSF	Granulocyte-colony stimulating factor
MMP-9	Matrix metalloproteinase-9
EGFR	Epidermal growth factor receptor
NTSR1	Neurotensin receptor 1
Keap1	Kelch-like epichlorohydrin-associated protein 1
STAT3	Signal transducer and activator of transcription 3
Sirt	Sirtuin
TLR4	Toll-like receptor 4
ATM	Ataxia telangiectasia mutated protein kinase
H2AX	H2A histone family member X
RAD51	Radiation-sensitive 51
XRCC1	X-ray repair cross-complementing protein 1
APAF1	Apoptotic peptidase activating factor 1
CASP3	Caspase 3
CASP9	Caspase 9
PD-L1	Programmed death-ligand 1
miR	microRNA
GSK-3β	Glycogen synthase kinase 3β
HDAC6	Histone deacetylase 6
IFIT2	Interferon-induced protein with tetratricopeptide repeats 2
NCOR2	Nuclear receptor corepressor 2
SerRS	Seryl-tRNA synthetase
HUVECs	Human umbilical vascular endothelial cells
TSN	Toosendanin
MET	Methionine
VEGFR2	Vascular endothelial growth factor receptor 2

References

Sung, H.; Ferlay, J.; Siegel, R.L.; Laversanne, M.; Soerjomataram, I.; Jemal, A.; Bray, F. Global Cancer Statistics 2020: GLOBOCAN Estimates of Incidence and Mortality Worldwide for 36 Cancers in 185 Countries. *CA Cancer J. Clin.* **2021**, *71*, 209–249. [CrossRef] [PubMed]

Prat, A.; Adamo, B.; Cheang, M.C.U.; Anders, C.K.; Carey, L.A.; Perou, C.M. Molecular Characterization of Basal-Like and Non-Basal-Like Triple-Negative Breast Cancer. *Oncologist* **2013**, *18*, 123–133. [CrossRef] [PubMed]

Garrido-Castro, A.C.; Lin, N.U.; Polyak, K. Insights into molecular classifications of triple-negative breast cancer: Improving patient selection for treatment. *Cancer Discov.* **2019**, *9*, 176–198. [CrossRef] [PubMed]

Dawson, J.C.; Serrels, A.; Stupack, D.G.; Schlaepfer, D.D.; Frame, M.C. Targeting FAK in anticancer combination therapies. *Nat. Rev. Cancer* **2021**, *21*, 313–324. [CrossRef]

Zhang, F.; Wang, Z.; Fan, Y.; Xu, Q.; Ji, W.; Tian, R.; Niu, R. Elevated STAT3 signaling-mediated upregulation of MMP-2/9 confers enhanced invasion ability in multidrug-resistant breast cancer cells. *Int. J. Mol. Sci.* **2015**, *16*, 24772–24790. [CrossRef]

Koudelka, Š.; Turánek, J. Liposomal paclitaxel formulations. *J. Control. Release* **2012**, *163*, 322–334. [CrossRef]

Samaan, T.M.A.; Samec, M.; Liskova, A.; Kubatka, P.; Büsselberg, D. Paclitaxel's mechanistic and clinical effects on breast cancer. *Biomolecules* **2019**, *9*, 789. [CrossRef]

Ferrari, P.; Scatena, C.; Ghilli, M.; Bargagna, I.; Lorenzini, G.; Nicolini, A. Molecular Mechanisms, Biomarkers and Emerging Therapies for Chemotherapy Resistant TNBC. *Int. J. Mol. Sci.* **2022**, *23*, 1665. [CrossRef]

Li, Y.; Zhang, H.; Merkher, Y.; Chen, L.; Liu, N.; Leonov, S.; Chen, Y. Recent advances in therapeutic strategies for triple-negative breast cancer. *J. Hematol. Oncol.* **2022**, *15*, 121. [CrossRef]

10. Keenan, T.E.; Tolaney, S.M. Role of immunotherapy in triple-negative breast cancer. *J. Natl. Compr. Cancer Netw.* **2020**, *18*, 479–48 [CrossRef]
11. Li, H.; Liu, Z.Y.; Wu, N.; Chen, Y.C.; Cheng, Q.; Wang, J. PARP inhibitor resistance: The underlying mechanisms and clinic implications. *Mol. Cancer* **2020**, *19*, 107. [CrossRef] [PubMed]
12. Yin, S.Y.; Wei, W.C.; Jian, F.Y.; Yang, N.S. Therapeutic applications of herbal medicines for cancer patients. *Evid.-Based Complemer Altern. Med.* **2013**, *2013*, 302426. [CrossRef] [PubMed]
13. Pan, S.-Y.; Litscher, G.; Gao, S.-H.; Zhou, S.-F.; Yu, Z.-L.; Chen, H.-Q.; Zhang, S.-F.; Tang, M.-K.; Sun, J.-N.; Ko, K.-M. Historic perspective of traditional indigenous medical practices: The current renaissance and conservation of herbal resources. *Evid.-Base Complement. Altern. Med.* **2014**, *2014*, 525340. [CrossRef]
14. Newman, D.J.; Cragg, G.M. Natural Products as Sources of New Drugs over the Nearly Four Decades from 01/1981 to 09/201 *J. Nat. Prod.* **2020**, *83*, 770–803. [CrossRef]
15. Tholl, D. Biosynthesis and biological functions of terpenoids in plants. *Adv Biochem Eng Biotechnol.* **2015**, *148*, 63–106. [CrossRe
16. El-Baba, C.; Baassiri, A.; Kiriako, G.; Dia, B.; Fadlallah, S.; Moodad, S.; Darwiche, N. Terpenoids' anti-cancer effects: Focus o autophagy. *Apoptosis* **2021**, *26*, 491–511. [CrossRef]
17. Zhao, M.; Tang, Y.; Xie, J.; Zhao, Z.; Cui, H. Meroterpenoids produced by fungi: Occurrence, structural diversity, biologic activities, and their molecular targets. *Eur. J. Med. Chem.* **2021**, *209*, 112860. [CrossRef]
18. Zeng, T.; Liu, Z.; Liu, H.; He, W.; Tang, X.; Xie, L.; Wu, R. Exploring Chemical and Biological Space of Terpenoids. *J. Chem. Ir Model.* **2019**, *59*, 3667–3678. [CrossRef]
19. Carneiro, B.A.; El-Deiry, W.S. Targeting apoptosis in cancer therapy. *Nat. Rev. Clin. Oncol.* **2020**, *17*, 395–417. [CrossRef]
20. Zhang, J.; Yang, F.; Mei, X.; Yang, R.; Lu, B.; Wang, Z.; Ji, L. Isotoosendanin inhibits triple-negative breast cancer metastasis b reducing mitochondrial fission and lamellipodia formation regulated by the Smad2/3-GOT2-MYH9 signaling axis. *Acta Pharmac Sin.* **2024**, *45*, 2672–2683. [CrossRef]
21. Xu, H.; Li, L.; Qu, L.; Tu, J.; Sun, X.; Liu, X.; Xu, K. Atractylenolide-1 affects glycolysis/gluconeogenesis by downregulating th expression of TPI1 and GPI to inhibit the proliferation and invasion of human triple-negative breast cancer cells. *Phytother. Re* **2023**, *37*, 820–833. [CrossRef] [PubMed]
22. Li, L.; Ji, Y.; Fan, J.; Li, F.; Li, Y.; Wu, M.; Cheng, H.; Xu, C. Demethylzeylasteral (T-96) inhibits triple-negative breast canc invasion by blocking the canonical and non-canonical TGF-β signaling pathways. *Naunyn-Schmiedeberg's Arch. Pharmacol.* **201** *392*, 593–603. [CrossRef] [PubMed]
23. Zhao, Y.; Fu, L.; Chen, J.; Zhou, J.; Tian, C.; Zhou, D.; Zhu, R. Eupalinolide O Induces Apoptosis in Human Triple-Negative Brea Cancer Cells via Modulating ROS Generation and Akt/p38 MAPK Signaling Pathway. *J. Oncol.* **2022**, *2022*, 8802453. [CrossRe [PubMed]
24. Yang, B.; Zhu, R.; Tian, S.; Wang, Y.; Lou, S.; Zhao, H. Jatamanvaltrate P induces cell cycle arrest, apoptosis and autophagy i human breast cancer cells in vitro and in vivo. *Biomed. Pharmacother.* **2017**, *89*, 1027–1036. [CrossRef]
25. Shen, Z.; Gu, Y.; Jiang, R.; Qian, H.; Li, S.; Xu, L.; Gu, W.; Zuo, Y. Antitumor Effect of Demethylzeylasteral (T-96) on Triple-Negativ Breast Cancer via LSD1-Mediate Epigenetic Mechanisms. *Anal. Cell. Pathol.* **2022**, *2022*, 2522597. [CrossRef]
26. Peng, F.; Wang, L.; Xiong, L.; Tang, H.; Du, J.; Peng, C. Maackiain Modulates miR-374a/GADD45A Axis to Inhibit Triple-Negativ Breast Cancer Initiation and Progression. *Front. Pharmacol.* **2022**, *13*, 806869. [CrossRef]
27. Yeo, S.K.; Ali, A.Y.; Hayward, O.A.; Turnham, D.; Jackson, T.; Bowen, I.D.; Clarkson, R. β-Bisabolene, a Sesquiterpene from th Essential Oil Extract of Opoponax (*Commiphora guidottii*), Exhibits Cytotoxicity in Breast Cancer Cell Lines. *Phytother. Res.* **201** *30*, 418–425. [CrossRef]
28. Lin, Z.; Yang, Y.; Liu, T.; Wu, Z.; Zhang, X.; Yang, J. Germacrone alleviates breast cancer-associated osteolysis by inhibitin osteoclastogenesis via inhibition of MAPK/NF-κB signaling pathways. *Phytother. Res.* **2024**, *38*, 2860–2874. [CrossRef]
29. Shahrestanaki, M.K.; Mirjani, A.; Ghanadian, M.; Aghaei, M. Cycloartane triterpenoid from *Euphorbia macrostegia* modulates E stress signaling pathways to induce apoptosis in MDA-MB231 and MCF-7 breast cancer cell lines. *Naunyn-Schmiedeberg's Arc Pharmacol.* **2023**, *396*, 1749–1758. [CrossRef]
30. Qin, J.-J.; Jin, H.-Z.; Huang, Y.; Zhang, S.-D.; Shan, L.; Voruganti, S.; Nag, S.; Wang, W.; Zhang, W.-D.; Zhang, R. Selectiv cytotoxicity, inhibition of cell cycle progression, and induction of apoptosis in human breast cancer cells by sesquiterpenoid from *Inula lineariifolia* Turcz. *Eur. J. Med. Chem.* **2013**, *68*, 473–481. [CrossRef]
31. Chen, M.; Zhang, M.; Lu, X.; Li, Y.; Lu, C. Diselenium-linked dimeric prodrug nanomedicine breaking the intracellular redo balance for triple-negative breast cancer targeted therapy. *Eur. J. Pharm. Biopharm.* **2023**, *193*, 16–27. [CrossRef] [PubMed]
32. Zhong, Z.; Chen, X.; Tan, W.; Xu, Z.; Zhou, K.; Wu, T.; Cui, L.; Wang, Y. Germacrone inhibits the proliferation of breast cancer ce lines by inducing cell cycle arrest and promoting apoptosis. *Eur. J. Pharmacol.* **2011**, *667*, 50–55. [CrossRef] [PubMed]
33. Duangmano, S.; Sae-lim, P.; Suksamrarn, A.; Domann, F.E.; Patmasiriwat, P. Cucurbitacin B inhibits human breast cancer ce proliferation through disruption of microtubule polymerization and nucleophosmin/B23 translocation. *BMC Complement. Alter Med.* **2012**, *12*, 185. [CrossRef]

Luo, W.W.; Zhao, W.W.; Lu, J.J.; Wang, Y.T.; Chen, X.P. Cucurbitacin B suppresses metastasis mediated by reactive oxygen species (ROS) via focal adhesion kinase (FAK) in breast cancer MDA-MB-231 cells. *Chin. J. Nat. Med.* **2018**, *16*, 10–19. [CrossRef]

Rakesh, R.; PriyaDharshini, L.C.; Sakthivel, K.M.; Rasmi, R.R. Role and regulation of autophagy in cancer. *Biochim. Biophys. Acta (BBA)—Mol. Basis Dis.* **2022**, *1868*, 166400. [CrossRef]

Yun, C.W.; Jeon, J.; Go, G.; Lee, J.H.; Lee, S.H. The dual role of autophagy in cancer development and a therapeutic strategy for cancer by targeting autophagy. *Int. J. Mol. Sci.* **2021**, *22*, 179. [CrossRef]

Zhang, J.-N.; Zhang, Z.; Huang, Z.-L.; Guo, Q.; Wu, Z.-Q.; Ke, C.; Lu, B.; Wang, Z.-T.; Ji, L.-L. Isotoosendanin exerts inhibition on triple-negative breast cancer through abrogating TGF-β-induced epithelial–mesenchymal transition via directly targeting TGFβR1. *Acta Pharm. Sin. B* **2023**, *13*, 2990–3007. [CrossRef]

Chen, Y.-M.; Xu, W.; Liu, Y.; Zhang, J.-H.; Yang, Y.-Y.; Wang, Z.-W.; Sun, D.-J.; Li, H.; Liu, B.; Chen, L.-X. Anomanolide C suppresses tumor progression and metastasis by ubiquitinating GPX4-driven autophagy-dependent ferroptosis in triple negative breast cancer. *Int. J. Biol. Sci.* **2023**, *19*, 2531–2550. [CrossRef]

Zhang, S.; Dong, Y.; Chen, X.; Tan, C.S.H.; Li, M.; Miao, K.; Lu, J.-H. Toosendanin, a late-stage autophagy inhibitor, sensitizes triple-negative breast cancer to irinotecan chemotherapy. *Chin. Med.* **2022**, *17*, 55. [CrossRef]

Lu, H.; Wu, Z.; Wang, Y.; Zhao, D.; Zhang, B.; Hong, M. Study on inhibition of Britannin on triple-negative breast carcinoma through degrading ZEB1 proteins. *Phytomedicine* **2022**, *104*, 154291. [CrossRef]

Zhang, M.; Meng, M.; Liu, Y.; Qi, J.; Zhao, Z.; Qiao, Y.; Hu, Y.; Lu, W.; Zhou, Z.; Xu, P.; et al. Triptonide effectively inhibits triple-negative breast cancer metastasis through concurrent degradation of Twist1 and Notch1 oncoproteins. *Breast Cancer Res.* **2021**, *23*, 116. [CrossRef] [PubMed]

Gao, B.; Chen, J.; Han, B.; Zhang, X.; Hao, J.; Giuliano, A.E.; Cui, Y.; Cui, X. Identification of triptonide as a therapeutic agent for triple negative breast cancer treatment. *Sci. Rep.* **2021**, *11*, 2408. [CrossRef] [PubMed]

Kim, D.; Wang, C.Y.; Hu, R.; Lee, J.Y.; Luu, T.-T.-T.; Park, H.-J.; Lee, S.K. Antitumor Activity of Vanicoside B Isolated from *Persicaria dissitiflora* by Targeting CDK8 in Triple-Negative Breast Cancer Cells. *J. Nat. Prod.* **2019**, *82*, 3140–3149. [CrossRef] [PubMed]

Huang, X.; Jia, Z.; Li, X.; Hu, Z.; Yu, X.; Xia, J. *Asiaticoside* hampers epithelial–mesenchymal transition by promoting PPARG expression and suppressing P2RX7-mediated TGF-β/Smad signaling in triple-negative breast cancer. *Phytother. Res.* **2023**, *37*, 1771–1786. [CrossRef]

Wang, H.; Wei, L.; Mao, D.; Che, X.; Ye, X.; Liu, Y. Combination of oxymatrine (Om) and astragaloside IV (As) enhances the infiltration and function of TILs in triple-negative breast cancer (TNBC). *Int. Immunopharmacol.* **2023**, *125*, 111026. [CrossRef]

Nagai, M.; Obata, Y.; Takahashi, D.; Hase, K. Fine-tuning of the mucosal barrier and metabolic systems using the diet-microbial metabolite axis. *Int. Immunopharmacol.* **2016**, *37*, 79–86. [CrossRef]

Gao, S.; Zhang, X.; Liu, J.; Ji, F.; Zhang, Z.; Meng, Q.; Zhang, Q.; Han, X.; Wu, H.; Yin, Y.; et al. Icariin Induces Triple-Negative Breast Cancer Cell Apoptosis and Suppresses Invasion by Inhibiting the JNK/c-Jun Signaling Pathway. *Drug Des. Dev. Ther.* **2023**, *17*, 821–836. [CrossRef]

Tang, Z.-H.; Li, T.; Gao, H.-W.; Sun, W.; Chen, X.-P.; Wang, Y.-T.; Lu, J.-J. Platycodin D from *Platycodonis Radix* enhances the anti-proliferative effects of doxorubicin on breast cancer MCF-7 and MDA-MB-231 cells. *Chin. Med.* **2014**, *9*, 16. [CrossRef]

Chun, J.; Kim, Y.S. Platycodin D inhibits migration, invasion, and growth of MDA-MB-231 human breast cancer cells via suppression of EGFR-mediated Akt and MAPK pathways. *Chem.-Biol. Interact.* **2013**, *205*, 212–221. [CrossRef]

Son, J.-A.; Lee, S.K.; Park, J.; Jung, M.J.; An, S.-E.; Yang, H.J.; Son, S.H.; Kim, K.R.; Park, K.-K.; Chung, W.-Y. Platycodin D Inhibits Vascular Endothelial Growth Factor-Induced Angiogenesis by Blocking the Activation of Mitogen-Activated Protein Kinases and the Production of Interleukin-8. *Am. J. Chin. Med.* **2022**, *50*, 1645–1661. [CrossRef]

Hajimehdipoor, H.; Tahmasvand, Z.; Nejad, F.G.; Maresca, M.; Rajabi, S. Rutin Promotes Proliferation and Orchestrates Epithelial–Mesenchymal Transition and Angiogenesis in MCF-7 and MDA-MB-231 Breast Cancer Cells. *Nutrients* **2023**, *15*, 2884. [CrossRef] [PubMed]

Mei, X.; Ouyang, H.; Zhang, H.; Jia, W.; Lu, B.; Zhang, J.; Ji, L. Scutellarin suppresses the metastasis of triple-negative breast cancer via targeting TNFα/TNFR2-RUNX1-triggered G-CSF expression in endothelial cells. *Biochem. Pharmacol.* **2023**, *217*, 115808. [CrossRef] [PubMed]

Mei, X.-Y.; Zhang, J.-N.; Jia, W.-Y.; Lu, B.; Wang, M.-N.; Zhang, T.-Y.; Ji, L.-L. Scutellarin suppresses triple-negative breast cancer metastasis by inhibiting TNFα-induced vascular endothelial barrier breakdown. *Acta Pharmacol. Sin.* **2022**, *43*, 2666–2677. [CrossRef]

Jiang, K.; Lu, Q.; Li, Q.; Ji, Y.; Chen, W.; Xue, X. Astragaloside IV inhibits breast cancer cell invasion by suppressing Vav3 mediated Rac1/MAPK signaling. *Int. Immunopharmacol.* **2017**, *42*, 195–202. [CrossRef]

Zhu, X.; Wang, K.; Chen, Y. Ophiopogonin D suppresses TGF-β1-mediated metastatic behavior of MDA-MB-231 breast carcinoma cells via regulating ITGB1/FAK/Src/AKT/β-catenin/MMP-9 signaling axis. *Toxicol. Vitr.* **2020**, *69*, 104973. [CrossRef]

Elsayed, H.E.; Ebrahim, H.Y.; Mohyeldin, M.M.; Siddique, A.B.; Kamal, A.M.; Haggag, E.G.; El Sayed, K.A. Rutin as A Novel c-Met Inhibitory Lead for The Control of Triple Negative Breast Malignancies. *Nutr. Cancer* **2017**, *69*, 1256–1271. [CrossRef]

57. Xia, J.; Zhang, S.; Zhang, R.; Wang, A.; Zhu, Y.; Dong, M.; Ma, S.; Hong, C.; Liu, S.; Wang, D.; et al. Targeting therapy and tumc microenvironment remodeling of triple-negative breast cancer by ginsenoside Rg3 based liposomes. *J. Nanobiotechnol.* **2022**, 2 414. [CrossRef]
58. Nakhjavani, M.; Palethorpe, H.M.; Tomita, Y.; Smith, E.; Price, T.J.; Yool, A.J.; Pei, J.V. Stereoselective anti-cancer activities c ginsenoside rg3 on triple negative breast cancer cell models. *Pharmaceuticals* **2019**, *12*, 117. [CrossRef]
59. Chen, J.C.; Chang, N.W.; Chung, J.G.; Chen, K.C. Saikosaponin-A induces apoptotic mechanism in human breast MDA-MB-23 and MCF-7 cancer cells. *Am. J. Chin. Med.* **2003**, *31*, 363–377. [CrossRef]
60. Xu, F.; Zang, J.; Chen, D.; Zhang, T.; Zhan, H.; Lu, M.; Zhuge, H. Neohesperidin induces cellular apoptosis in human breas adenocarcinoma MDA-MB-231 cells via activating the Bcl-2/Bax-mediated signaling pathway. *Nat. Prod. Commun.* **2012**, 1475–1478. [CrossRef]
61. Tandrasasmita, O.M.; Lee, J.S.; Baek, S.H.; Tjandrawinata, R.R. Induction of cellular apoptosis in human breast cancer b DLBS1425, a *Phaleria macrocarpa* compound extract, via downregulation of PI3-kinase/AKT pathway. *Cancer Biol. Ther.* **2010**, *1* 814–823. [CrossRef]
62. Wu, H.; Wei, G.; Luo, L.; Li, L.; Gao, Y.; Tan, X.; Wang, S.; Chang, H.; Liu, Y.; Wei, Y.; et al. Ginsenoside Rg3 nanoparticle with permeation enhancing based chitosan derivatives were encapsulated with doxorubicin by thermosensitive hydrogel an anti-cancer evaluation of peritumoral hydrogel injection combined with PD-L1 antibody. *Biomater. Res.* **2022**, *26*, 77. [CrossRef
63. Kim, K.H.; Kim, J.Y.; Kwak, J.H.; Pyo, S. Different anticancer effects of Saxifragifolin A on estrogen receptor-positive and estroge receptor-negative breast cancer cells. *Phytomedicine* **2015**, *22*, 820–828. [CrossRef] [PubMed]
64. Kim, B.-M.; Kim, D.-H.; Park, J.-H.; Na, H.-K.; Surh, Y.-J. Ginsenoside Rg 3 Induces Apoptosis of Human Breast Cance (MDA-MB-231) Cells. *J. Cancer Prev.* **2013**, *18*, 177–185. [CrossRef] [PubMed]
65. Kim, B.-M.; Kim, D.-H.; Park, J.-H.; Surh, Y.-J.; Na, H.-K. Ginsenoside Rg3 Inhibits Constitutive Activation of NF-κB Signalin in Human Breast Cancer (MDA-MB-231) Cells: ERK and Akt as Potential Upstream Targets. *J. Cancer Prev.* **2014**, *19*, 23–3 [CrossRef] [PubMed]
66. Liu, S.; Huang, J.; Gao, F.; Yin, Z.; Zhang, R. Ginsenoside RG1 augments doxorubicin-induced apoptotic cell death in MDA-MB-23 breast cancer cell lines. *J. Biochem. Mol. Toxicol.* **2022**, *36*, e22945. [CrossRef]
67. Gao, Y.; Chu, S.; Shao, Q.; Zhang, M.; Xia, C.; Wang, Y.; Li, Y.; Lou, Y.; Huang, H.; Chen, N. Antioxidant activities of ginsenosid Rg1 against cisplatin-induced hepatic injury through Nrf2 signaling pathway in mice. *Free Radic. Res.* **2017**, *51*, 1–13. [CrossRef
68. Liu, P.; Liu, Y.; Chen, L.; Fan, Z.; Luo, Y.; Cui, Y. Anemoside A3 Inhibits Macrophage M2-Like Polarization to Prevent Triple Negative Breast Cancer Metastasis. *Molecules* **2023**, *28*, 1611. [CrossRef]
69. Chan, R.Y.K.; Chen, W.F.; Dong, A.; Guo, D.; Wong, M.S. Estrogen-like activity of ginsenoside Rg1 derived from *Panax notoginsen* *J. Clin. Endocrinol. Metab.* **2002**, *87*, 3691–3695. [CrossRef]
70. Fu, R.; Zhang, L.; Li, Y.; Li, B.; Ming, Y.; Li, Z.; Xing, H.; Chen, J. Saikosaponin D inhibits autophagosome-lysosome fusion anc induces autophagy-independent apoptosis in MDA-MB-231 breast cancer cells. *Mol. Med. Rep.* **2020**, *22*, 1026–1034. [CrossRef]
71. Ma, H.; Yue, G.G.; Lee, J.K.-M.; Gao, S.; Yuen, K.-K.; Cheng, W.; Li, X.; Lau, C.B. Scutellarin, a flavonoid compound fron *Scutellaria barbata*, suppresses growth of breast cancer stem cells in vitro and in tumor-bearing mice. *Phytomedicine* **2024**, *12* 155418. [CrossRef] [PubMed]
72. Foss, K.; Przybyłowicz, K.E.; Sawicki, T. Antioxidant Activity and Profile of Phenolic Compounds in Selected Herbal Plants *Plant Foods Hum. Nutr.* **2022**, *77*, 383–389. [CrossRef] [PubMed]
73. Shi, L.; Zhao, W.; Yang, Z.; Subbiah, V.; Suleria, H.A.R. Extraction and characterization of phenolic compounds and their potentia antioxidant activities. *Environ. Sci. Pollut. Res.* **2022**, *29*, 81112–81129. [CrossRef]
74. Singh, B.; Singh, J.P.; Kaur, A.; Singh, N. Phenolic composition, antioxidant potential and health benefits of citrus peel. *Food Res Int.* **2020**, *132*, 109114. [CrossRef]
75. Jung, K.; Everson, R.J.C.; Joshi, B.; Bulsara, P.A.; Upasani, R.; Clarke, M.J. Structure-function relationship of phenolic antioxidants in topical skin health products. *Int. J. Cosmet. Sci.* **2017**, *39*, 217–223. [CrossRef]
76. Peng, F.; Xiong, L.; Peng, C. (-)-Sativan Inhibits Tumor Development and Regulates miR-200c/PD-L1 in Triple Negative Breas Cancer Cells. *Front. Pharmacol.* **2020**, *11*, 251. [CrossRef]
77. Peng, F.; Tang, H.; Du, J.; Chen, J.; Peng, C. Isoliquiritigenin Suppresses EMT-Induced Metastasis in Triple-Negative Breast Cance through miR-200c/C-JUN/β-Catenin. *Am. J. Chin. Med.* **2021**, *49*, 505–523. [CrossRef]
78. Pai, J.T.; Chen, X.H.; Leu, Y.L.; Weng, M.S. Propolin G-Suppressed Epithelial-to-Mesenchymal Transition in Triple-Negative Breas Cancer Cells via Glycogen Synthase Kinase 3β-Mediated Snail and HDAC6-Regulated Vimentin Degradation. *Int. J. Mol. Sci* **2022**, *23*, 1672. [CrossRef]
79. Koh, S.Y.; Moon, J.Y.; Unno, T.; Cho, S.K. Baicalein suppresses stem cell-like characteristics in radio-and chemoresistant MDA MB-231 human breast cancer cells through up-regulation of IFIT2. *Nutrients* **2019**, *11*, 624. [CrossRef]

Ebrahim, H.Y.; Elsayed, H.E.; Mohyeldin, M.M.; Akl, M.R.; Bhattacharjee, J.; Egbert, S.; El Sayed, K.A. Norstictic Acid Inhibits Breast Cancer Cell Proliferation, Migration, Invasion, and In Vivo Invasive Growth Through Targeting C-Met. *Phytother. Res.* **2016**, *30*, 557–566. [CrossRef]

Yu, S.; Wang, Z.; Su, Z.; Song, J.; Zhou, L.; Sun, Q.; Liu, S.; Li, S.; Li, Y.; Wang, M.; et al. Gigantol inhibits Wnt/β-catenin signaling and exhibits anticancer activity in breast cancer cells. *BMC Complement. Med. Ther.* **2018**, *18*, 59. [CrossRef] [PubMed]

Zou, G.; Zhang, X.; Wang, L.; Li, X.; Xie, T.; Zhao, J.; Yan, J.; Wang, L.; Ye, H.; Jiao, S.; et al. Herb-sourced emodin inhibits angiogenesis of breast cancer by targeting VEGFA transcription. *Theranostics* **2020**, *10*, 6839–6853. [CrossRef] [PubMed]

Abdellatef, A.A.; Fathy, M.; Mohammed, A.E.-S.I.; Bakr, M.S.A.; Ahmed, A.H.; Abbass, H.S.; El-Desoky, A.H.; Morita, H.; Nikaido, T.; Hayakawa, Y. Inhibition of cell-intrinsic NF-κB activity and metastatic abilities of breast cancer by aloe-emodin and emodic-acid isolated from *Asphodelus microcarpus*. *J. Nat. Med.* **2021**, *75*, 840–853. [CrossRef] [PubMed]

Yu, Y.; Cai, W.; Pei, C.G.; Shao, Y. Rhamnazin, a novel inhibitor of VEGFR2 signaling with potent antiangiogenic activity and antitumor efficacy. *Biochem. Biophys. Res. Commun.* **2015**, *458*, 913–919. [CrossRef]

Yeon Park, J.; Young Kim, H.; Shibamoto, T.; Su Jang, T.; Cheon Lee, S.; Suk Shim, J.; Hahm, D.-H.; Lee, H.-J.; Lee, S.; Sung Kang, K. Beneficial effects of a medicinal herb, *Cirsium japonicum* var. *maackii*, extract and its major component, cirsimaritin on breast cancer metastasis in MDA-MB-231 breast cancer cells. *Bioorg. Med. Chem. Lett.* **2017**, *27*, 3968–3973. [CrossRef]

Shi, Y.; Wu, Y.; Li, F.; Zhang, Y.; Hua, C.; Yang, J.; Zheng, J.; Chen, L.; Wei, Z.; Yue, H.; et al. Identifying the anti-metastasis effect of Anhydroicaritin on breast cancer: Coupling network pharmacology with experimental validation. *J. Ethnopharmacol.* **2022**, *293*, 115326. [CrossRef]

Sun, C.; Wang, T.; Wang, C.; Zhu, Z.; Wang, X.; Xu, J.; An, H. The Protective Effect of Kaempferol Against Ischemia/Reperfusion Injury Through Activating SIRT3 to Inhibit Oxidative Stress. *Braz. J. Cardiovasc. Surg.* **2022**, *37*, 335–342. [CrossRef]

Maharjan, S.; Lee, M.G.; Kim, S.Y.; Lee, K.S.; Nam, K.S. Morin Sensitizes MDA-MB-231 Triple-Negative Breast Cancer Cells to Doxorubicin Cytotoxicity by Suppressing FOXM1 and Attenuating EGFR/STAT3 Signaling Pathways. *Pharmaceuticals* **2023**, *16*, 672. [CrossRef]

Hua, F.; Xiao, Y.-Y.; Qu, X.-H.; Li, S.-S.; Zhang, K.; Zhou, C.; He, J.-L.; Zhu, Y.; Wan, Y.-Y.; Jiang, L.-P.; et al. Baicalein sensitizes triple negative breast cancer MDA-MB-231 cells to doxorubicin via autophagy-mediated down-regulation of CDK1. *Mol. Cell. Biochem.* **2023**, *478*, 1519–1531. [CrossRef]

Liu, Q.; Li, J.; Pu, G.; Zhang, F.; Liu, H.; Zhang, Y. Co-delivery of baicalein and doxorubicin by hyaluronic acid decorated nanostructured lipid carriers for breast cancer therapy. *Drug Deliv.* **2016**, *23*, 1364–1368. [CrossRef]

Pérez-Gutiérrez, L.; Ferrara, N. Biology and therapeutic targeting of vascular endothelial growth factor A. *Nat. Rev. Mol. Cell Biol.* **2023**, *24*, 816–834. [CrossRef] [PubMed]

Takahashi, H.; Shibuya, M. The vascular endothelial growth factor (VEGF)/VEGF receptor system and its role under physiological and pathological conditions. *Clin. Sci.* **2005**, *109*, 227–241. [CrossRef] [PubMed]

Du, Y.; Khan, M.; Fang, N.; Ma, F.; Du, H.; Tan, Z.; Wang, H.; Yin, S.; Wei, X. Berberine Attenuates Cell Motility via Inhibiting Inflammation-Mediated Lysyl Hydroxylase-2 and Glycolysis. *Front. Pharmacol.* **2022**, *13*, 856777. [CrossRef] [PubMed]

Xing, Z.; Zhang, Y.; Zhang, X.; Yang, Y.; Ma, Y.; Pang, D. Fangchinoline induces g1 arrest in breast cancer cells through cell-cycle regulation. *Phytother. Res.* **2013**, *27*, 1790–1794. [CrossRef]

Wang, C.-D.; Yuan, C.-F.; Bu, Y.-Q.; Wu, X.-M.; Wan, J.-Y.; Zhang, L.; Hu, N.; Liu, X.-J.; Zu, Y.; Liu, G.-L.; et al. Fangchinoline inhibits cell proliferation via Akt/GSK-3beta/cyclin D1 signaling and induces apoptosis in MDA-MB-231 breast cancer cells. *Asian Pac. J. Cancer Prev.* **2014**, *15*, 769–773. [CrossRef]

Wei, S.; Zhang, Y.; Ma, X.; Yao, Y.; Zhou, Q.; Zhang, W.; Zhou, C.; Zhuang, J. MAT as a promising therapeutic strategy against triple-negative breast cancer via inhibiting PI3K/AKT pathway. *Sci. Rep.* **2023**, *13*, 12351. [CrossRef]

Guo, Q.; Yu, Y.; Tang, W.; Zhou, S.; Lv, X. Matrine exerts an anti-tumor effect via regulating HN1 in triple breast cancer both in vitro and in vivo. *Chem. Biol. Drug Des.* **2023**, *102*, 1469–1477. [CrossRef]

Guzmán, E.A.; Peterson, T.A.; Wright, A.E. The Marine Natural Compound Dragmacidin D Selectively Induces Apoptosis in Triple-Negative Breast Cancer Spheroids. *Mar. Drugs* **2023**, *21*, 642. [CrossRef]

Peng, F.; Xiong, L.; Xie, X.; Tang, H.; Huang, R.; Peng, C. Isoliquiritigenin Derivative Regulates miR-374a/BAX Axis to Suppress Triple-Negative Breast Cancer Tumorigenesis and Development. *Front. Pharmacol.* **2020**, *11*, 378. [CrossRef]

0. Maharjan, S.; Kwon, Y.S.; Lee, M.G.; Lee, K.S.; Nam, K.S. Cell cycle arrest-mediated cell death by morin in MDA-MB-231 triple-negative breast cancer cells. *Pharmacol. Rep.* **2021**, *73*, 1315–1327. [CrossRef]

1. Jin, H.; Lee, W.S.; Eun, S.Y.; Jung, J.H.; Park, H.-S.; Kim, G.; Choi, Y.H.; Ryu, C.H.; Jung, J.M.; Hong, S.C.; et al. Morin, a flavonoid from Moraceae, suppresses growth and invasion of the highly metastatic breast cancer cell line MDA-MB-231 partly through suppression of the Akt pathway. *Int. J. Oncol.* **2014**, *45*, 1629–1637. [CrossRef] [PubMed]

2. Yan, W.; Ma, X.; Zhao, X.; Zhang, S. Baicalein induces apoptosis and autophagy of breast cancer cells via inhibiting PI3K/AKT pathway in vivo and vitro. *Drug Des. Dev. Ther.* **2018**, *12*, 3961–3972. [CrossRef] [PubMed]

103. Yao, P.; Yao, P.; Ku, X.; Yang, J. Harmine suppresses the malignant phenotypes and PI3K activity in breast cancer. *Anticancer Dru* **2023**, *34*, 373–383. [CrossRef] [PubMed]
104. Al-Sayed, E.; Ke, T.-Y.; Hwang, T.-L.; Chen, S.-R.; Korinek, M.; Chen, S.-L.; Cheng, Y.-B. Cytotoxic and Anti-Inflammatory Effec of Lignans and Diterpenes from Cupressus Macrocarpa. *Bioorg. Med. Chem. Lett.* **2020**, *30*, 127127. [CrossRef]
105. Bommareddy, A.; Knapp, K.; Nemeth, A.; Steigerwalt, J.; Landis, T.; Vanwert, A.L.; Gorijavolu, H.P.; Dwivedi, C. Alpha-Santal a Component of Sandalwood Oil Inhibits Migration of Breast Cancer Cells by Targeting the β-Catenin Pathway. *Anticancer R* **2018**, *38*, 4475–4480. [CrossRef]
106. Mbaveng, A.T.; Bitchagno, G.T.M.; Kuete, V.; Tane, P.; Efferth, T. Cytotoxicity of ungeremine towards multi-factorial drug resista cancer cells and induction of apoptosis, ferroptosis, necroptosis and autophagy. *Phytomedicine* **2019**, *60*, 152832. [CrossRef]
107. Zhang, Y.; Gao, X.; Liu, C.; Wang, M.; Zhang, R.; Sun, J.; Liu, Y. Design, synthesis and in vitro evaluation of fangchinolir derivatives as potential anticancer agents. *Bioorg. Chem.* **2020**, *94*, 103431. [CrossRef]
108. He, K.; Gao, J.L. Protopine inhibits heterotypic cell adhesion in MDA-MB-231 cells through down-regulation of multi-adhesi factors. *Afr. J. Tradit. Complement. Altern. Med. AJTCAM/Afr. Netw. Ethnomed.* **2014**, *11*, 415–424. [CrossRef]
109. Shaala, L.A.; Youssef, D.T.A.; Sulaiman, M.; Behery, F.A.; Foudah, A.I.; El Sayed, K.A. Subereamolline A as a potent breast canc migration, invasion and proliferation inhibitor and bioactive dibrominated alkaloids from the red sea sponge *Pseudoceratir arabica*. *Mar. Drugs* **2012**, *10*, 2492–2508. [CrossRef]
110. Zhang, Y.; Chen, J.; Mi, D.; Ling, J.; Li, H.; He, P.; Liu, N.; Chen, Q.; Chen, Y.; Huang, L. Discovery of YH677 as a canc stemness inhibitor that suppresses triple-negative breast cancer growth and metastasis by regulating the TGFβ signaling pathwa *Cancer Lett.* **2023**, *560*, 216142. [CrossRef]
111. Qi, M.; Liu, X.; Zhou, Y.; Wang, H.; Zhao, Y.; Ren, J.; Xiang, J. Berberine inhibits mda-mb-231 cells as an agonist of g protein-couple estrogen receptor 1. *Int. J. Mol. Sci.* **2021**, *22*, 11466. [CrossRef] [PubMed]
112. Mabeta, P.; Steenkamp, V. The VEGF/VEGFR Axis Revisited: Implications for Cancer Therapy. *Int. J. Mol. Sci.* **2022**, *23*, 1558 [CrossRef] [PubMed]
113. Li, Y.L.; Zhao, H.; Ren, X.B. Relationship of VEGF/VEGFR with immune and cancer cells: Staggering or forward? *Cancer Bic Med.* **2016**, *13*, 206–214. [CrossRef]
114. Dakowicz, D.; Zajkowska, M.; Mroczko, B. Relationship between VEGF Family Members, Their Receptors and Cell Death in th Neoplastic Transformation of Colorectal Cancer. *Int. J. Mol. Sci.* **2022**, *23*, 3375. [CrossRef] [PubMed]
115. Kang, Y.; Li, H.; Liu, Y.; Li, Z. Regulation of VEGF-A expression and VEGF-A-targeted therapy in malignant tumors. *J. Cancer Re Clin. Oncol.* **2024**, *150*, 221. [CrossRef]
116. Elebiyo, T.C.; Rotimi, D.; Evbuomwan, I.O.; Maimako, R.F.; Iyobhebhe, M.; Ojo, O.A.; Oluba, O.M.; Adeyemi, O.S. Reassessin vascular endothelial growth factor (VEGF) in anti-angiogenic cancer therapy. *Cancer Treat. Res. Commun.* **2022**, *32*, 10062 [CrossRef]
117. Peng, W.; Zhang, H.; Yin, M.; Kong, D.; Kang, L.; Teng, X.; Wang, J.; Chu, Z.; Sun, Y.; Long, P.; et al. Combined Inhibition of PI3 and STAT3 signaling effectively inhibits bladder cancer growth. *Oncogenesis* **2024**, *13*, 29. [CrossRef]
118. Glaviano, A.; Foo, A.S.C.; Lam, H.Y.; Yap, K.C.H.; Jacot, W.; Jones, R.H.; Eng, H.; Nair, M.G.; Makvan-di, P.; Geoerger, B.; et a PI3K/AKT/mTOR signaling transduction pathway and targeted therapies in cancer. *Mol. Cancer* **2023**, *22*, 138. [CrossRef]
119. Liu, R.; Chen, Y.; Liu, G.; Li, C.; Song, Y.; Cao, Z.; Li, W.; Hu, J.; Lu, C.; Liu, Y. PI3K/AKT pathway as a key link modulates th multidrug resistance of cancers. *Cell Death Dis.* **2020**, *11*, 797. [CrossRef]
120. He, Y.; Sun, M.M.; Zhang, G.G.; Yang, J.; Chen, K.S.; Xu, W.W.; Li, B. Targeting PI3K/Akt signal transduction for cancer therap *Signal Transduct. Target. Ther.* **2021**, *6*, 425. [CrossRef]
121. Davies, M.A. The role of the PI3K-AKT pathway in melanoma. *Cancer J.* **2012**, *18*, 142–147. [CrossRef] [PubMed]
122. Liu, J.; Xiao, Q.; Xiao, J.; Niu, C.; Li, Y.; Zhang, X.; Zhou, Z.; Shu, G.; Yin, G. Wnt/β-catenin signalling: Function, biologica mechanisms, and therapeutic opportunities. *Signal Transduct. Target. Ther.* **2022**, *7*, 3. [CrossRef] [PubMed]
123. Yu, F.; Yu, C.; Li, F.; Zuo, Y.; Wang, Y.; Yao, L.; Wu, C.; Wang, C.; Ye, L. Wnt/β-catenin signaling in cancers and targeted therapie *Signal Transduct. Target. Ther.* **2021**, *6*, 307. [CrossRef]
124. Cargnello, M.; Roux, P.P. Activation and Function of the MAPKs and Their Substrates, the MAPK-Activated Protein Kinase *Microbiol. Mol. Biol. Rev.* **2011**, *75*, 50–83. [CrossRef]
125. Rovida, E.; Tusa, I. Targeting MAPK in Cancer 2.0. *Int. J. Mol. Sci.* **2022**, *23*, 5702. [CrossRef]
126. Ramanan, M.; Sinha, S.; Sudarshan, K.; Aidhen, I.S.; Doble, M. Inhibition of the Enzymes in the Leukotriene and Prostaglandi Pathways in Inflammation by 3-Aryl Isocoumarins. *Eur. J. Med. Chem.* **2016**, *124*, 428–434. [CrossRef]
127. Sudarshan, K.; Boda, A.K.; Dogra, S.; Bose, I.; Yadav, P.N.; Aidhen, I.S. Discovery of an Isocoumarin Analogue That Modulate Neuronal Functions via Neurotrophin Receptor TrkB. *Bioorg. Med. Chem. Lett.* **2019**, *29*, 585–590. [CrossRef]
128. Han, H.S.; Vikas, P.; Costa, R.L.B.; Jahan, N.; Taye, A.; Stringer-Reasor, E.M. Early-Stage Triple-Negative Breast Cancer Journey Beginning, End, and Everything in Between. *Am. Soc. Clin. Oncol. Educ. Book* **2023**, *43*, e390464. [CrossRef]

29. Pusztai, L.; Yau, C.; Wolf, D.M.; Han, H.S.; Du, L.; Wallace, A.M.; String-Reasor, E.; Boughey, J.C.; Chien, A.J.; Elias, A.D.; et al. Durvalumab with Olaparib and Paclitaxel for High-Risk HER2-Negative Stage II/III Breast Cancer: Results from the Adaptively Randomized I-SPY2 Trial. *Cancer Cell* **2021**, *39*, 989–998.e5. [CrossRef]
30. LaFargue, C.J.; Dal Molin, G.Z.; Sood, A.K.; Coleman, R.L. Exploring and Comparing Adverse Events between PARP Inhibitors. *Lancet Oncol.* **2019**, *20*, e15–e28. [CrossRef]
31. Jin, Y.; Huang, Z.; Li, L.; Yang, Y.; Wang, C.; Wang, Z.; Ji, L. Quercetin Attenuates Toosendanin-Induced Hepatotoxicity through Inducing the Nrf2/GCL/GSH Antioxidant Signaling Pathway. *Acta Pharmacol. Sin.* **2019**, *40*, 75–85. [CrossRef]
32. Lu, X.; Ji, C.; Tong, W.; Lian, X.; Wu, Y.; Fan, X.; Gao, Y. Integrated Analysis of microRNA and mRNA Expression Profiles Highlights the Complex and Dynamic Behavior of Toosendanin-Induced Liver Injury in Mice. *Sci. Rep.* **2016**, *6*, 34225. [CrossRef]
33. Thant, M.T.; Bhummaphan, N.; Wuttiin, J.; Puttipanyalears, C.; Chaichompoo, W.; Rojsitthisak, P.; Punpreuk, Y.; Böttcher, C.; Likhitwitayawuid, K.; Sritularak, B. New Phenolic Glycosides from *Coelogyne fuscescens* Lindl. Var. *brunnea* and Their Cytotoxicity against Human Breast Cancer Cells. *ACS Omega* **2024**, *9*, 7679–7691. [CrossRef] [PubMed]
34. Chakravarti, B.; Maurya, R.; Siddiqui, J.A.; Bid, H.K.; Rajendran, S.M.; Yadav, P.P.; Konwar, R. In Vitro Anti-Breast Cancer Activity of Ethanolic Extract of Wrightia Tomentosa: Role of pro-Apoptotic Effects of Oleanolic Acid and Urosolic Acid. *J. Ethnopharmacol.* **2012**, *142*, 72–79. [CrossRef]
35. Chen, D.-L.; Ma, G.-X.; Yang, E.-L.; Yang, Y.; Wang, C.-H.; Sun, Z.-C.; Liang, H.-Q.; Xu, X.-D.; Wei, J.-H. Cadinane-Type Sesquiterpenoid Dimeric Diastereomers Hibisceusones A-C from Infected Stems of Hibiscus Tiliaceus with Cytotoxic Activity against Triple-Negative Breast Cancer Cells. *Bioorganic Chem.* **2022**, *127*, 105982. [CrossRef] [PubMed]
36. Cheng, L.; Xia, T.-S.; Shi, L.; Xu, L.; Chen, W.; Zhu, Y.; Ding, Q. D Rhamnose β-Hederin Inhibits Migration and Invasion of Human Breast Cancer Cell Line MDA-MB-231. *Biochem. Biophys. Res.* **2018**, *495*, 775–780. [CrossRef] [PubMed]
37. Chun, J.; Li, R.-J.; Cheng, M.-S.; Kim, Y.S. Alantolactone Selectively Suppresses STAT3 Activation and Exhibits Potent Anticancer Activity in MDA-MB-231 Cells. *Cancer Lett.* **2015**, *357*, 393–403. [CrossRef]
38. Chun, J.; Song, K.; Kim, Y.S. Sesquiterpene Lactones-enriched Fraction of *Inula helenium* L. Induces Apoptosis through Inhibition of Signal Transducers and Activators of Transcription 3 Signaling Pathway in MDA-MB-231 Breast Cancer Cells. *Phytother. Res.* **2018**, *32*, 2501–2509. [CrossRef]
39. Güney Eskiler, G.; Deveci Özkan, A.; Kaleli, S.; Bilir, C. Inhibition of TLR4/TRIF/IRF3 Signaling Pathway by Curcumin in Breast Cancer Cells. *J. Pharm. Pharm. Sci.* **2019**, *22*, 281–291. [CrossRef]
40. Dawood, M.; Ooko, E.; Efferth, T. Collateral Sensitivity of Parthenolide via NF-κB and HIF-α Inhibition and Epigenetic Changes in Drug-Resistant Cancer Cell Lines. *Front. Pharmacol.* **2019**, *10*, 542. [CrossRef]
41. Gou, L.; Yue, G.G.-L.; Lee, J.K.-M.; Puno, P.T.; Lau, C.B.-S. Natural Product Eriocalyxin B Suppressed Triple Negative Breast Cancer Metastasis Both in Vitro and in Vivo. *Biochem. Pharmacol.* **2023**, *210*, 115491. [CrossRef] [PubMed]
42. Jang, H.; Ko, H.; Song, K.; Kim, Y. A Sesquiterpenoid from Farfarae Flos Induces Apoptosis of MDA-MB-231 Human Breast Cancer Cells through Inhibition of JAK–STAT3 Signaling. *Biomolecules* **2019**, *9*, 278. [CrossRef] [PubMed]
43. Liu, L.-Y.; Yan, Z.; Kang, J.; Chen, R.-Y.; Yu, D.-Q. Three New Triterpenoids from *Ganoderma theaecolum*. *J. Asian Nat. Prod. Res.* **2017**, *19*, 847–853. [CrossRef]
44. Park, J.-H.; Yoon, J.; Park, B. Pomolic Acid Suppresses HIF1α/VEGF-Mediated Angiogenesis by Targeting P38-MAPK and mTOR Signaling Cascades. *Phytomedicine* **2016**, *23*, 1716–1726. [CrossRef]
45. Qiao, A.; Wang, Y.; Xiang, L.; Wang, C.; He, X. A Novel Triterpenoid Isolated from Apple Functions as an Anti-Mammary Tumor Agent via a Mitochondrial and Caspase-Independent Apoptosis Pathway. *J. Agric. Food Chem.* **2015**, *63*, 185–191. [CrossRef]
46. Sathya, S.; Sudhagar, S.; Vidhya Priya, M.; Bharathi Raja, R.; Muthusamy, V.S.; Niranjali Devaraj, S.; Lakshmi, B.S. 3β-Hydroxylup-20(29)-Ene-27,28-Dioic Acid Dimethyl Ester, a Novel Natural Product from Plumbago Zeylanica Inhibits the Proliferation and Migration of MDA-MB-231 Cells. *Chem. Biol. Interact.* **2010**, *188*, 412–420. [CrossRef]
47. Shin, M.; Lee, B.-M.; Kim, O.; Tran, H.N.K.; Lee, S.; Hwangbo, C.; Min, B.-S.; Lee, J.-H. Triterpenoids from *Ziziphus jujuba* Induce Apoptotic Cell Death in Human Cancer Cells through Mitochondrial Reactive Oxygen Species Production. *Food Funct.* **2018**, *9*, 3895–3905. [CrossRef]
48. Huang, W.-C.; Yen, J.-H.; Sung, Y.-W.; Tung, S.-L.; Chen, P.-M.; Chu, P.-Y.; Shih, Y.-C.; Chi, H.-C.; Huang, Y.-C.; Huang, S.-J.; et al. Novel Function of THEMIS2 in the Enhancement of Cancer Stemness and Chemoresistance by Releasing PTP1B from MET. *Oncogene* **2022**, *41*, 997–1010. [CrossRef]
49. Yen, J.-H.; Huang, W.-C.; Lin, S.-C.; Huang, Y.-W.; Chio, W.-T.; Tsay, G.J.; Hung, M.-C.; Huang, S.-T. Metabolic Remodeling in Tumor-Associated Macrophages Contributing to Antitumor Activity of Cryptotanshinone by Regulating TRAF6-ASK1 Axis. *Mol. Ther. Oncolytics* **2022**, *26*, 158–174. [CrossRef]
50. Xia, Y.; Zhang, Y.; Chen, L.; Min, L.; Huang, D.; Zhang, Y.; Li, C.; Li, Z. Suppression of Migration and Invasion by Taraxerol in the Triple-Negative Breast Cancer Cell Line MDA-MB-231 via the ERK/Slug Axis. *PLoS ONE* **2023**, *18*, e0291693. [CrossRef]

151. Kashyap, A.; Umar, S.; Dev J R, A.; Prasad, C. Dihydrotanshinone-I Modulates Epithelial Mesenchymal Transition (EMT) Thereb Impairing Migration and Clonogenicity of Triple Negative Breast Cancer Cells. *Asian Pac. J. Cancer Prev.* **2021**, *22*, 2177–218 [CrossRef] [PubMed]
152. Aumsuwan, P.; Khan, S.I.; Khan, I.A.; Ali, Z.; Avula, B.; Walker, L.A.; Shariat-Madar, Z.; Helferich, W.G.; Katzenellenbogen, B.S Dasmahapatra, A.K. The Anticancer Potential of Steroidal Saponin, Dioscin, Isolated from Wild Yam (*Dioscorea villosa*) Roc Extract in Invasive Human Breast Cancer Cell Line MDA-MB-231 In Vitro. *Arch. Biochem. Biophys.* **2016**, *591*, 98–110. [CrossRe [PubMed]
153. Christodoulou, P.; Boutsikos, P.; Neophytou, C.M.; Kyriakou, T.-C.; Christodoulou, M.-I.; Papageorgis, P.; Stephanou, A.; Patrikio I. Amygdalin as a Chemoprotective Agent in Co-Treatment with Cisplatin. *Front. Pharmacol.* **2022**, *13*, 1013692. [CrossRe [PubMed]
154. He, Z.; Chen, H.; Li, G.; Zhu, H.; Gao, Y.; Zhang, L.; Sun, J. Diosgenin Inhibits the Migration of Human Breast Cancer MDA-MB-23 Cells by Suppressing Vav2 Activity. *Phytomedicine* **2014**, *21*, 871–876. [CrossRef]
155. Ho, D.V.; Hoang, H.N.T.; Vo, H.Q.; Nguyen, K.V.; Pham, T.V.; Le, A.T.; Van Phan, K.; Nguyen, H.M.; Morita, H.; Nguye H.T. Three New Steroidal Saponins from *Aspidistra letreae* Plants and Their Cytotoxic Activities. *J. Nat. Med.* **2020**, *74*, 591–59 [CrossRef]
156. Kwak, J.H.; Park, J.Y.; Lee, D.; Kwak, J.Y.; Park, E.H.; Kim, K.H.; Park, H.-J.; Kim, H.Y.; Jang, H.J.; Ham, J.; et al. Inhibitory Effec of *Ginseng sapogenins* on the Proliferation of Triple Negative Breast Cancer MDA-MB-231 Cells. *Bioorg. Med. Chem. Lett.* **2014**, 2 5409–5412. [CrossRef]
157. Pederson, P.J.; Cai, S.; Carver, C.; Powell, D.R.; Risinger, A.L.; Grkovic, T.; O'Keefe, B.R.; Mooberry, S.L.; Cichewicz, R.H Triple-Negative Breast Cancer Cells Exhibit Differential Sensitivity to Cardenolides from *Calotropis gigantea*. *J. Nat. Prod.* **2020**, *8* 2269–2280. [CrossRef]
158. Xu, H.; Li, Y.; Han, B.; Li, Z.; Wang, B.; Jiang, P.; Zhang, J.; Ma, W.; Zhou, D.; Li, X.; et al. Anti-Breast-Cancer Activity Exerte by β-Sitosterol-D-Glucoside from Sweet Potato via Upregulation of MicroRNA-10a and via the PI3K–Akt Signaling Pathwa *J. Agric. Food Chem.* **2018**, *66*, 9704–9718. [CrossRef]
159. Pei, S.; Yang, X.; Wang, H.; Zhang, H.; Zhou, B.; Zhang, D.; Lin, D. Plantamajoside, a Potential Anti-Tumor Herbal Medicin Inhibits Breast Cancer Growth and Pulmonary Metastasis by Decreasing the Activity of Matrix Metalloproteinase-9 and -2. *BM Cancer* **2015**, *15*, 965. [CrossRef]
160. Tan, H.; Li, M.; Han, L.; Zhao, Y.; Zhang, X. Gypensapogenin I Suppresses Cell Proliferation in Triple-Negative Breast Cancer Vi Triggering the Closure of AKT/GSK3β/β-Catenin and Notch-1 Signaling Pathways. *J. Agric. Food Chem.* **2022**, *70*, 5438–544 [CrossRef]
161. Zhou, Y.; Fu, X.; Guan, Y.; Gong, M.; He, K.; Huang, B. 1,3-Dicaffeoylquinic Acid Targeting 14-3-3 Tau Suppresses Human Breas Cancer Cell Proliferation and Metastasis through IL6/JAK2/PI3K Pathway. *Bioorg. Med. Chem. Lett.* **2020**, *172*, 113752. [CrossRef [PubMed]
162. Yang, J.; Su, J.-C.; Lei, X.-P.; Huang, X.-J.; Zhang, D.-M.; Ye, W.-C.; Wang, Y. Acylphloroglucinol Derivatives from the Leaves o *Syzygium samarangense* and Their Cytotoxic Activities. *Fitoterapia* **2018**, *129*, 1–6. [CrossRef] [PubMed]
163. Sunilkumar, D.; Drishya, G.; Chandrasekharan, A.; Shaji, S.K.; Bose, C.; Jossart, J.; Perry, J.J.P.; Mishra, N.; Kumar, G.B.; Nai B.G. Oxyresveratrol Drives Caspase-Independent Apoptosis-like Cell Death in MDA-MB-231 Breast Cancer Cells through th Induction of ROS. *Biochem. Pharmacol.* **2020**, *173*, 113724. [CrossRef]
164. Sun, L.-P.; Chen, A.-L.; Hung, H.-C.; Chien, Y.-H.; Huang, J.-S.; Huang, C.-Y.; Chen, Y.-W.; Chen, C.-N. Chrysin: A Histon Deacetylase 8 Inhibitor with Anticancer Activity and a Suitable Candidate for the Standardization of Chinese Propolis. *J. Agri Food Chem.* **2012**, *60*, 11748–11758. [CrossRef]
165. Nguyen, P.H.; Sharma, G.; Dao, T.T.; Uddin, M.N.; Kang, K.W.; Ndinteh, D.T.; Mbafor, J.T.; Oh, W.K. New Prenylated Isoflavonoid as Protein Tyrosine Phosphatase 1B (PTP1B) Inhibitors from *Erythrina addisoniae*. *Bioorg. Med. Chem.* **2012**, *20*, 6459–646 [CrossRef]
166. Liu, F.; Lu, W.-J.; Li, N.-P.; Liu, J.-W.; He, J.; Ye, W.-C.; Wang, L. Four New Cinnamoyl-Phloroglucinols from the Leaves o *Xanthostemon chrysanthus*. *Fitoterapia* **2018**, *128*, 93–96. [CrossRef]
167. Lee, M.-F.; Pan, M.-H.; Chiou, Y.-S.; Cheng, A.-C.; Huang, H. Resveratrol Modulates MED28 (Magicin/EG-1) Expression an Inhibits Epidermal Growth Factor (EGF)-Induced Migration in MDA-MB-231 Human Breast Cancer Cells. *J. Agric. Food Chem* **2011**, *59*, 11853–11861. [CrossRef]
168. Lee, S.-J.; Hong, S.; Yoo, S.-H.; Kim, G.-W. Cyanidin-3-O-Sambubioside from Acanthopanax Sessiliflorus Fruit Inhibits Metastasi by Downregulating MMP-9 in Breast Cancer Cells MDA-MB-231. *Planta Med.* **2013**, *79*, 1636–1640. [CrossRef]
169. Gao, Y.; Tollefsbol, T.O. Combinational Proanthocyanidins and Resveratrol Synergistically Inhibit Human Breast Cancer Cell and Impact Epigenetic–Mediating Machinery. *Int. J. Mol. Sci.* **2018**, *19*, 2204. [CrossRef]

). El Gaafary, M.; Saber, F.R.; Mahrous, E.A.; Ashour, R.M.; Okba, M.M.; Jin, L.; Lang, S.J.; Schmiech, M.; Simmet, T.; Syrovets, T. The Phloroglucinol Calcitrinone A, a Novel Mitochondria-Targeting Agent, Induces Cell Death in Breast Cancer Cells. *Food Chem. Toxicol.* **2022**, *162*, 112896. [CrossRef]

1. Bajpai, V.K.; Alam, M.B.; Quan, K.T.; Choi, H.-J.; An, H.; Ju, M.-K.; Lee, S.-H.; Huh, Y.S.; Han, Y.-K.; Na, M. Cytotoxic Properties of the Anthraquinone Derivatives Isolated from the Roots of *Rubia philippinensis*. *BMC Complement. Altern. Med.* **2018**, *18*, 200. [CrossRef] [PubMed]

2. Arunasree, K.M. Anti-Proliferative Effects of Carvacrol on a Human Metastatic Breast Cancer Cell Line, MDA-MB 231. *Phytomedicine* **2010**, *17*, 581–588. [CrossRef]

3. Bello, I.; Smimmo, M.; d'Emmanuele Di Villa Bianca, R.; Bucci, M.; Cirino, G.; Panza, E.; Brancaleone, V. Erucin, an H2S-Releasing Isothiocyanate, Exerts Anticancer Effects in Human Triple-Negative Breast Cancer Cells Triggering Autophagy-Dependent Apoptotic Cell Death. *Int. J. Mol. Sci.* **2023**, *24*, 6764. [CrossRef] [PubMed]

4. Chang, F.-R.; Yen, C.-T.; El-Shazly, M.; Yu, C.-Y.; Yen, M.-H.; Cheng, Y.-B.; Chen, S.-L.; Wu, Y.-C. Spirostanoids with 1,4-Dien-3-One or 3β,7α-Diol-5,6-Ene Moieties from *Solanum violaceum*. *Bioorg. Med. Chem. Lett.* **2013**, *23*, 2738–2742. [CrossRef] [PubMed]

5. Li, X.; Wu, Q.; Xie, Y.; Ding, Y.; Du, W.W.; Sdiri, M.; Yang, B.B. Ergosterol Purified from Medicinal Mushroom *Amauroderma rude* Inhibits Cancer Growth in vitro and in vivo by up-Regulating Multiple Tumor Suppressors. *Oncotarget* **2015**, *6*, 17832–17846. [CrossRef]

6. Yu, T.-J.; Shiau, J.-P.; Tang, J.-Y.; Farooqi, A.A.; Cheng, Y.-B.; Hou, M.-F.; Yen, C.-H.; Chang, H.-W. Physapruin A Exerts Endoplasmic Reticulum Stress to Trigger Breast Cancer Cell Apoptosis via Oxidative Stress. *Int. J. Mol. Sci.* **2023**, *24*, 8853. [CrossRef]

7. Wang, H.-C.; Hu, H.-H.; Chang, F.-R.; Tsai, J.-Y.; Kuo, C.-Y.; Wu, Y.-C.; Wu, C.-C. Different Effects of 4β-Hydroxywithanolide E and Withaferin A, Two Withanolides from Solanaceae Plants, on the Akt Signaling Pathway in Human Breast Cancer Cells. *Phytomedicine* **2019**, *53*, 213–222. [CrossRef]

8. Xiang, K.; Li, C.; Li, M.-X.; Song, Z.-R.; Ma, X.-X.; Sun, D.-J.; Li, H.; Chen, L.-X. Withanolides Isolated from Tubocapsicum Anomalum and Their Antiproliferative Activity. *Bioorganic Chem.* **2021**, *110*, 104809. [CrossRef]

9. Ghosh, K.; De, S.; Das, S.; Mukherjee, S.; Sengupta Bandyopadhyay, S. Withaferin A Induces ROS-Mediated Paraptosis in Human Breast Cancer Cell-Lines MCF-7 and MDA-MB-231. *PLoS ONE* **2016**, *11*, e0168488. [CrossRef]

0. Grossmann, M.E.; Mizuno, N.K.; Schuster, T.; Cleary, M.P. Punicic Acid Is an Omega-5 Fatty Acid Capable of Inhibiting Breast Cancer Proliferation. *Int. J. Mol. Sci.* **2010**, *36*, 421–426.

1. Guzmán, E.A.; Pitts, T.P.; Winder, P.L.; Wright, A.E. The Marine Natural Product Furospinulosin 1 Induces Apoptosis in MDA-MB-231 Triple Negative Breast Cancer Cell Spheroids, But Not in Cells Grown Traditionally with Longer Treatment. *Mar Drugs.* **2021**, *19*, 249. [CrossRef] [PubMed]

2. Khaw-on, P.; Pompimon, W.; Banjerdpongchai, R. Goniothalamin Induces Necroptosis and Anoikis in Human Invasive Breast Cancer MDA-MB-231 Cells. *Int. J. Mol. Sci.* **2019**, *20*, 3953. [CrossRef] [PubMed]

3. Mi, C.; Ma, J.; Wang, K.S.; Wang, Z.; Li, M.Y.; Li, J.B.; Li, X.; Piao, L.X.; Xu, G.H.; Jin, X. Amorfrutin A Inhibits TNF-α Induced JAK/STAT Signaling, Cell Survival and Proliferation of Human Cancer Cells. *Int. J. Mol. Sci.* **2017**, *39*, 338–347. [CrossRef] [PubMed]

molecules

MDPI

Review

A Comprehensive Review of the Phytochemistry, Pharmacology, and Other Applications of Euphorbiae Humifusae Herba

Jian Xiao [1,†], Hong Gu [1,†], Jiaqing Zhang [1,†], Yuqian Xue [1], Yunyi Chen [1], Weizhong Zhu [1], Hong Zhang [2], Boyi Fan [1,*] and Wenli Wang [1,*]

[1] School of Pharmacy, Nantong University, 9 Seyuan Road, Nantong 226019, China; 15902504768@163.com (J.X.); 15190810916@163.com (H.G.); 15380330612@163.com (J.Z.); 15651305816@163.com (Y.X.); 19502239732@163.com (Y.C.); zhuwz@ntu.edu.cn (W.Z.)

[2] The Second Affiliated Hospital of Zhejiang Chinese Medical University, Hangzhou 310053, China; zhanghong_16@126.com

* Correspondence: fanboyi@ntu.edu.cn (B.F.); wangwenli16@126.com (W.W.)

† These authors contributed equally to this work.

Abstract: Euphorbiae Humifusae Herba (EHH) is a globally distributed plant commonly utilized in traditional Chinese medicine (TCM) and health food within China. The dried aerial parts of EHH are well-recognized as health foods for the management of diarrhea and other intestinal diseases. Known for its therapeutic effects, such as heat-clearing, detoxification, blood cooling, hemostasis, dampness, elimination, and jaundice relief, EHH has yielded 197 bioactive compounds identified to date, including terpenoids, flavonoids, phenolic acids, tannins, alkaloids, sterols, lactones, coumarins, and other constituents, with flavonoids and terpenoids, highlighting its wide-ranging pharmacological properties and medicinal significance. Despite its popularity in research, limited systematic and comprehensive information has been provided on the EHH. Therefore, we provide an in-depth overview of EHH, covering its botanical characteristics, traditional uses, phytochemical composition, pharmacological properties, and additional applications. Furthermore, it addresses the current challenges and limitations in modern research on EHH, offering critical perspectives to guide future studies in this field.

Keywords: Euphorbiae Humifusae Herba; pharmacological effects; chemical constituents; traditional use

1. Introduction

Euphorbiae Humifusae Herba (EHH) is the whole herb of *Euphorbia humifusa* Willd. or *Euphorbia maculata* L. in the Euphorbia family. Known by several names, including "Dijincao" (DJC), "Dijin", and "Bandijin", EHH is widely distributed across various provinces in China, with the exception of Hainan province. It holds significant importance in traditional Chinese medicine (TCM), where it has been used therapeutically for centuries [1,2].

Within TCM, EHH is traditionally valued for its effects in clearing heat and toxins, cooling blood, stopping bleeding, dispelling dampness, and reducing yellowing. Modern applications in Chinese medicine extend to its anti-inflammatory, hemostatic, liver-protective, and blood glucose-lowering properties [3]. Beyond its medicinal uses, EHH is consumed as a health food in China, and its extract is popular as a flavorful beverage in Korea [1,4,5]. These wide-ranging applications are attributed to its diverse chemical profile. To date, about 197 compounds have been identified in EHH, representing a broad spectrum of chemical groups, including terpenoids, flavonoids, phenolic acids and tannins, alkaloids, sterols, lactones, and coumarins, among others.

Currently, the Pharmacopeia of People's Republic of China (2020) [2] lists four formulations containing EHH: Xiao'er Xieting Keli, Changyanning Pian, Changyanning Tangjiang, and Jidesheng Sheyao Pian. Notably, three of these formulations are primarily used to treat conditions such as diarrhea, dysentery, and both acute and chronic gastroenteritis, underscoring the significance of EHH as an essential component in TCM.

In recent years, extensive studies have examined the formulations, diverse compositions, pharmacological properties, and applications of EHH. Accordingly, this article aims to present a thorough and systematic review encompassing the botanical characteristics, ethnopharmacological uses, phytochemistry, pharmacological effects, and potential applications of EHH. This review is intended to serve as a valuable resource for advancing research and applications in the fields of medicine and chemistry.

2. Materials and Methods

The study design followed the guidelines of the PRISMA 2020 statement: an updated guideline for reporting systematic reviews [6].

2.1. Search Strategy

Research data were sourced from reputable references, including the Pharmacopoeia of China and the Flora of China, as well as various online databases such as Scopus, CAS SciFinder, Web of Science, CNKI, and Wanfang Data. The databases were queried for articles published up to December 2024. The search terms utilized included "Euphorbiae Humifusae Herba," "*Euphorbia humifusa* Willd", "*Euphorbia humifusa*", "*E.humifusa*", "*Euphorbia maculata* L.", "*Euphorbia maculata*", "*E. maculata* L.", "*Euphorbia maculata* ", "Dijincao" (DJC), "Dijin", and "Bandijin" (Figure 1).

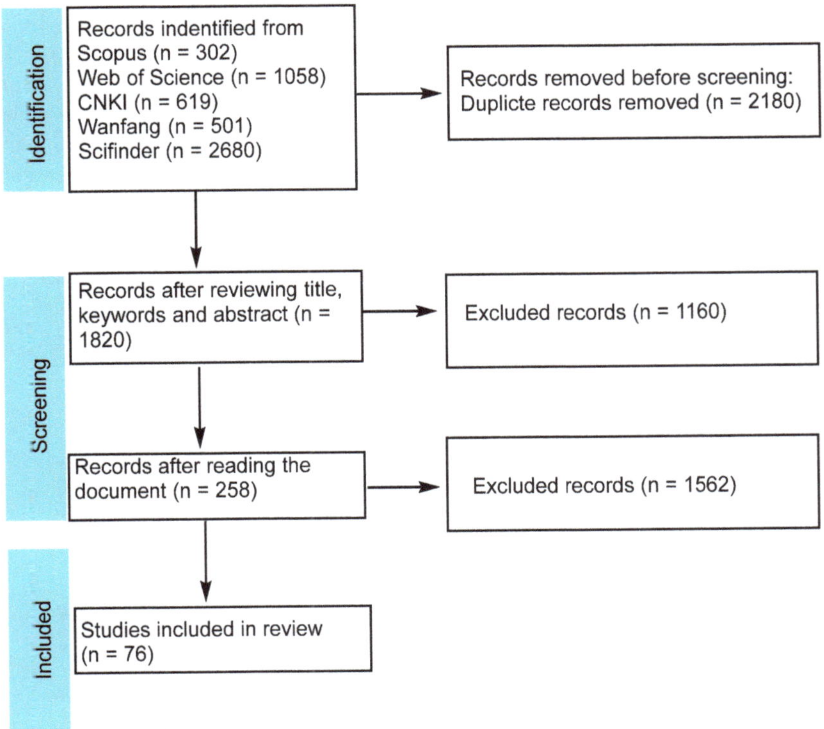

Figure 1. Search and selection of published articles.

2.2. Exclusion Criteria

The inclusion and exclusion criteria of evidence found in databases are listed below. Inclusion criteria included the following:

1. Studies carried out in vitro, in animal clinical studies and used as food
2. Studies utilizing the fresh, dried or extracts of EHH
3. Phytochemical studies on EHH

Exclusion criteria included the following:

1. Duplication of data
2. The title or the abstract does not meet the inclusion criteria
3. The document does not meet the inclusion criteria

3. Botany

EHH is an annual herbaceous plant characterized by slender, unbranched roots measuring approximately 10–18 cm in length and 2–3 mm in diameter. The stem is prostrate, branching above the base, with tips that sometimes incline upward. The base is often tinged red or pinkish-red, reaching 20–30 cm in length and 1–3 mm in diameter, and bears soft or sparse hairs. The leaves are opposite, rectangular to elliptical, 5–10 mm in length and 3–6 mm in width, with a blunt, rounded tip and a slightly oblique, tapering base. They frequently display fine serrations on the upper edge, occasionally with a central oblong purple spot. The upper leaf surface is green, while the underside is light green, sometimes pinkish-red, with sparse fur on both sides. The petioles are very short, measuring 1–2 mm.

Inflorescences appear solitary in the leaf axils, with a short stalk of 1–3 mm at the base. The gyro-shaped bracts are about 1 mm in both height and diameter, featuring four-lobed edges. The lobes are triangular, with four rectangular glands, each edged with white or pinkish-red appendages. Several male flowers extend outward, approximately matching the bract edge length, along with one outward-extending female flower with an ovary stalk reaching the bract edge. The capsule is triangular-ovate, about 2 mm long and 2.2 mm in diameter (Figure 2). Upon maturation, it divides into three fruit valves, leaving the style intact. Seeds are triangular-ovoid, approximately 1.3 mm long and 0.9 mm wide, gray, lacking transverse grooves or mounds on each edge. The flowering and fruiting period extends from May to October [7].

EHH is distributed throughout China, except in Hainan province, thriving in wild areas, roadsides, fields, sand dunes, beaches, and mountain slopes, especially in the regions north of the Yangtze River. It is also widely found in the temperate zone of Eurasia.

Figure 2. EHH plant photo. (**A**): *E. maculata* plant actual field photo; (**B**): *E. humifusa* Willd plant actual field photo; (**C**): The morphology of *E. humifusa* Willd (cited from http://www.iplant.cn (accessed on 28 December 2024)); (**D**): *E. maculata*; (**E**): The tea of EHH. (**F**): Geographical distribution of *E. humifusa* (Data from the Global Biodiversity Information Facility). (**G**): Geographical distribution of *E. maculata*.

4. Traditional Uses

EHH has a rich history of use in TCM in China. Its medicinal properties were first documented in ancient texts such as "Jia You Ben Cao", "Zheng Lei Ben Cao", and the "Compendium of Materia Medica", where the renowned physician. Li Shizhen described EHH as effective in treating abscesses, malignant sores, bleeding from cuts, and bloody dysentery, as well as for dispersing blood, stopping bleeding, and promoting urination. Other traditional texts, including "Jing Yan Fang", "Qian Kun Sheng Yi", and "Shi Yi De Xiao Fang", also mention their therapeutic applications. Today, EHH remains commonly employed TCM. Classified in TCM as having a pungent and neutral taste, it primarily affects the liver and large intestine to promote heat-clearing, detoxification, blood cooling, hemostasis, dampness clearing, and yellowing reduction [2].

Beyond its role in TCM, EHH is used in Uyghur and Mongolian medicine, which it is applied to treat fungal infections like tinea (hand, body, foot, and versicolor) and psoriasis and is valued for cooling the blood and supporting liver health [8–10]. Additionally, EEH is also used in food preservation and other applications. Recent studies have shown that the ethanol extract can extend the freshness of strawberries, and its formulation, Dijincaokeli, is used in veterinary medicine to treat bacterial infections in chickens [11].

5. Phytochemistry

To date, the literature reports the identification and isolation of 197 distinct compounds from EHH, representing a broad spectrum of chemical classes, including terpenoids, flavonoids, phenolic acids, tannins, alkaloids, sterols, lactones, coumarins, and other constituents. Among these, phenolic acids and tannins are the most prevalent, followed by flavonoids and terpenoids, while alkaloids, sterols, lactones, and coumarins are present in smaller quantities (Figure 3).

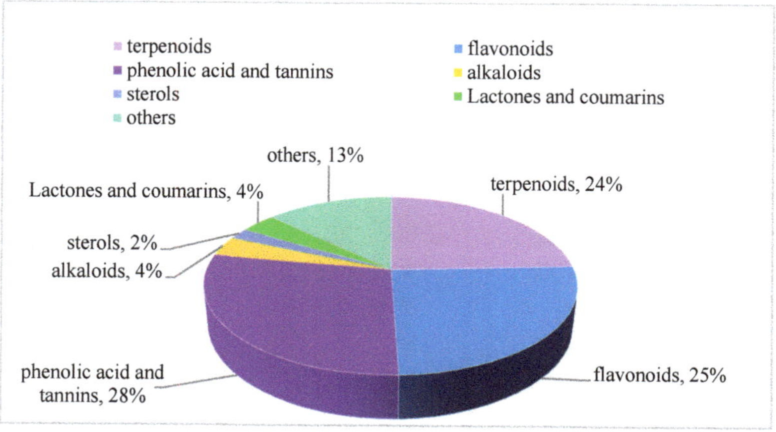

Figure 3. The proportion of chemical components in EHH.

5.1. Triterpenoids

Triterpenes represent one of the primary active components in EHH and are categorized into four distinct types. Of these, triterpenoid compounds are more numerous, totaling 47 identified compounds (Figure 4, Table 1). Moreover, 17 triterpenes have been isolated from *E. maculata*, including such as: (3*S*,4*S*,7*S*,9*R*)-4-methyl-3,7-dihydroxy-7(8→9) *abeo*-lanost-24(28)-en-8-one (**1**) and 24-hydroperoxylanost-7,25-dien-3*β*-ol (**2**), 3-hydroxycycloart-25-ene-24-hydroperoxide (**3**), 3*β*-hydroxy-26-nor-9,19-cyclolanost-23-en-25-one (**4**), cycloart-23en-3*β*,25-diol (**5**), cycloeucalenol (**6**), (23*E*)-3*β*,25-

dihydroxytirucalla-7,23-diene (**7**), (23*Z*)-3*β*, 25-dihydroxy-tirucalla-7,23-diene (**8**), obtusifoliol (**9**), 4*α*,l4*α*-dimethyl-5*α*-ergosta-7, 9(11), 24(28)-trien-3*β*-ol (**10**), gramisterol (**11**), urs-12-ene-3*β*, 11*α*-diol (**12**), neoilexonol (**13**), 12-oleanene-3*β*,11*β*-diol (**14**), (3*β*,15*α*,16*α*)-15,16-epoxy,olean-12-en-3-ol (**15**), multiflorenol (**16**), and lupeol (**17**) [12]. Additionally, 10 triterpenoids have been identified from *E. humifusa*, including 3,4-seco-lupa-4(23),20(29)-dien-24-hydroxy-3-oic acid (**18**), lup-20(29)-ene-3,30-diol (**19**), 24(*R*)-3,4-secocycloart-4(29),25-dien-24-hydroxy-3-oic acid (**20**), 24(*S*)-3,4-seco-cycloart-4(29),25-dien-24-hydroxy-3-oic acid (**21**), 23(*Z*)-cycloart-23-en-3,25-diol (**22**), 24(*S*)-cycloart-25-en-3,24-diol (**23**), 23(*E*)-cycloart-23-en-25-ethoxy-3-ol (**24**), 3-hydroxy-4,14-dimethyl-5-ergosta-8,24(28)-dien-7-one (**25**), 3-hydroxy-4,14-dimethyl-5-ergosta-8,24(28)-dien-7,11-dione (**26**), 3-hydroxy-4,14-dimethyl-5-ergosta-7,9(11),24(28)-trien (**27**) [13]. Further isolated from *E. humifusa* were (3*S*,5*R*,7*R*,8*S*,10*S*,13*S*,14*S*,17*R*,18*R*,21*R*)−7*α*,8*α*-epoxyfern-9(11)-en-3*β*-ol (**28**), (3*R*,7*R*,8*S*,9*S*,13*S*,14*S*,17*R*,18*R*,21*R*)−7*α*,8*α*-epoxyadian-5(10)-en-3*α*-ol (**29**), fern-8(9)-en-3*β*-ol (**30**), and 17*β*,21*β*-epoxyhopan-3*β*-ol (**31**) [14]. Additional compounds include spiromaculatols A–C, spiropedroxodiol, and spiroinonotsuoxodiol (**32–36**), euphomaculatoids A–E, 3*β*, 7*α*-dihydroxy-4*α*,14*α*-dimethyl-5*α*-ergosta-8,24 (28)-dien-11-one, and 3*β*-hydroxy-4*α*,14*α*-dimethyl-5*α*-ergosta-8,24 (28)-diene-7,11-dione (**37–43**), euphomaculatoids F-H (**44–46**). and 3*β*,11*β*-3,11-dihydroxylanosta-8,24-dien-7-one (**47**), derived from *E. maculate* [15].

Table 1. Triterpenoids in EHH.

NO.	Name	Molecular Formula	Molecular Weight	Source	References
1	(3*S*,4*S*,7*S*,9*R*)-4-methyl-3,7-dihydroxy-7(8→9) *abeo*-lanost-24(28)-en-8-one	$C_{30}H_{50}O_3$	458.73	*E. maculata*	[12]
2	24-hydroperoxylanost-7,25-dien-3*β*-ol	$C_{30}H_{50}O_3$	458.73	*E. maculata*	[12]
3	cycloart-23*Z*-en-3*β*, 25-diol	$C_{30}H_{50}O_2$	442.73	*E. maculata*	[12]
4	3*β*-hydroxy-26-nor-9,19-cyclolanost-23-en-25-one	$C_{29}H_{46}O_2$	426.69	*E. maculata*	[12]
5	cycloart-23en-3*β*,25-diol	$C_{30}H_{50}O_2$	442.73	*E. maculata*	[12]
6	cycloeucalenol	$C_{30}H_{50}O$	426.73	*E. maculata*	[12]
7	(23*E*)-3*β*,25-dihydroxytirucalla-7,23-diene	$C_{30}H_{50}O_2$	442.73	*E. maculata*	[12]
8	(23*Z*)-3*β*, 25-dihydroxy-tirucalla-7,23-diene	$C_{30}H_{50}O_2$	442.73	*E. maculata*	[12]
9	obtusifoliol	$C_{30}H_{50}O$	426.73	*E. maculata*	[12]
10	4*α*, l4*α*-dimethyl-5*α*-ergosta-7,9(11), 24(28)-trien-3β-ol	$C_{30}H_{48}O$	424.71	*E. maculata*	[12]
11	gramisterol	$C_{30}H_{50}O$	426.73	*E. maculata*	[12]
12	urs-12-ene-3*β*, 11*α*-diol	$C_{30}H_{50}O_2$	442.73	*E. maculata*	[12]
13	neoilexonol	$C_{30}H_{48}O_2$	440.71	*E. maculata*	[12]
14	12-oleanene-3*β*,11*β*-diol	$C_{30}H_{50}O_2$	442.73	*E. maculata*	[12]
15	(3*β*,15*α*, 16*α*)-15,16-epoxy, olean-12-en-3-ol	$C_{30}H_{48}O_2$	440.71	*E. maculata*	[12]
16	multiflorenol	$C_{30}H_{50}O$	426.73	*E. maculata*	[12]
17	lupeol	$C_{30}H_{50}O$	426.73	*E. maculata*	[12]
18	3,4-seco-lupa-4(23),20(29)-dien-24-hydroxy-3-oic acid	$C_{30}H_{48}O_3$	456.71	*E. humifusa*	[13]
19	lup-20(29)-ene-3,30-diol	$C_{30}H_{50}O_2$	442.73	*E. humifusa*	[13]
20	24(*R*)-3,4-secocycloart-4(29),25-dien-24-hydroxy-3-oic acid	$C_{30}H_{48}O_3$	456.71	*E. humifusa*	[13]
21	24(*S*)-3,4-seco-cycloart-4(29),25-dien-24-hydroxy-3-oic acid	$C_{30}H_{48}O_3$	456.71	*E. humifusa*	[13]
22	23(*Z*)-cycloart-23-en-3,25-diol	$C_{30}H_{50}O_2$	442.73	*E. humifusa*	[13]
23	24(*S*)-cycloart-25-en-3,24-diol	$C_{30}H_{50}O_2$	442.73	*E. humifusa*	[13]
24	23(*E*)-cycloart-23-en-25-ethoxy-3-ol	$C_{32}H_{54}O_2$	470.78	*E. humifusa*	[13]
25	3-hydroxy-4,14-dimethyl-5-ergosta-8,24(28)-dien-7-one	$C_{30}H_{48}O_2$	440.71	*E. humifusa*	[13]
26	3-hydroxy-4,14-dimethyl-5-ergosta-8,24(28)-dien-7,11-dione	$C_{30}H_{46}O_3$	454.70	*E. humifusa*	[13]

Table 1. *Cont.*

NO.	Name	Molecular Formula	Molecular Weight	Source	References
27	3-hydroxy-4,14-dimethyl-5-ergosta-7,9(11),24(28)-trien	$C_{30}H_{48}O$	424.71	*E. humifusa*	[13]
28	(3*S*,5*R*,7*R*,8*S*,10*S*,13*S*,14*S*,17*R*,18*R*,21*R*)−7*α*,8*α*-epoxyfern-9(11)-en-3*β*-ol	$C_{30}H_{48}O_2$	440.71	*E. humifusa*	[14]
29	(3*R*,7*R*,8*S*,9*S*,13*S*,14*S*,17*R*,18*R*,21*R*)−7*α*,8*α*-epoxyadian-5(10)-en-3*α*-ol	$C_{30}H_{48}O_2$	440.71	*E. humifusa*	[14]
30	fern-8(9)-en-3*β*-ol	$C_{30}H_{50}O$	426.73	*E. humifusa*	[14]
31	17*β*,21*β*-epoxyhopan-3*β*-ol	$C_{30}H_{50}O_2$	442.73	*E. humifusa*	[14]
32	spiromaculatol A	$C_{31}H_{52}O_3$	472.75	*E. maculate*	[15]
33	spiromaculatol B	$C_{31}H_{52}O_3$	472.75	*E. maculate*	[15]
34	spiromaculatol C	$C_{30}H_{50}O_3$	458.73	*E. maculate*	[15]
35	spiropedroxodiol	$C_{30}H_{50}O_3$	458.73	*E. maculate*	[15]
36	spiroinonotsuoxodiol	$C_{30}H_{50}O_3$	458.73	*E. maculate*	[15]
37	euphomaculatoid A	$C_{30}H_{44}O_4$	468.68	*E. maculate*	[15]
38	euphomaculatoid B	$C_{30}H_{46}O_4$	470.69	*E. maculate*	[15]
39	euphomaculatoid C	$C_{30}H_{46}O_4$	470.69	*E. maculate*	[15]
40	euphomaculatoid D	$C_{30}H_{46}O_3$	454.70	*E. maculate*	[15]
41	euphomaculatoid E	$C_{30}H_{46}O_3$	454.70	*E. maculate*	[15]
42	3*β*, 7*α*-dihydroxy-4*α*,14*α*-dimethyl-5*α*-ergosta-8,24 (28)-dien-11-one	$C_{30}H_{48}O_3$	456.71	*E. maculate*	[15]
43	3*β*-hydroxy-4*α*,14*α*-dimethyl-5*α*-ergosta-8,24 (28)-diene-7,11-dione	$C_{30}H_{46}O_3$	454.70	*E. maculate*	[15]
44	euphomaculatoid F	$C_{29}H_{44}O_4$	456.67	*E. maculate*	[15]
45	euphomaculatoid G	$C_{29}H_{42}O_4$	454.65	*E. maculate*	[15]
46	euphomaculatoid H	$C_{30}H_{48}O_3$	456.71	*E. maculate*	[15]
47	3*β*,11*β*-3,11-dihydroxylanosta-8,24-dien-7-one	$C_{30}H_{48}O_3$	456.71	*E. maculate*	[15]

5.2. Flavonoids

The specific flavonoid composition in EHH includes various subclasses such as flavonoids, flavonols, flavanones, and dihydrochalcone, as well as their glycosides (Figure 5, Table 2) [3,4,16–19]. Ying Tian and colleagues have isolated 11 flavonoid compounds from *E. humifusa*, including luteolin (**48**), apigenin (**49**), luteolin-7-*O*-(6″-*O*-transferuloyl)-*β*-*D*-glucopyranoside (**50**), luteolin-7-*O*-(6″-*O*-coumaroyl)-*β*-*D*-glucopyranoside (**51**), apigenin-7-*O*-*β*-*D*-lutinoside (**52**), apigenin-7-*O*-*β*-*D*-apiofuranosyl(1→2)-*β*-*D*-glucopyranoside (**53**), 6,8-di-*C*-*β*-*D*-glucopyranosyl apigenin (**54**), apigenin-7-*O*-(6″-*O*-galloyl)-*β*-*D*-glucopyranoside (**55**), quercetin-3-*O*-*β*-*D*-galactoside (**56**), quercetin-3-*O*-*β*-*D*-glucopyranoside (**57**), and hesperidin (**58**) [15]. Xiaoying Wang has isolated quercetin-7-*O*-*β*-*D*-glucopyranoside (**59**), quercetin-3-*O*-*α*-*L*-rhamnosyl(1→6)-*β*-*D*-galactoside (**60**), and kaempferol-3-*O*-*β*-*D*-glucopyranoside (**61**) from *E. humifusa* [20]. Using polyamide chromatography, Rongzhi Li has identified quercetin (**62**) and kaempferol (**63**) from *E. humifusa* [21]. In 2008, Deng et al. found that isorhoifolin (**64**), kaempferol-3-*O*-*α*-*L*-arabinoforanoside (**65**) [22]. Runhui Liu and colleagues have employed Sephadex LH-20 purification to isolate apigenin-7-*O*-glucoside (**66**), luteolin-7-*O*-glucoside (**67**), and quercetin-3-*O*-arabinoside (**68**) from *E. humifusa* [23]. Additional flavonoids identified in *E. humifusa* including luteolin-7-*O*-*β*-*D*-glucopyranoside (**69**), apigenin-7-*O*-*β*-*D*-glucopyranoside (**70**), quercetin-3-*O*-(2″,3″-di-*O*-galloyl)-*β*-*D*-glucopyranoside (**71**), quercetin-3-*O*-(6″-*O*-galloyl)-glucopyranoside (**72**), quercetin-3-*O*-{rhamnosyl-(1→6)-[xylosyl-(1→2)]-galactoside} (**73**), quercetin-3-*O*-*β*-*D*-xylosyl-(1→2)-*β*-*D*-glucopyranoside (**74**), nicotiflorin (**75**), kaempferol-3-*O*-(6″-*O*-galloyl)-*β*-*D*-galactoside (**76**), isomyricitrin (**77**) and phlorizin (**78**) [24]. Ahn et al. isolated quercetin-3-*O*-*α*-*L*-arabinofuranoside (**79**) from *E. humifusa* [3]. Four flavonoids, apigenin (**49**), quercetin (**62**), kaempferol (**63**), 8-hydroxyluteolin (**80**) and one prenylated chalcone paratocarpin E (**81**), were obtained from *E. humifusa* by Gao et al. [25]. Four flavonoid glycosides, quercetin-3-*O*-apiosyl (1–2) galactoside (**82**), quercetin-3-*O*-glucoside (**83**), astragalin (**84**), and vitexin

(**85**), were isolated from *E. humifusa*. Isoquercitrin 6″-*O*-gallate (**86**), astragalin 6″-*O*-gallate (**87**), avicularin (**88**), juglanin (**89**), isoquercitrin (**90**), and hyperoside (**91**) were isolated from *E. maculata* by Nugroho [18]. Lastly, Luyen et al. have isolated 12 additional flavonoids from *E. humifusa*, including astragalin 2″,3″-*O*-digallate (**92**), astragalin 2″-*O*-gallate (**93**), isoquercitin 2″-*O*-gallate (**94**), quercitrin 2″-*O*-gallate (**95**), phlorizin (**78**), quercetin (**62**), hyperin (**96**), isoquercitrin (**90**), rutin (**97**), isomyricitrin (**77**), nicotiflorin (**75**), and astragalin (**84**) [4].

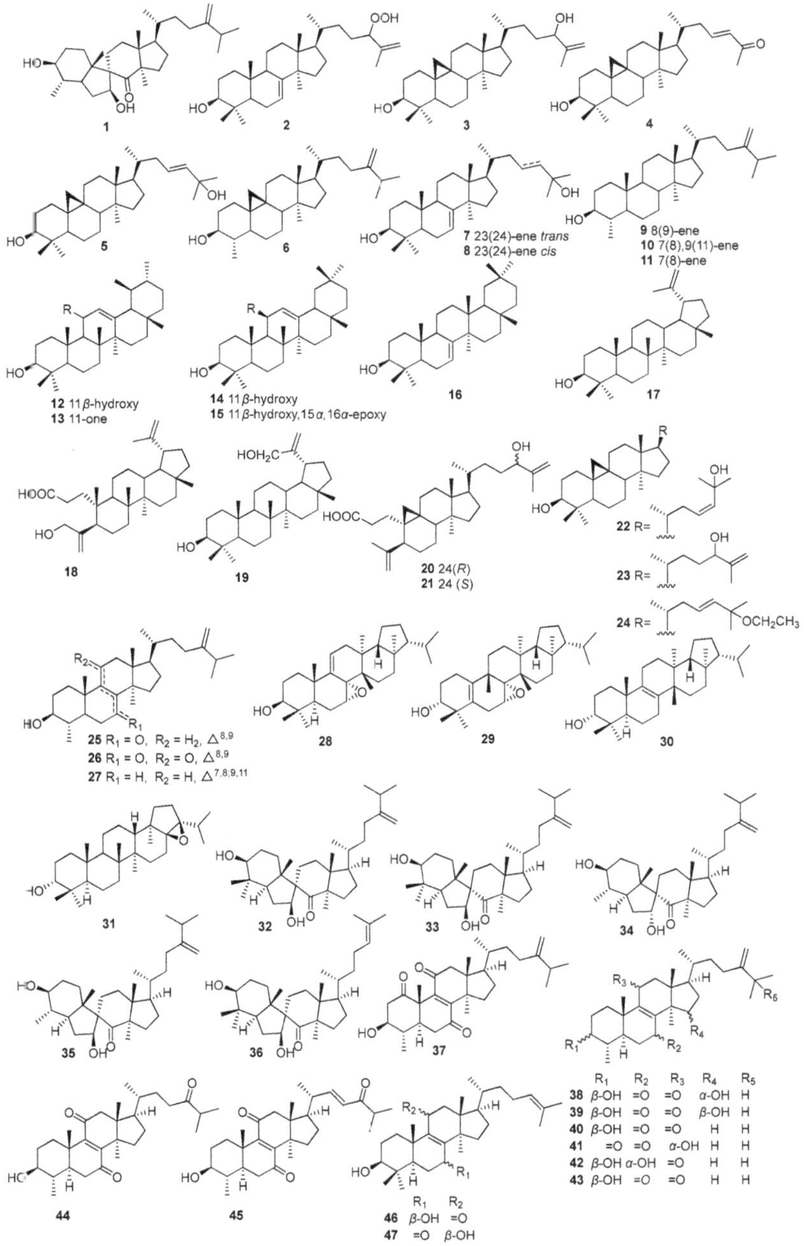

Figure 4. The structures of triterpenoids in EHH.

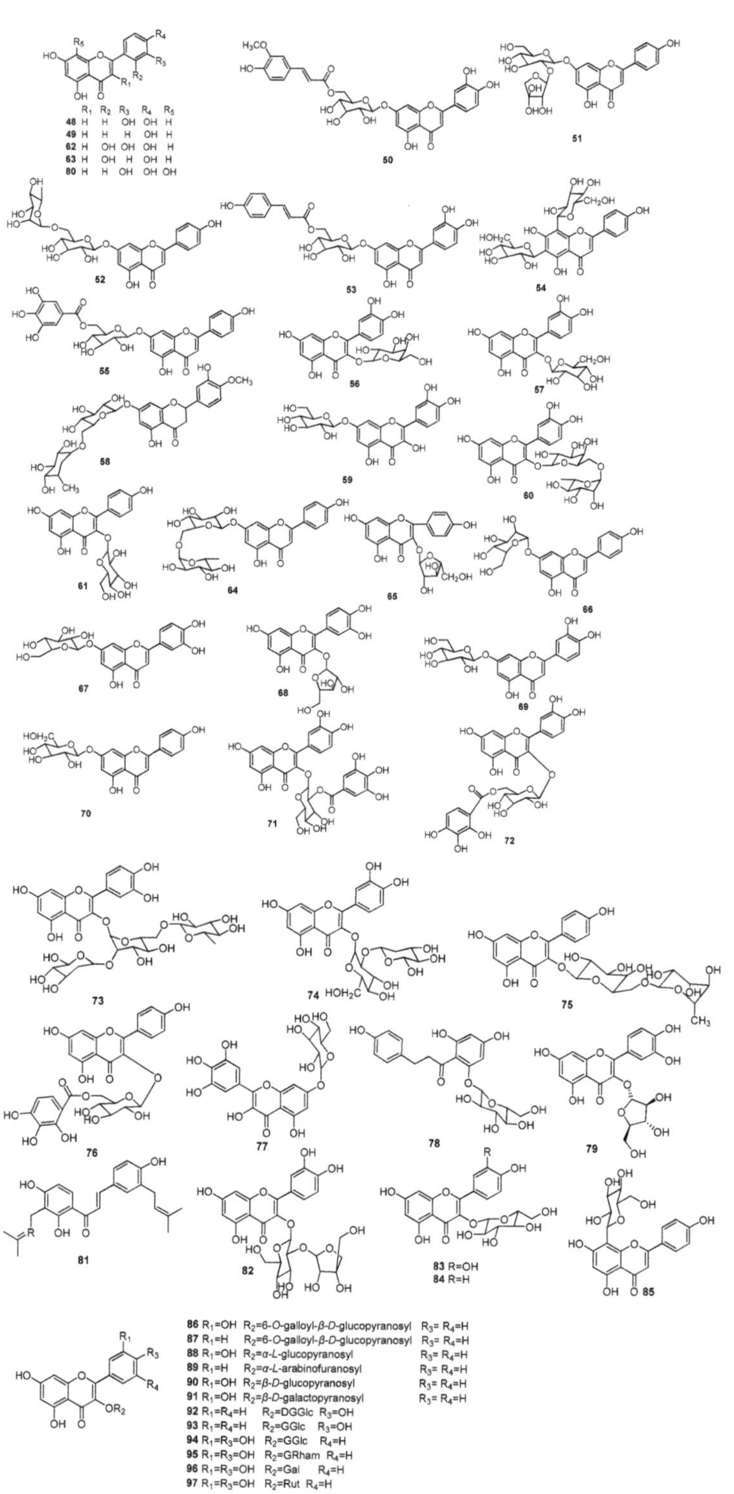

Figure 5. The structures of flavonoids in EHH (DGGlc: 2,3-digalloylglucosyl, GGlc: 2-galloylglucosyl, GRham: 2-galloylrhamosyl; Gal: galactosyl; Rut: rutinosyl).

Table 2. Flavonoids in EHH.

NO.	Name	Molecular Formula	Molecular Weight	Source	References
48	luteolin	$C_{15}H_{10}O_6$	286.24	*E. humifusa*	[16,25]
49	apigenin	$C_{15}H_{10}O_5$	270.24	*E. humifusa*	[16]
50	luteolin-7-*O*-(6″-*O*-trans-feruloyl)-*β*-*D*-glucopyranoside	$C_{31}H_{28}O_{14}$	624.55	*E. humifusa*	[16]
51	luteolin-7-*O*-(6″-*O*-coumaroyl)-*β*-*D*-glucopyranoside	$C_{25}H_{26}O_{14}$	550.47	*E. humifusa*	[16]
52	apigenin-7-*O*-*β*-*D*-lutinoside	$C_{27}H_{30}O_{14}$	578.52	*E. humifusa*	[16]
53	apigenin-7-*O*-*β*-*D*-apiofuranosyl(1→2)-*β*-*D*-glucopyranoside	$C_{30}H_{26}O_{13}$	594.14	*E. humifusa*	[16]
54	6,8-di-*C*-*β*-*D*-glucopyranosyl apigenin	$C_{27}H_{30}O_{15}$	594.52	*E. humifusa*	[16]
55	apigenin-7-*O*-(6″-*O*-galloyl)-*β*-*D*-glucopyranoside	$C_{28}H_{24}O_{14}$	584.49	*E. humifusa*	[16]
56	quercetin-3-*O*-*β*-*D*-galactoside	$C_{21}H_{20}O_{12}$	464.38	*E. humifusa*	[16]
57	quercetin-3-*O*-*β*-*D*-glucopyranoside	$C_{21}H_{20}O_{12}$	464.38	*E. humifusa*	[16]
58	hesperidin	$C_{29}H_{36}O_{14}$	608.59	*E. humifusa*	[16]
59	quercetin-7-*O*-*β*-*D*-glucopyranoside	$C_{21}H_{20}O_{12}$	464.38	*E. humifusa*	[20]
60	quercetin-3-*O*-*α*-*L*-rhamnosyl(1→6)-*β*-*D*-galactoside	$C_{27}H_{30}O_{16}$	610.52	*E. humifusa*	[20]
61	kaempferol-3-*O*-*β*-*D*-glucopyranoside	$C_{21}H_{20}O_{11}$	448.38	*E. humifusa*	[20]
62	quercetin	$C_{15}H_{10}O_7$	302.24	*E. humifusa*	[20]
63	kaempferol	$C_{15}H_{10}O_6$	286.24	*E. humifusa*	[21]
64	isorhoifolin	$C_{27}H_{30}O_{14}$	578.52	*E. humifusa*	[22]
65	kaempferol-3-*O*-*α*-*L*-arabinoforanoside	$C_{20}H_{18}O_{10}$	418.35	*E. humifusa*	[22]
66	apigenin-7-*O*-glucoside	$C_{21}H_{20}O_{10}$	432.38	*E. humifusa*	[23]
67	luteolin-7-*O*-glucoside	$C_{21}H_{20}O_{11}$	448.38	*E. humifusa*	[23]
68	quercetin-3-*O*-arabinoside	$C_{20}H_{18}O_{11}$	434.35	*E. humifusa*	[22]
69	luteolin-7-*O*-*β*-*D*-glucopyranoside	$C_{21}H_{20}O_{11}$	448.38	*E. humifusa*	[24]
70	apigenin-7-*O*-*β*-*D*-glucopyranoside	$C_{21}H_{20}O_{10}$	432.38	*E. humifusa*	[24]
71	quercetin-3-*O*-(2″,3″-di-*O*-galloyl)-*β*-*D*-glucopyranoside	$C_{28}H_{24}O_{16}$	616.48	*E. humifusa*	[24]
72	quercetin-3-*O*-(6″-*O*-galloyl)-glucopyranoside	$C_{28}H_{24}O_{16}$	616.48	*E. humifusa*	[24]
73	quercetin-3-*O*-{rhamnosyl-(1→6)-[xylosyl-(1→2)]-galactoside}	$C_{32}H_{38}O_{20}$	742.64	*E. humifusa*	[24]
74	quercetin-3-*O*-*β*-*D*-xylosyl-(1→2)-*β*-*D*-glucopyranoside	$C_{26}H_{28}O_{16}$	596.49	*E. humifusa*	[24]
75	nicotiflorin	$C_{27}H_{30}O_{15}$	594.52	*E. humifusa*	[24]
76	kaempferol-3-*O*-(6″-*O*-galloyl)-*β*-*D*-galactoside	$C_{28}H_{24}O_{15}$	600.49	*E. humifusa*	[24]
77	isomyricitrin	$C_{21}H_{20}O_{13}$	480.38	*E. humifusa*	[24]
78	phlorizin	$C_{21}H_{24}O_{10}$	436.41	*E. humifusa*	[24]
79	quercetin-3-*O*-*α*-*L*-arabinofuranoside	$C_{20}H_{18}O_{11}$	434.35	*E. humifusa*	[3]
80	8-hydroxyluteolin	$C_{15}H_{10}O_7$	302.24	*E. humifusa*	[25]
81	paratocarpin E	$C_{25}H_{28}O_4$	392.50	*E. humifusa*	[25]
82	quercetin-3-*O*-apiosyl (1–2) galactoside	$C_{26}H_{28}O_{16}$	596.49	*E. humifusa*	[19]
83	quercetin-3-*O*-glucoside	$C_{21}H_{20}O_{11}$	448.38	*E. humifusa*	[19]
84	astragalin	$C_{21}H_{20}O_{12}$	464.38	*E. humifusa*	[3,19]
85	vitexin	$C_{21}H_{20}O_{10}$	432.38	*E. humifusa*	[19]
86	isoquercitrin 6″-*O*-gallate	$C_{28}H_{24}O_{16}$	616.48	*E. maculata*	[18]
87	astragalin 6″-*O*-gallate	$C_{28}H_{24}O_{15}$	600.48	*E. maculata*	[18]
88	avicularin	$C_{20}H_{18}O_{11}$	434.35	*E. maculata*	[18]
89	juglanin	$C_{20}H_{18}O_{10}$	418.35	*E. maculata*	[18]
90	isoquercitrin	$C_{21}H_{20}O_{12}$	464.38	*E. maculata*	[18]
91	hyperoside	$C_{21}H_{20}O_{12}$	464.38	*E. maculata*	[18]
92	astragalin 2″,3″-*O*-digallate	$C_{36}H_{30}O_{19}$	766.62	*E. humifusa*	[4]
93	astragalin 2″-*O*-gallate	$C_{29}H_{26}O_{15}$	614.51	*E. humifusa*	[4]
94	isoquercitin 2″-*O*-gallate	$C_{29}H_{26}O_{16}$	630.51	*E. humifusa*	[4]
95	quercitrin 2″-*O*-gallate	$C_{29}H_{26}O_{15}$	614.51	*E. humifusa*	[4]
96	hyperin	$C_{21}H_{20}O_{12}$	464.38	*E. humifusa*	[4]
97	rutin	$C_{27}H_{30}O_{16}$	616.49	*E. humifusa*	[4]

5.3. Phenolic Acids and Tannins

The 55 phenolic constituents of EHH include simple phenolic acids and ellagitannins (Figure 6, Table 3). Nugroho et al. have isolated gallic acid (**98**) and methyl gallate (**99**) from *E. humifusa* [18]. In 1994, Yoshida identified several tannins and phenolics, including 1,3,6-tri-*O*-galloyl-2-*O*-brevifolincarboxyl-*β*-*D*-glucose (**100**), ellimagrandin I (**101**), and excoecarianin (**102**) from a 70% acetone extract of *E. humifusa* using Ttoyptearl HW-40, MCI-gel, CHP-20P, among others [26]. Tian et al. have isolated ethyl gallate (**103**), phyllanthussin E methyl ester (**104**), humifusaone (**105**), and valoneaic aciddilactone (**106**) from *E. humifusa* in 2010 [16]. Yoshiaki and colleagues have isolated eumaculin E (**107**) from *E. maculata* [27]. In 2014, Luyen and colleagues fractionated several tannins and derivatives, including dehydropicrorhiza acid methyl diester (**108**), methylsyringin (**109**), methylconiferin (**110**), syringin (**111**), and sphaerophyside SC (**112**) from the methanolic extract of *E. humifusa* [4]. According to the literature, Isao isolated 1,2,3-tri-*O*-galloyl-*β*-*D*-glucose (**113**) from the aerial parts of *E. maculata* [28]. In 1997, Amakura et al. extracted granatin B (**114**) from an acetone extract of *E. maculata* [27]. Takashi et al. have isolated ellagic acid-4-*O*-*β*-*D*-glucopyranoside (**115**) from the above-ground parts of *E. humifusa* [26]. Deng et al. have obtained 3,3′-di-*O*-methyl ellagic acid-4-*O*-*β*-*D*-glucopyranoside (**116**) using silica gel column chromatography and Sephadex LH-20 from *E. humifusa* [22]. Luyen et al. have isolated brevifolin carboxylic acid (**117**) [4], and Tian and colleagues have obtained methyl brevifolin carboxylate (**118**), ethyl brevifolin carboxylate (**119**), sanguisorbicacid dilactone (**120**), 7″-ethyl-sanguisorbic acid dilactone (**121**), 1-*O*-methyl-6-*O*-p-digalloyl-*α*-*D*-glucopyranoside (**122**), 1-*O*-ethyl-6-*O*-p-digalloyl-*α*-*D*-glucopyranoside (**123**), chebulanin (**124**), and furosin (**125**) from *E. humifusa* [16]. Yoshida et al. have obtained 1,3,4,6-tetra-*O*-galloyl-*β*-*D*-glucose (**126**), euphormisin M3 (**127**), 1,2,4,6-tetra-*O*-galloyl-*β*-*D*-glucose (**128**), 1,2,6-tri-*O*-galloyl-*β*-*D*-glucose (**129**), 1,3,4,6-tetra-*O*-galloyl-*β*-*D*-glucose (**130**), 2,4,6-tri-*O*-galloyl-*D*-glucose (**131**), 3,4,6-tri-*O*-galloyl-*D*-glucose (**132**), and 1,3,6-tri-*O*-galloyl-*D*-glucose (**133**) were also extracted from 70% acetone extract of *E. humifusa* [26]. Using toypyearl HW-40 and Sephadex LH-20, Isao et al. have isolated euphormisin M2 (**134**), eumaculin A (**135**), geraniin (**136**), euphorbin B (**137**), euphorbin A (**138**), chebulagic acid (**139**), mallotusinin (**140**), corilagin (**141**), 1-*O*-ethyl-3,6-*O*-(*R*)-hexahydroxydiphenoyl-(1C4)-*β*-*D*-glucose (**142**), 1,3,4,6-tetra-*O*-galloyl-*β*-*D*-glucose (**143**), tercatain (**144**), eumaculin B (**145**), and eumaculin D (**146**) from *E. maculata* [28]. Other compounds have been isolated, including protocatechuic acid (**147**), *cis*-caffeic acid (**148**), *trans*-caffeic acid (**149**) [29], rosmarinic (**150**) [30], euphorbinoside (**151**), and benzyl *β*-D-ribofuranoside (**152**) [4] were isolated from *E. humifusa*.

Table 3. Phenolic acids and tannins in EHH.

NO.	Name	Molecular Formula	Molecular Weight	Source	References
98	gallic acid	$C_7H_6O_5$	170.12	*E. humifusa*	[18]
99	methyl gallate	$C_8H_8O_5$	184.15	*E. humifusa*	[18]
100	1,3,6-tri-*O*-galloyl-2-*O*-brevifolincarboxyl-*β*-*D*-glucose	$C_{40}H_{30}O_{25}$	910.66	*E. humifusa*	[26]
101	tellimagrandin I	$C_{35}H_{28}O_{21}$	784.11	*E. humifusa*	[26]
102	excoecarianin	$C_{82}H_{56}O_{53}$	1888.17	*E. humifusa*	[26]
103	ethyl gallate	$C_9H_{10}O_5$	198.17	*E. humifusa*	[16]
104	phyllanthussin E methyl ester	$C_{14}H_{10}O_8$	306.4	*E. humifusa*	[16]
105	humifusaone	$C_{12}H_{10}O_7$	266.21	*E. humifusa*	[16]
106	valoneaic aciddilactone	$C_{21}H_{10}O_{13}$	470.30	*E. humifusa*	[27]

Table 3. *Cont.*

NO.	Name	Molecular Formula	Molecular Weight	Source	References
107	eumaculin E	$C_{41}H_{32}O_{28}$	972.68	*. maculata*	[16]
108	dehydropicrorhiza acid methyl diester	$C_{16}H_{12}O_{10}$	364.26	*E. humifusa*	[4]
109	methylsyringin	$C_{18}H_{26}O_{9}$	386.16	*E. humifusa*	[4]
110	syringin	$C_{17}H_{24}O_{8}$	356.15	*E. humifusa*	[4]
111	methylconiferin	$C_{17}H_{24}O_{9}$	372.14	*E. humifusa*	[4]
112	sphaerophyside SC	$C_{16}H_{22}O_{7}$	326.35	*E. humifusa*	[4]
113	1,2,3-tri-*O*-galloyl-*β*-*D*-glucose	$C_{27}H_{24}O_{18}$	636.47	*E. maculata*	[28]
114	granatin B	$C_{42}H_{30}O_{26}$	950.68	*E. maculata*	[27]
115	ellagic acid-4-*O*-*β*-*D*-glucopyranoside	$C_{20}H_{16}O_{13}$	464.34	*E. humifusa*	[26]
116	3,3′-di-*O*-methyl ellagic acid-4-*O*-*β*-*D*-glucopyranoside	$C_{22}H_{20}O_{13}$	492.39	*E. humifusa*	[22]
117	brevifolin carboxylic acid	$C_{12}H_{8}O_{6}$	248.19	*E. humifusa*	[4]
118	methyl brevifolin carboxylate	$C_{14}H_{10}O_{8}$	306.23	*E. humifusa*	[16]
119	ethyl brevifolin carboxylate	$C_{15}H_{12}O_{8}$	320.25	*E. humifusa*	[16]
120	sanguisorbicacid dilactone	$C_{21}H_{10}O_{13}$	470.30	*E. humifusa*	[16]
121	7′′-ethyl-sanguisorbic acid dilactone	$C_{23}H_{14}O_{13}$	498.35	*E. humifusa*	[16]
122	1-*O*-methyl-6-*O*-p-digalloyl-*α*-*D*-glucopyranoside	$C_{21}H_{22}O_{12}$	466.40	*E. humifusa*	[12]
123	1-*O*-ethyl-6-*O*-p-digalloyl-*α*-*D*-glucopyranoside	$C_{22}H_{24}O_{12}$	480.42	*E. humifusa*	[16]
124	chebulanin	$C_{27}H_{24}O_{19}$	652.47	*E. humifusa*	[16]
125	furosin	$C_{27}H_{22}O_{19}$	650.45	*. humifusa*	[16]
126	1,3,4,6-tetra-*O*-galloyl-*β*-*D*-glucose	$C_{27}H_{24}O_{18}$	636.47	*E. humifusa*	[26]
127	Euphormisin M3	$C_{41}H_{32}O_{26}$	940.68	*E. humifusa*	[26]
128	1,2,4,6-tetra-*O*-galloyl-*β*-*D*-glucose	$C_{34}H_{28}O_{22}$	788.58	*E. humifusa*	[26]
129	1,2,6-tri-*O*-galloyl-*β*-*D*-glucose	$C_{27}H_{24}O_{18}$	636.47	*E. humifusa*	[26]
130	1,3,4,6-tetra-*O*-galloyl-*β*-*D*-glucose	$C_{34}H_{28}O_{22}$	788.58	*E. humifusa*	[26]
131	2,4,6-tri-*O*-galloyl-*D*-glucose	$C_{27}H_{24}O_{18}$	636.47	*E. humifusa*	[26]
132	3,4,6-tri-*O*-galloyl-*D*-glucose	$C_{27}H_{24}O_{18}$	636.47	*E. humifusa*	[26]
133	1,3,6-tri-*O*-galloyl-*D*-glucose	$C_{27}H_{24}O_{18}$	636.47	*E. humifusa*	[26]
134	euphormisin M2	$C_{41}H_{30}O_{25}$	922.67	*E. maculata*	[28]
135	eumaculinA	$C_{69}H_{54}O_{43}$	1571.15	*E. maculata*	[28]
136	geraniin	$C_{41}H_{28}O_{27}$	952.65	*E. maculata*	[28]
137	euphorbin B	$C_{83}H_{60}O_{52}$	1889.34	*E. maculata*	[28]
138	euphorbin A	$C_{125}H_{92}O_{77}$	2826.03	*E. maculata*	[28]
139	chebulagic acid	$C_{41}H_{30}O_{27}$	954.66	*E. maculata*	[28]
140	mallotusinin	$C_{41}H_{26}O_{25}$	918.63	*E. maculata*	[27]
141	corilagin	$C_{27}H_{22}O_{18}$	634.46	*E. maculata*	[27]
142	1-*O*-ethyl-3,6-*O*-(*R*)-hexahydroxydiphenoyl-(1C4)-*β*-*D*-glucose	$C_{22}H_{22}O_{14}$	510.40	*E. maculata*	[27]
143	1,3,4,6-tetra-*O*-galloyl-*β*-*D*-glucose	$C_{41}H_{28}O_{27}$	952.65	*E. maculata*	[27]
144	tercatain	$C_{34}H_{26}O_{22}$	786.56	*E. maculata*	[27]
145	eumaculin B	$C_{69}H_{54}O_{43}$	1571.15	*E. maculata*	[27]
146	eumaculin D	$C_{69}H_{54}O_{43}$	1571.15	*E. maculata*	[27]
147	protocatechuic acid	$C_{7}H_{6}O_{4}$	154.12	*E. humifusa*	[29]
148	*cis*-caffeic acid	$C_{9}H_{8}O_{4}$	180.16	*E. humifusa*	[29]
149	*trans*-caffeic acid	$C_{9}H_{8}O_{4}$	180.16	*E. humifusa*	[29]
150	rosmarinic	$C_{18}H_{16}O_{8}$	360.32	*E. humifusa*	[30]
151	euphorbinoside	$C_{27}H_{32}O_{14}$	580.54	*E. humifusa*	[4]
152	benzyl *β*-D-ribofuranoside	$C_{12}H_{16}O_{5}$	240.26	*E. humifusa*	[4]

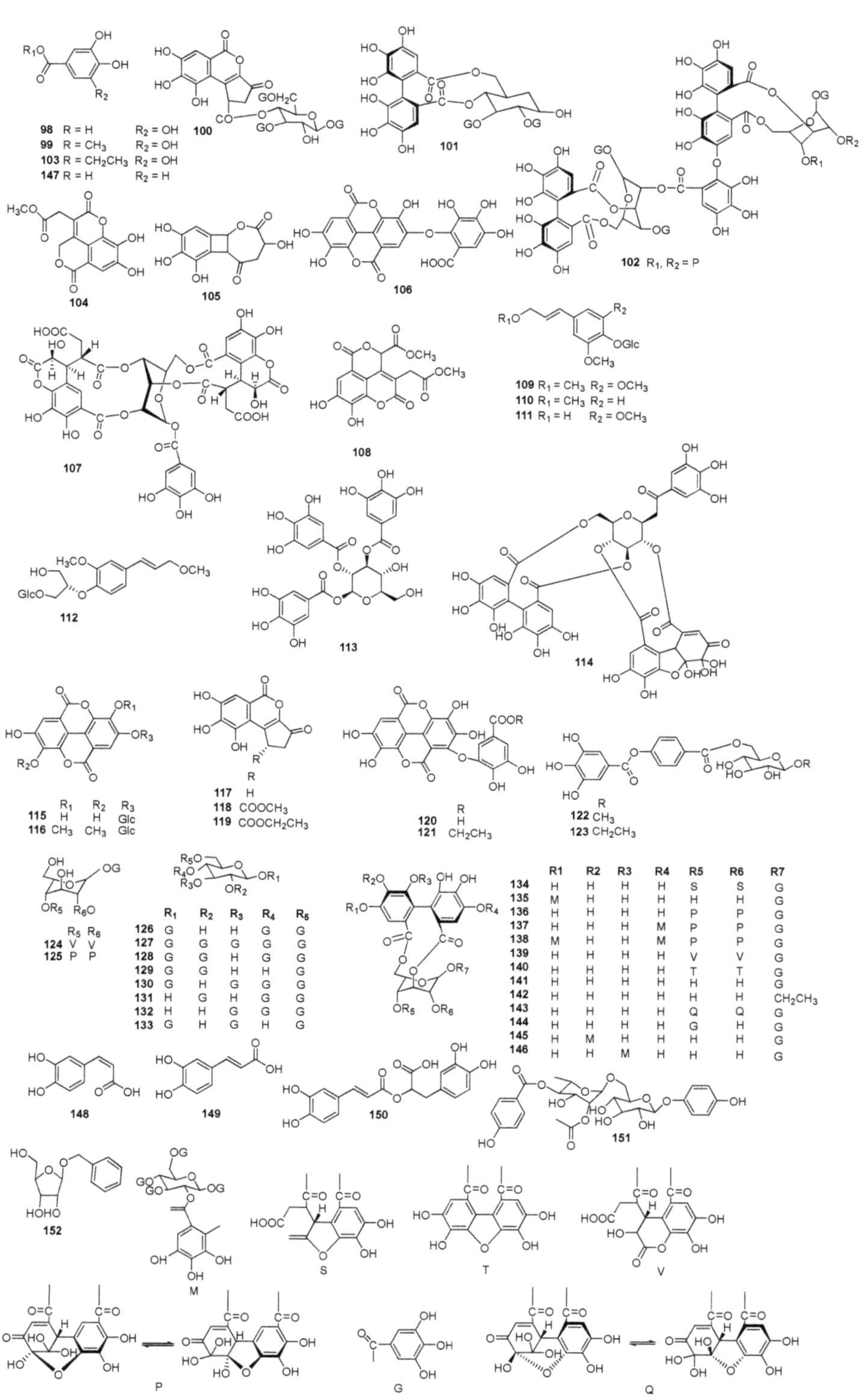

Figure 6. The structures of phenolic acids and tannins in EHH.

5.4. Alkaloids

Deng et al. have separated four alkaloids from *E. humifusa*: 5-β-methoxy-4β-hydroxy-3-methylene-α-pyrrolidinone (**153**), 5-β-methoxy-4α-hydroxy-3-methylene-α-pyrrolidinone (**154**), 5β-butoxy-4α-hydroxy-3-methylene-α-pyrrolidinone (**155**), and 3-(2-hydroxyethyl)-5-(1-*O*-glucopyranosyloxy)-indole (**156**) [22] (Figure 7, Table 4). Tian et al. have separated 1-(2′,3′,4′,5′-tetrahydroxypentyl)-6,7-dimethyl-guinoxaline-2,3-(1H,4H)-dione (**157**), from *E. humifusa* [16]. Two other alkaloids have also been separated from *E. humifusa* called uinoxadione (**158**) [26] and (−)-neoechinulin A (**159**) [29].

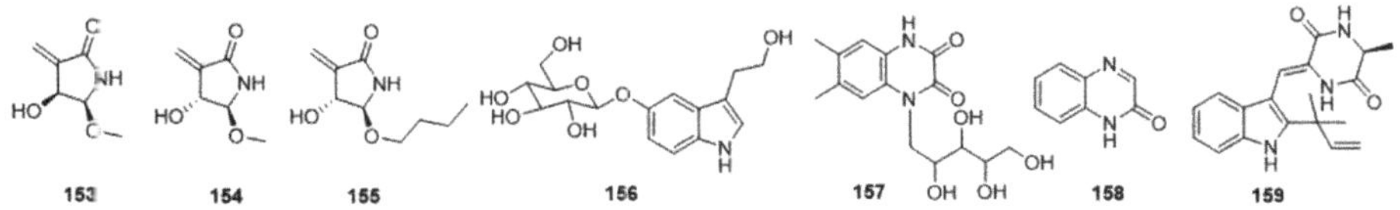

Figure 7. The structures of alkaloids in EHH.

Table 4. The alkaloids in EHH.

NO.	Name	Molecular Formula	Molecular Weight	Source	References
153	5-β-methoxy-4β-hydroxy-3-methylene-α-pyrrolidinone	$C_6H_9NO_3$	143.14	*E. humifusa*	[22]
154	5-β-methoxy-4α-hydroxy-3-methylene-α-pyrrolidinone	$C_6H_9NO_3$	143.14	*E. humifusa*	[22]
155	5β-butoxy-4α-hydroxy-3-methylene-α-pyrrolidinone	$C_9H_{15}NO_3$	185.22	*E. humifusa*	[22]
156	3-(2-hydroxyethyl)-5-(1-*O*-glucopyranosyloxy)-indole	$C_{16}H_{21}NO_7$	339.34	*E. humifusa*	[22]
157	1-(2′,3′,4′,5′-tetrahydroxypentyl)-6,7-dimethyl-guinoxaline-2,3-(1H,4H)-dione	$C_{15}H_{20}N_2O_6$	324.33	*E. humifusa*	[16]
158	uinoxadione	$C_8H_6N_2O$	146.15	*E. humifusa*	[26]
159	(−)-neoechinulin A	$C_{19}H_{21}N_3O_2$	323.40	*E. humifusa*	[29]

5.5. Sterols

The sterols identified from *E. humifusa,* including β-sitosterol (**160**), β-daucosterol (**161**), stigmaster-5-ene-3-*O*-(6-linoyl-114yl)-β-*D*-glucopyranoside (**162**) and 7β-hydroxy-sitosterol (**163**) [24] (Figure 8, Table 5).

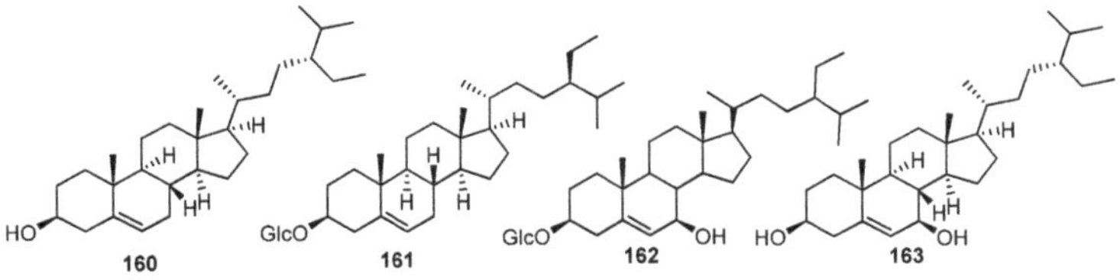

Figure 8. The structures of sterols in EHH.

Table 5. The sterols in EHH.

NO.	Name	Molecular Formula	Molecular Weight	Source	References
160	β-sitosterol	$C_{29}H_{50}O$	414.72	*E. humifusa*	[24]
161	β-daucosterol	$C_{35}H_{60}O_6$	576.85	*E. humifusa*	[24]
162	stigmaster-5-ene-3-*O*-(6-linoyl-114yl)-β-*D*-glucopyranoside	$C_{35}H_{60}O_7$	592.85	*E. humifusa*	[24]
163	7β-hydroxy-sitosterol	$C_{29}H_{50}O_2$	430.72	*E. humifusa*	[24]

5.6. Lactones and Coumarins

Several lactones and coumarins have been isolated from *E. humifusa*, including scopoletin (**164**) [24], umbelliferone (**165**) [23], 7-methoxy-6-hydroxyl-coumarin (**166**) [13], esculetin (**167**), 5-methoxyscopoletin (**168**), isofraxidin (**169**) [29], and ethyl brevifolincarboxylate (**170**) [13] (Figure 9, Table 6).

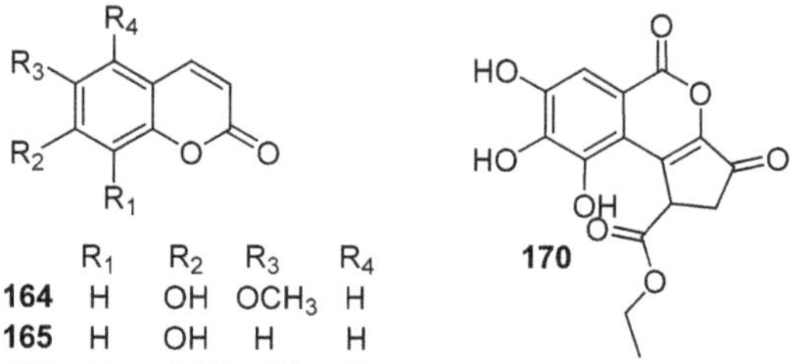

	R_1	R_2	R_3	R_4
164	H	OH	OCH_3	H
165	H	OH	H	H
166	H	OCH_3	OH	H
167	H	OH	OH	H
168	H	OH	OCH_3	OCH_3
169	OCH_3	OH	OCH_3	H

Figure 9. The structures of lactones and coumarins in EHH.

Table 6. The lactones and coumarins in EHH.

NO.	Name	Molecular Formula	Molecular Weight	Source	References
164	scopoletin	$C_{10}H_8O_4$	192.17	*E. humifusa*	[24]
165	umbelliferone	$C_9H_6O_3$	162.14	*E. humifusa*	[23]
166	7-methoxy-6-hydroxyl-coumarin	$C_{10}H_8O_4$	192.17	*E. humifusa*	[13]
167	esculetin	$C_9H_6O_4$	178.14	*E. humifusa*	[29]
168	5-methoxyscopoletin	$C_{11}H_{10}O_5$	222.20	*E. humifusa*	[29]
169	isofraxidin	$C_{11}H_{10}O_5$	222.20	*E. humifusa*	[29]
170	ethyl brevifolincarboxylate	$C_{15}H_{12}O_8$	320.25	*E. humifusa*	[13]

5.7. Other Compounds

Previous analyzing the volatile components of EHH have identified ketones (32.33%), acids (25.32%), and esters (14.55%) as their primary constituents [31] (Figure 10, Table 7). Additional compounds identified in this plant include α-pyrone (**171**), γ-pyrone (**172**), 2-methoxy-4-vinylphenol (**173**), desogestrel (**174**), α-ionone (**175**), dihydroactinidiolide (**176**), lauric acid (**177**), isononyl phthalate (**178**), methyl hexadecanoate (**179**), dibutyl phthalate (**180**), palmitic acid (**181**), methyl linoleate (**182**), 9,12,15-octadecatrienoic acid, methyl ester (**183**), phytol (**184**), linoleic acid (**185**), and α-linolenic acid (**186**). From the whole plant of *E. humifusa*, four fatty glycosides have been isolated: humionoactosides A (**187**), (2*S*)-3-*O*-octadeca-9*Z*,12*Z*,15*Z*-trienoylglyceryl-*O*-β-D-galactopyranoside (**188**), ingerglycolipid A (**189**), and 6′-*O*-linolenoylsucrose (**190**) [32]. Other compounds obtained (5*Z*)-Nonenoic acid (**191**), (5*Z*)-undecenoic acid (**192**), corchoionol C (**193**), vomifoliol (**194**), (−)-phaseic acid (**195**), and isololiiolide (**196**) were also isolated from *E. humifusa* [29]. (4*S*)-α-terpineol 8-*O*-[α-L-arabinopyranosyl-(1→6)-β-D-glucopyranoside] (**197**) was isolated from *E. humifusa* [4]. *E. humifusa* also contains inorganic trace elements such as K, Mg, Ca, Na, Fe, Mn, Zn, and Cu [33].

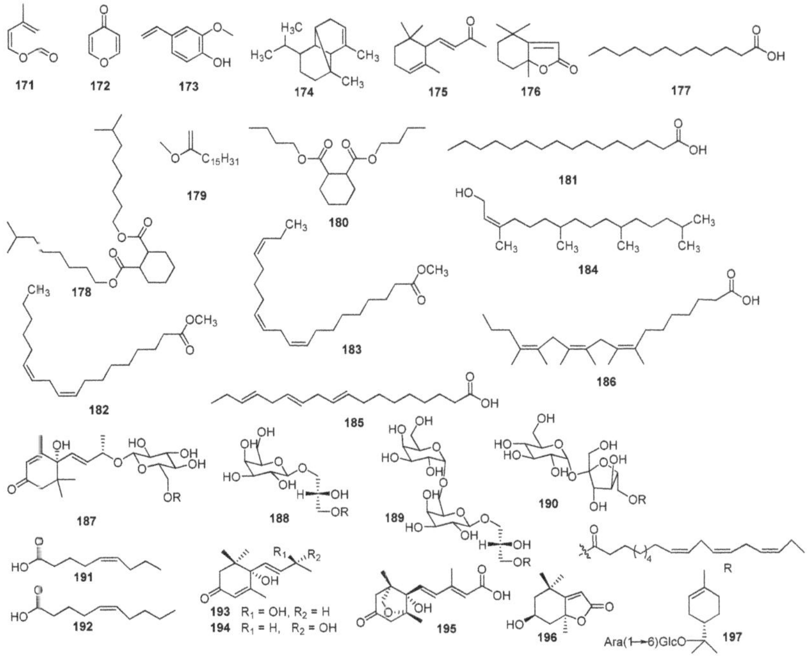

Figure 10. The structure of other compounds in EHH.

Tab_e 7. Other compounds in EHH.

NO.	Name	Molecular Formula	Molecular Weight	Source	References
171	α-pyrone	$C_6H_6O_2$	110.04	*E. humifusa*	[31]
172	γ-pyrone	$C_5H_4O_2$	96.09	*E. humifusa*	[31]
173	2-methoxy-4-vinylphenol	$C_9H_{10}O_2$	150.18	*E. humifusa*	[31]
174	desogestrel	$C_{15}H_{24}$	204.36	*E. humifusa*	[31]
175	α-ionone	$C_{13}H_{20}O$	192.30	*E. humifusa*	[31]
176	dihydroactinidiolide	$C_{11}H_{16}O_2$	180.25	*E. humifusa*	[31]
177	lauric acid	$C_{12}H_{24}O_2$	200.32	*E. humifusa*	[31]
178	isononyl phthalate	$C_{26}H_{48}O_4$	424.67	*E. humifusa*	[31]
179	methyl hexadecanoate	$C_{18}H_{36}O$	268.49	*E. humifusa*	[31]
]180	dibutyl phthalate	$C_{16}H_{28}O_4$	284.40	*E. humifusa*	[31]
181	palmitic acid	$C_{16}H_{32}O_2$	256.43	*E. humifusa*	[31]
182	methyl linoleate	$C_{19}H_{34}O_2$	294.48	*E. humifusa*	[31]
183	9,12,15-octadecatrienoic acid, methyl ester	$C_{21}H_{36}O_2$	320.52	*E. humifusa*	[31]
184	phytol	$C_{20}H_{40}O$	296.54	*E. humifusa*	[31]
185	linoleic acid	$C_{18}H_{30}O_2$	278.44	*E. humifusa*	[31]
186	α-linolenic acid	$C_{24}H_{42}O_2$	362.60	*E. humifusa*	[31]
187	humionoactoside A	$C_{38}H_{60}O_9$	660.89	*E. humifusa*	[32]
188	(2*S*)-3-*O*-octadeca-9*Z*,12*Z*,15*Z*-trienoylglyceryl-*O*-β-D-galactopyranoside	$C_{28}H_{48}O_9$	528.68	*E. humifusa*	[32]
189	gingerglycolipid A	$C_{34}H_{58}O_{14}$	690.82	*E. humifusa*	[32]
190	6′-*O*-linolenoylsucrose	$C_{31}H_{52}O_{12}$	616.75	*E. humifusa*	[32]
191	(5*Z*)-nonenoic acid	$C_9H_{16}O_2$	156.23	*E. humifusa*	[29]
192	(5*Z*)-undecenoic acid	$C_{10}H_{18}O_2$	170.25	*E. humifusa*	[29]
193	corchoionol C	$C_{13}H_{20}O_3$	224.30	*E. humifusa*	[29]
194	vomifoliol	$C_{13}H_{20}O_3$	224.30	*E. humifusa*	[29]
195	(−)-phaseic acid	$C_{15}H_{20}O_5$	280.32	*E. humifusa*	[29]
196	isololiiolide	$C_{11}H_{16}O_3$	196.25	*E. humifusa*	[29]
197	(4*S*)-α-terpineol 8-*O*-[α-L-arabinopyranosyl-(1→6)-β-D-glucopyranoside]	$C_{21}H_{36}O_{10}$	448.51	*E. humifusa*	[4]

6. Pharmacological Effects

EHH has a sweet taste and is considered relatively mild in its medicinal action. Traditionally, EHH is used to clear heat and detoxify, cool the blood, stop bleeding, alleviate dampness and jaundice, and treat diarrhea. Recent research has revealed that EHH possesses multiple pharmacological activities, including hypoglycemic, antioxidant, antidiarrhea, antibacterial, antiviral, anti-allergic (for skin conditions), anti-rheumatoid arthritis, antiemetic, detoxifying, anticancer, and chronic urticaria treatments. These pharmacological effects align with its traditional therapeutic applications (Figure 11).

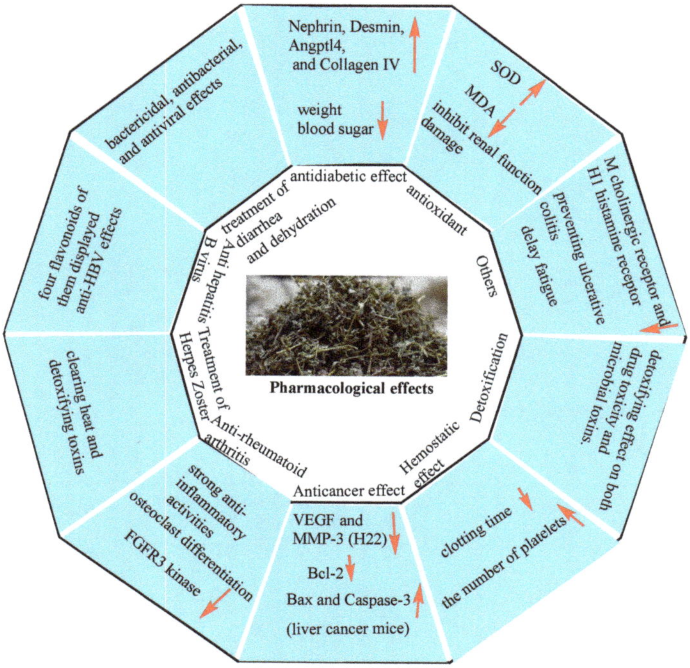

Figure 11. EHH has a wide range of pharmacological activities.

6.1. *Hypoglycemic Effect (Diabetes and Diabetic Nephropathy)*

Diabetic kidney disease (DKD), a prevalent microvascular complication of diabetes, poses a serious health threat and is the primary cause of end-stage renal disease. In a 2023 study, Li et al. investigated the effects of *E. humifusa* on DKD by dividing specific pathogen-free (SPF) rats into three groups: a normal group, a DKD group, and a DKD group treated with *E. humifusa*. Findings have revealed that the expression levels of Nephrin, Desmin, Angptl4, and Collagen IV in podocytes are significantly higher in the *E. humifusa*-treated DKD group than in the untreated DKD group. This increase inhibits podocyte transdifferentiation, resulting in reduced proteinuria and improved renal function in DKD rats, indicating that EHH holds therapeutic potential for diabetic nephropathy [34].

Diabetes mellitus (DM) arises from a combination of genetic and environmental factors, leading to insufficient insulin secretion, insulin resistance, and consequential metabolic disturbances in glucose, protein, fat, water, and electrolyte balance. Type II diabetes constitutes the majority of cases [35]. In a study using KK-Ay mice on a high-fat diet, the mice are separated into several groups: a metformin group, a model group, low-dose and high-dose *E. humifusa* groups, and a control group regular diet. A few weeks later, enzyme-linked immunosorbent assay (ELISA) is conducted to measure serum insulin, TNFα, IL-6, adiponectin, and leptin levels across the groups. The results show that *E. humifusa* effectively reduces body weight and fasting blood glucose in mice, alleviating pancreatic islets stress, thereby demonstrating a notable hypoglycemic effect [36,37]. Previous studies further support that polysaccharides, saponins, flavonoids, alkaloids, and phenols in EHH contribute to its blood glucose-lowering effects [38]. Collectively, these findings underscore *E. humifusa* as a promising agent for blood sugar regulation.

6.2. Antioxidant Properties

The total flavonoids from EHH have shown potential as feed additives in broiler chicken diets, enhancing immune function, antioxidant capacity, and intestinal microecological balance, thereby improving overall production performance [39]. Additionally, *E. humifusa* has been found to increase superoxide dismutase (SOD) activity in mice, a critical enzyme in oxidative and antioxidative processes [40]. It also inhibits malondialdehyde (MDA) production, thus protecting cell membranes from oxidative damage.

In studies involving a rat model of renal ischemia-reperfusion injury, treatment with *E. humifusa* significantly decreases MDA levels and increases SOD activity compared to untreated groups, indicating its ability to mitigate renal function damage caused by ischemia-reperfusion group through antioxidative free radical effects [41]. Further research has demonstrated a marked reduction in plasma lipid peroxides following treatment with *E. humifusa* water extract, indicating its potential for reducing oxidative stress [42]. In aging mice, the total flavonoids of *E. humifusa* increase telomerase content and SOD activity in testicular and brain tissues while reducing MDA levels, contributing to its antioxidant properties [43]. In summary, these studies affirm the potent antioxidant effects of EHH.

6.3. Treatment of Diarrhea and Dehydration

Fungal, viral, and bacterial infections are among the most common causes of diarrhea, and *E. humifusa* contains various active compounds with bactericidal, antibacterial, and antiviral properties. Domestic research has shown that [15] flavonoids, tannins, and phenolic acids in *E. humifusa* are the primary active ingredients for treating diarrhea. Studies indicate that *E. humifusa* is effective in treating recurrent ulcerative colitis [44] and has an inhibitory effect on small intestine peristalsis, making it highly effective in managing diarrhea and dehydration [45].

The extract of *E. humifusa* has demonstrated strong antimicrobial activity in vitro. For instance, it showed a significant inhibitory effect against 20 common pathogenic bacteria, while concentrations of 0.005–1.25 mg/mL exhibit bactericidal effects. Notably, compounds such as gallic acid and quercetin isolated from *E. humifusa* have also shown potent antibacterial activity [46]. The ethanol extract of *E. humifusa* has proven effective against dysentery bacteria, typhoid and paratyphoid bacteria, *Proteus* species, pathogenic *Escherichia coli*, and others [47].

In TCM, EHH is traditionally used for heat-clearing and detoxification, exhibiting detoxifying effects against various microbial toxins. Its antiviral effect may stem from its ability to alter the ultrastructure of endotoxins, rendering them non-toxic [48].

6.4. Anti-Hepatitis B Virus (HBV) Effects

HBV has been a major global health threat since the 1960s, with significant implications for human life and health. According to the World Health Organization (WHO), about 2 million people die from liver disease every year worldwide, including about 1 million deaths from complications of liver cirrhosis. China accounts for about 11 percent of the total deaths from live cirrhosis globally. According to data from 2015, China has the largest number of patients with chronic hepatitis B (CHB), about 74 million. In the development of

CHB, compensated cirrhosis, about 4–12% of patients develop decompensated cirrhosis every year [49]. However, it is necessary to find the active constituent from a natural source.

In research conducted by Tian, 38 compounds have been isolated from *E. humifusa*, with four flavonoids demonstrating anti-HBV effects in vitro. This marks the first discovery of natural products with significant anti-HBV activity in vitro, presenting promising candidates for anti-HBV drug development [16].

6.5. Treatment of Herpes Zoster (HZ)

HZ, an acute skin disease caused by the varicella-zoster virus (VZV), arises when it invades the body [50]. In TCM, HZ is thought to result from damp-heat and fire-toxin imbalance. EHH is traditionally used to dispel external toxins through clear-heat and detoxifying actions, as well as to alleviate damp-heat.

The mechanism of HZ is linked to damp-heat accumulation and stagnation of heat in the qi and blood, causing congestion in blood vessels; *E. humifusa* is believed to regulate the qi clear heat and promote diuresis, helping to relieve these symptoms. According to Ben Cao Hui Yan, EHH is used to cool the blood and disperse blood, detoxify, and stop dysentery. It is particularly effective in clearing bleeding veins and detoxifying sores.

Modern pharmacological studies support these traditional uses, showing that some compounds isolated from *E. humifusa* exhibit anti-inflammatory properties, potentially contributing to its heat-clearing and detoxifying effects [4]. These findings suggest that EHH may offer therapeutic benefits in managing HZ.

6.6. Anti-Rheumatoid Arthritis (RA)

RA is a prevalent chronic autoimmune disease that affects multiple systems and is associated with prolonged disability. While RA can manifest at any age, it is most commonly diagnosed in young adults, with a prevalence rate reaching between 0.2% and 0.4%, significantly diminishing patients' quality of life. Extensive studies have investigated the chemical constituents of *E. humifusa*, with particular emphasis on quercetin and kaempferol for their therapeutic effects on RA [51]. Quercetin, a natural flavonoid isolated from EHH, primarily alleviates RA symptoms by inhibiting processes like angiogenesis, synovial hyperplasia, inflammatory factor infiltration, and neutrophil extracellular trap formation. The study analyzed the therapeutic role of quercetin in collagen-induced arthritis in C57BL/6 mice. The animals were allocated into five groups that were subjected to the following treatments: negative (untreated) control, positive control (arthritis-induced), arthritis + methotrexate, arthritis + quercetin, and arthritis + methotrexate + quercetin [52]. Additionally, it modulates the balance between ruptured/osteoblasts and influences Th17/regulatory T cell dynamics, involving mediators such as TNF-α, MCP-1, IL-6, IL-1β, IL-17, IL-10, NF-κB, CXCL1, CXCL5, LTB4, TGF-β, as well as various immune cells like neutrophils and macrophages [52–58]. Kaempferol, another flavonoid isolated from EHH, on the other hand, exerts anti-RA activity mainly through inhibition of FGFR3 kinase activity. Firstly, The MTT assay to investigate the cytotoxicity of kaempferol resulted in no cytotoxicity in murine CD^{4+} T-cells up to 25 μM of kaempferol. Next, the research investigated whether kaempferol suppressed inflammation and joint destruction in an experimental RA murine model. One group of mice was intraperitoneally injected with 2 mg/kg of kaempferol three times a week after type II collagen-boosting immunization, and the other group was only injected with the vehicle [59]. The results showed that it could reduce osteoclast differentiation both in vivo and ex vivo. and downregulate osteoclast markers such as tartrate-resistant acid phosphatase, integrin β3, and MMP9 [59]. Furthermore, kaempferol demonstrates potent anti-inflammatory effects by modulating pathways including MAPK, PKC, and PI3K, as well as inflammatory mediators IL-1, IL-8, IP-10, PGE2, IL-2, TNF-α,

NF-κB, and AP-1 [60,61], which are critical in RA pathogenesis. These findings suggest that *E. humifusa* may possess potential anti-RA properties.

6.7. Anticancer Effects

EHH has demonstrated inhibitory effects on liver cancer cell growth, potentially due to decreased expression of VEGF and MMP-3 in H22 tumor-bearing mice. In research conducted by Zou, mice are divided into five groups: model control, positive control, and three treatment groups receiving (high 264 mg/kg·d), medium (132 mg/kg·d), and low (66 mg/kg·d) doses of EHH. The treatment is administered over 14 days, after which tumor weight and size are measured, and histopathological examination of tumor tissues is conducted under an optical microscope. Immunohistochemistry reveals significantly reduced VEGF protein expression in the high-dose group (0.160 ± 0.004) compared to the model control group (0.228 ± 0.020), $t = 5.011$, $p < 0.001$. Similarly, MMP-3 expression is markedly lower across all EHH dose groups (0.316 ± 0.062, 0.303 ± 0.057, and 0.302 ± 0.058) than in the control group (t = 6.322, 6.845, and 6.534, $p < 0.001$) [62,63].

Further research suggested that EHH inhibits liver cancer cell growth by enhancing antioxidative capacity, reducing Bcl-2 expression, and upregulating Bax and Caspase-3 expression in transplanted liver tumors [64]. Another proposed mechanism involves the suppression of tumor angiogenesis, potentially through activation of the NF-κB/VEGF signaling pathway [65]. In a separate study by Geng, administration of a high dose of DJC (water extract of DJC) significantly increased p19ARF in tumor tissues [66]. Xie's study has reported a total effectiveness rate of 91% in treating 290 lung cancer patients with *E. humifusa* [49], while Jiang's research has indicated that *E. humifusa* can inhibit Hela cell proliferation, potentially by inducing apoptosis [67]. Collectively, these findings underscore the substantial anticancer potential of *E. humifusa*, suggesting it may play a valuable role in cancer therapy.

6.8. Hemostatic Effects

Research on the hemostatic properties of *E. maculate* has demonstrated a marked ability to significantly reduce clotting time, thereby showing a strong hemostatic effect [68]. Similarly, studies on *E. humifusa* have revealed that it can swiftly elevate platelet counts, contributing to effective hemostasis [69]. Given these properties, *E. humifusa* shows potential as a treatment for various bleeding disorders, dental bleeding, acute hemorrhagic necrotizing enteritis, and other acute bleeding conditions.

6.9. Detoxification

E. humifusa has been shown to mitigate the severe organ damage induced by the chemical hexachlorocyclohexane, effectively protecting the heart, liver, spleen, kidneys, and other organs in animals, demonstrating a greater protective effect than vitamin C [70]. Additionally, microbial toxins, which lead to various symptoms and organ damage during infections, are notably countered by DJC tinctures at 100%, 50%, and 25% concentrations, exhibiting a strong "neutralizing" effect on diphtheria toxin. These findings suggest that this herb possesses significant detoxifying properties, addressing both chemical and microbial toxin-induced toxicity.

6.10. Others

The extract of *E. humifusa* has shown effectiveness in treating chronic urticaria, enhancing therapeutic outcomes, accelerating symptom relief, and demonstrating minimal adverse reactions, establishing it as a safe and reliable option. Additionally, the alcoholic extract significantly inhibits the contraction amplitude of rabbit small intestine smooth muscle, likely involving the activation of α-adrenergic receptors and inhibition of M cholinergic and H_1 histamine receptors [45]. Research by Ma has indicated that EHH offers protective effects on ischemia-reperfusion in myocardial tissues [71]. Chang's study explored the effect of *E. humifusa* polysaccharide on preventing ulcerative colitis through modulation of the intestinal microbiota [72].

Furthermore, *E. humifusa* polysaccharides are found to improve metabolic processes, enhance endurance, boost lactate dehydrogenase activity and glycogen levels, accelerate lactic acid clearance, reduce lactic acid buildup, inhibit excessive blood urea nitrogen levels post-exercise, delay fatigue, and enhance the body's resilience to physical stress [73]. Analgesic properties of *E. humifusa* are also demonstrated in mice through a hotplate pain induction method [74]. Additionally, *E. humifusa* total polysaccharides have shown potential in preventing and treating animal diseases [75]. EHH is used as food in some countries, and toxicological studies have shown no significant toxicity from chronic use in rats, supporting its safety profile [5,76].

7. Summary and Perspectives

EHH has a long history of medicinal use and is showing encouraging results across diverse therapeutic areas, from treating diarrhea and dehydration to enhancing drug efficacy in combination therapies. Pharmacological research has uncovered its antidiabetic, antioxidant, antiviral, anti-rheumatoid arthritis, anticancer, and immunomodulatory properties, along with benefits for fatigue reduction, cardiovascular health, and animal disease prevention. Phytochemical analyses have identified numerous active compounds within EHH, including terpenoids, flavonoids, phenolic acids, alkaloids, steroids, and polysaccharides.

Before advancing its applications, comprehensive investigations into the chemical composition and pharmacological efficacy of EHH are essential. Although EHH has begun to be cultivated in certain areas, medicinal varieties in China are still largely sourced from the wild. The plant's growth in diverse regions and climates results in variations in its quality and medicinal potency. To fully harness EHH's pharmacological benefits, it is crucial to identify key bioactive markers and assess samples from multiple sources. Additionally, standardized cultivation bases should be established, with quality assessments conducted through advanced techniques such as NMR spectroscopy and HPLC-MS. By rigorously controlling market quality, TCM standards can be upheld, ensuring consistent therapeutic effects.

Moreover, 197 chemical constituents have been isolated from EHH, with terpenoids, flavonoids, and phenolic acids identified as particularly abundant. Future pharmacological research should continue focusing on these primary components while also employing more sophisticated methods and technologies to explore the effects of other components. Expanding this pharmacological understanding will provide a robust scientific foundation for clinical applications and broaden the therapeutic potential of EHH.

Historically, EHH has been valued for traditional medicinal uses, including its ability to clear heat and toxins, cool the blood and stop bleeding, eliminate dampness, and alleviate jaundice. In recent years, research and clinical applications have primarily focused on its effectiveness in treating diarrhea and dehydration. However, further exploration is required to understand other traditional effects, such as blood glucose reduction and anti-fatigue properties, by clarifying the underlying mechanisms and pharmacological actions to support broader clinical applications. This investigation will be crucial in validating EHH's traditional uses.

As a plant with a rich history in traditional medicine, EHH has been widely employed for wellness and illness management. In this review, we examine the phytochemistry, pharmacological properties, and contemporary applications of EHH, aiming to deepen understanding of ongoing research and identify areas for further advancement. We also critically discuss potential future research directions for EHH, highlighting the need to investigate its traditional therapeutic effects and explore new therapeutic applications. These insights will provide strong scientific support for expanding EHH's role in clinical interventions, enabling its use in a wider range of pharmacological and therapeutic contexts.

To deepen our understanding of EHH and advance its research, several key questions remain to be addressed. Firstly, although the value of EHH has gained increasing recognition, the quantity and quality of wild-harvested resources remain inconsistent. Thus, the development of artificial cultivation methods, along with the selection of disease-resistant varieties, is essential to improve both yield and quality. Secondly, while numerous studies have examined the phytochemistry of EHH, more focused attention is needed to lay the groundwork for future therapeutic applications.

Furthermore, given EHH's extensive use in traditional medicine, systematic pharmacological studies, including an investigation into its pharmacodynamics and molecular mechanisms, are crucial to support its clinical application as both a TCM herb and a folk remedy in China. EHH represents a valuable resource. Future research should prioritize understanding its active compounds, mechanisms of action and the principles of its use in classical formulations, providing a solid theoretical foundation for broader applications and potential developments across various medical fields.

Author Contributions: Conceptualization, J.X., H.G., J.Z., B.F. and W.W.; methodology, Y.X.; software, Y.C. validation, J.X., H.G. and J.Z.; formal analysis, J.X.; investigation, H.Z.; resources, H.G.; data curation, J.Z.; writing—original draft preparation, J.X., B.F. and W.W.; writing—review and editing, W.Z.; visualization, Y.X.; supervision, W.W.; project administration, B.F.; funding acquisition, H.Z., B.F. and W.W. All authors have read and agreed to the published version of the manuscript.

Funding: This work was supported by the National Natural Science Foundation of China (32100317 and 32370419), National College Students Innovation and Entrepreneurship Training Program (202410304061Z), and the Zhejiang Province Traditional Chinese Medicine Science and Technology Project (2023ZL455).

Institutional Review Board Statement: Not applicable.

Informed Consent Statement: Not applicable.

Data Availability Statement: No new data were created or analyzed in this study.

Conflicts of Interest: The authors declare no conflicts of interest.

Abbreviations

AP-1	Activator protein 1
CHB	Chronic hepatitis B
CNKI	China National Knowledge Infrastructure
CXCL1	(C-X-C motif) ligand 1 protein
CXCL5	(C-X-C motif) ligand 5 protein
DGGlc	2,3-Digalloyglucosyl
DJC	Dijincao
DKD	Diabetic kidney disease
DM	Diabetes mellitus
EHH	Euphorbiae Humifusae Herba
ELISA	Enzyme-linked immunosorbent assay
FvGFR3	Fibroblast growth factor receptors
Gal	Galactosyl
GGlc	2-Galloyglucosyl

GRham	2-Galloylrhamosyl
HBV	Hepatitis B virus
IL-1	Interleukin-1
IL-10	Interleukin-10
IL-17	Interleukin-17
IL-1β	Interleukin-1β
IL-2	Interleukin-2
IL-6	Interleukin-6
IL-8	Interleukin-8
IP-10	Inducible Protein 10
LTB4	Leukotriene B4
MAPK	Mitogen-activated protein kinase
MCP-1	Monocyte chemoattractant protein-1
MDA	Malondialdehyde
MMP9	Matrix metalloproteinase 9
NF-κB	Nuclear factor-κB
PGE2	Prostaglandin E2
PI3K	Phosphatidylinositol 3-kinase
PKC	Protein kinase C
RA	Rheumatoid arthritis
Rut	Rutinosyl
SOD	Superoxide dismutase
SPF	Specific pathogen-free
TCM	Traditional Chinese medicine
TGF-β	Transforming growth factor-β
Th17	T helper cell 17
TNF-α	Tumor necrosis factor-α
VZV	Varicella-zoster virus
WHO	World Health Organization

References

1. Wang, L.; Huo, Z.T.; Xu, W.B.; Zhou, P.N.; Nan, W.X.; Guo, H.J.; Zhang, Q.W.; Yang, P.; Alolga, R.N.; Yin, X.J.; et al. Comparativ plastomes of eight subgenus *Chamaesyce* plants and system authentication of Euphorbiae Humifusae Herba. *Food Chem.* **2024**, *44* 139039. [CrossRef] [PubMed]
2. Chinese Pharmacopoeia Commission. *Pharmacopoeia of the People's Republic of China*; China Medical Science Publisher: Beijing China, 2020; Volume 1, p. 131.
3. Ahn, S.; Jung, H.; Jung, Y.; Lee, J.; Shin, S.Y.; Yoongho, L.; Lee, Y.H. Identification of the active components inhibiting th expression of matrix metallopeptidase-9 by TNFα in ethyl acetate extract of *Euphorbia humifusa* Willd. *J. Appl. Biol. Chem.* **2019**, *6* 367–374. [CrossRef]
4. Luyen, B.T.T.; Tai, B.H.; Thao, N.P.; Eun, K.J.; Cha, J.Y.; Xin, M.J.; Lee, Y.M.; Kim, Y.H. Anti-inflammatory components of *Euphorbi humifusa* Willd. *Bioorg. Med. Chem. Lett.* **2014**, *24*, 1895–1900. [CrossRef]
5. Xiang, N.; Zhao, J.B.; Chang, S.Q.; Li, S.S.; Liu, S.W.; Wang, C. In vitro fecal fermentation of *Euphorbia humifusa*-derive polysaccharides and their protective effect against ulcerative colitis in mice. *Foods* **2023**, *12*, 751. [CrossRef]
6. Page, M.J.; Mckenzie, J.E.; Bossuyt, P.M.; Boutron, I.; Hoffmann, T.C.; Mulrow, C.D.; Shamseer, L.; Tetzlaff, J.M.; Akl, E.A Brennan, S.E.; et al. The PRISMA 2020 statement: An undated guideline for reporting systematic reviews. *Rev. Esp. Cardiol.* **2021** *74*, 790–799. [CrossRef]
7. Board, E. *Flora of China*; Chinese Academy of Sciences: Beijing, China, 1994.
8. Li, Z.J.; Gulnar, D.; Silafu, A. Research progress on the Uighur herb *Euphorbia humifusa*. *J. Med. Pharm. Chin. Minor.* **2008**, *8*, 15–21
9. Tan, X.; Li, X.; Wang, Q.; Shi, M.; Yang, J.; Cao, Y. HPLC fingerprint spectrum of Uighur herb Euphorbiae humifusae. *Chin. Wil Plant Resour.* **2011**, *30*, 47–50.
10. An, H.; Gulnar, D.; Li, Z.J.; Silafu, A. Active component of *Euphorbia humifusa* antifungal activity and mechanism. *Chin. Pharm Bull.* **2010**, *26*, 1162–1165.
11. Yang, X.Z.; Gao, Y.C.; Zhang, X.H.; He, X.L. Study on one step granulation process for *Euphorbia humifusa* granular. *J. Tradit. Chin Vet. Med.* **2016**, *35*, 54–56.

Gao, Y.; Sun, L.L.; Tang, M.Y.; Feng, B.M.; Pei, Y.H.; Yasukawa, K. Triterpenoids from *Euphorbia maculata* and their anti-inflammatory effects. *Molecules* **2018**, *23*, 2112. [CrossRef]

Pei, Y.G.; Wu, Q.X.; Shi, Y.P. Triterpenoids and other constituents from *Euphorbia humifusa*. *J. Chin. Chem. Soc.* **2007**, *54*, 1565–1572. [CrossRef]

Huang, P.Z.; Yang, H.Y.; Zhang, L.J.; Sun, Y.; Li, Y.; Gao, K. Anti-inflammatory activities of triterpenoids from *Euphorbia humifusa* willd. *Phytochem. Lett.* **2023**, *54*, 129–132. [CrossRef]

Xia, R.F.; Su, J.C.; Yu, J.; Zha, H.J.; Wu, J.L.; Fu, X.N.; Cai, Q.; Wan, L.S. Anti-inflammatory lanostane triterpenoids with rearranged spirobi[indene] scaffold and their biogenetically related analogues from *Euphorbia maculata*. *Phytochemistry* **2023**, *211*, 113682. [CrossRef]

Tian, Y.; Sun, L.M.; Liu, X.Q.; Li, B.; Wang, Q.; Dong, J.X. Anti-HBV active flavone glucosides from *Euphorbia humifusa* Willd. *Fitoterapia* **2010**, *81*, 799–802. [CrossRef] [PubMed]

Luyen, B.T.T.; Tai, B.H.; Thao, N.P.; Lee, S.H.; Jang, H.D.; Lee, Y.M.; Kim, Y.H. Evaluation of the anti-osteoporosis and antioxidant activities of phenolic compounds from *Euphorbia maculata*. *J. Korean Soc. Appl. Biol. Chem.* **2014**, *57*, 573–579. [CrossRef]

Nugroho, A.; Rhim, T.J.; Choi, M.Y.; Choi, J.S.; Kim, Y.C.; Kim, M.S.; Park, H.J. Simultaneous analysis and peroxynitrite-scavenging activity of galloylated flavonoid glycosides and ellagic acid in *Euphorbia supina*. *Arch. Pharmacal Res.* **2014**, *37*, 890–898. [CrossRef]

Rakotondrabe, T.F.; Fan, M.; Guo, M. Exploring potential antidiabetic and anti-inflammatory flavonoids from *Euphorbia humifusa* with an integrated strategy. *Front. Pharmacol.* **2022**, *13*, 980945. [CrossRef]

Wang, X.Y. Studies on the Chemical Constituents and Quality Control of *Euphorbia humifusa*. Master's Thesis, Shenyang Pharmaceutical University, Shenyang, China, 2007.

Li, R.Z.; He, Y.C.; Liu, H.; Wang, B. The separation and identification of flavonoids in the *Euphorbia humifusa*. *J. Beijing Med. Coll.* **1983**, *15*, 72–74.

Deng, F.; Tang, N.; Xu, J.; Shi, Y.H.; Zhao, M.; Zhang, J.S. New α-pyrrolidinonoids and glycosides from *Euphorbia humifusa*. *J. Asian Nat. Prod. Res.* **2008**, *10*, 531–539. [CrossRef]

Liu, R.H.; Wang, H.B.; Kong, L.Y. Studies on chemical constituents of *Euphorbia humifusa*. *Chin. Tradit. Herb. Drugs* **2001**, *32*, 107–108.

Li, H. Study on the Material Basis of Anti-Dermatophytes of Euphorbiae Humifusae Herba. Master's Thesis, Peking Union Medical College, Beijing, China, 2017.

Gao, S.Y.; Sun, D.J.; Wang, G.; Zhang, J.; Jiang, Y.N.; Li, G.Y.; Zhang, K.; Wang, L.; Huang, J.; Chen, L.X. Growth inhibitory effect of paratocarpin E, a prenylated chalcone isolated from *Euphorbia humifusa* Wild., by induction of autophagy and apoptosis in human breast cancer cells. *Bioorg. Chem.* **2016**, *69*, 121–128. [CrossRef] [PubMed]

Yoshida, T.; Amakura, Y.; Liu, Y.Z. Tannins and related polyphenols of Euphorbiaceous plants. XI. three new hydrolyzable tannins and a polyphenol glucoside from *Euphorbia humifusa*. *Chem. Pharm. Bull.* **1994**, *42*, 1803–1807. [CrossRef] [PubMed]

Yoshiaki, A.; Keita, K.; Tsutomu, H.; Takuo, O.; Takashi, Y.; Isao, A.; Takeshi, S.; Sansei, N. Four new hydrolyzable tannins and an acylated flavonol glycoside from *Euphorbia maculata*. *Can. J. Chem.* **1997**, *75*, 727–733.

Isao, A.; Tsutomu, H.; Yoko, N.; Sugaya, T.; Nishibe, S.; Yoshida, T.; Okuda, T. Tannins and related polyphenols of Euphorbiaceous Plants. VIII. Eumaculin A and Eusupinin A, and accompanying polypbnols from *Euphorbia maculata* L. and *E. Supina* Rafin. *Chem. Pharm. Bull.* **1991**, *39*, 881–883.

Wu, X.; Ma, G.L.; Chen, H.W.; Zhao, Z.Y.; Zhu, Z.P.; Xiong, J.; Yang, G.X.; Hu, J.F. Antibacterial and antibiofilm efficacy of the preferred fractions and compounds from *Euphorbia humifusa* (herba Euphorbiae humifusae) against *Staphylococcus aureus*. *J. Ethnopharmacol.* **2023**, *306*, 116177. [CrossRef]

Tian, S.G.; Liu, W.X.; Liu, F.; Zhang, X.; Upur, H. Development and validation of a high-performance liquid chromatography-diode array detector method for the simultaneous determination of six phenolic compounds in abnormal savda munziq decoction. *Pharmacogn. Mag.* **2015**, *11*, 157–162.

Zhang, W.; Lu, Y.; Gu, X.Z.; Kang, W.Y. Volatile compounds from *Euphorbia humifusa*. *Chin. J. Exp. Tradit. Med. Formulae* **2012**, *18*, 66–68.

Yuan, Y.; Rouzimaimait, R.; Hasan, A.; Zang, D.; Aisa, H.A. A new anti-vitiligo unsaturated fatty glycoside from *Euphorbia humifusa* Willd. *Nat. Prod. Res.* **2022**, *37*, 2225–2231. [CrossRef]

Xiong, W.; Cai, M.Z.; Wang, H.; Yu, R.P.; Cheng, C.X. Analysis and comparison of trace elements of herba Euphorbiae humifusae in different periods by microwave digestion-atomic absorption spectroscopy. *Spectrosc. Spect. Anal.* **2010**, *30*, 1975–1978.

Li, X.Y.; Tian, Y.F.; Liu, C.; Chen, S.; Fan, F.Y.; Zhang, C.; Wu, X. Effect of *Euphorbiae humifusa* on expression of glomerular podocyte-related proteins in diabetickidney disease rats. *Chin. J. Comp. Med.* **2023**, *33*, 105–113.

Xu, M.Y. *Diabetes Mellitus*; Shanghai Scientific & Technical Publishers: Shanghai, China, 2010.

Wang, L.L.; Fu, H.; Li, W.W.; Song, F.J.; Song, Y.X.; Yu, Q.; Liu, G.X.; Wang, X.M. Study of effect of Euphorbia Humifuse herb on alleviating insulin resistance in type 2 diabetic model KK-Ay mice. *J. Chin. Mater. Med.* **2015**, *40*, 1994–1998.

37. Wang, X.M.; Gao, Z.G.; Zhou, Z.Y. Effect of Euphorbiae Humifusae herba on pancreatic islet beta cells on oxidative stress an apoptosis in diabetic mice. *Chin. J. Exp. Tradit. Med. Formulae* **2015**, *21*, 120–123.
38. Wang, Y.; Zhang, X. Research progress on α-glucosidase inhibitors. *Strait Pharm. J.* **2009**, *21*, 4–6.
39. Xi, L.; Qin, X.X.; Du, L.X.; Cui, Y.Y.; Li, Z.Q.; Zhang, C.M. Effects of total flavonoids from *Euphorbia humifusa* on grow performance, immune function, antioxidant capacity and cecal microbiota of Ross 308 Broilers. *China Poult.* **2023**, *8*, 1732–173
40. Chen, F.X.; Chen, W.Y.; Gong, X.C.; Zhong, X.H. Study on the antioxidant effect of Dijincao on different tissues of mice. *Heilongjiar Anim. Sci. Veter. Med.* **2008**, *6*, 91–92.
41. Zhang, Z.T.; Wu, R.Q.; Zhang, W.; Zhang, J. Experimental study on the antioxidant effect of Dijincao and its protective effect o renal ischemia-reperfusion injury. *Chin. J. Tradit. Med. Sci. Technol.* **1999**, *6*, 892–905.
42. Li, B.S.; Wu, R.N.; Zhang, X.Y.; Ba, G.N.; Sun, F.X.; Wang, Z.M.; Zhao, Z.X. Effects of five traditional Mongolian medicines o scavenging free radical. *Shizhen J. Tradit. Chin. Med.* **1997**, *8*, 308–309.
43. Cao, R.Z.; Zhang, W.; She, J.K.; Qiao, W.; Xue, Y. Effects of total flavonoids from Dijincao on serum aging indexes in aging mod mice. *Chin. J. Gerontol.* **2008**, *6*, 562–563.
44. Dong, X.L. Experience of Cha Ansheng in treating recurrent ulcerative colitis. *J. Anhui TCM Coll.* **2013**, *32*, 53–54.
45. Xiang, D.B.; Yang, Y.; Liu, X.J.; Shi, S.C. Study on influences of ethanol extract from *Euphorbia humifusa* (EEEH) on smooth musc of isolated small intestine in rabbits and its mechanism. *Chin. Hosp. Pharm. J.* **2015**, *35*, 1371–1375.
46. Wang, T.T.; Wen, J.F.; Jin, S.N. Research progress on chemical constituents and pharmacological effects of *Euphorbia humifusa*. *Taishan Med. Coll.* **2012**, *8*, 628–631.
47. Zhou, H.H.; Yan, Y.B.; Wang, F. Research progress of Pueraria Mirifica and Dijincao in the treatment of diarrhea dehydratio *Electr. J. Clin. Med. Liter* **2015**, *2*, 4061.
48. Xie, J.Z.; Gao, Y.M. Treatment of 290 cases of lung cancer with "Lung Dan" series of Chinese patent medicines. *Chin. J. Civil Me* **1996**, *8*, 222–223.
49. Dong, J.H.; Liu, C.C.; Zhang, M.M.; Yu, H.L.; Zhao, D.; Bai, X.; Zheng, M.; Liu, Y.; Ji, J.C.; Li, R.; et al. Prediction modelling fc gastroesophageal variceal bleeding in patients with chronic hepatitis B using four-dimensional flow MRI. *J. Clin. Exp. Hepatc* **2025**, *15*, 102403. [CrossRef] [PubMed]
50. Geng, W.W.; Qiao, L.P.; Wang, J. Professor Tong Xiaolin's experience in the treatment of herpes zoster. *Jilin J. Chin. Med.* **2022**, 4 1132–1160.
51. Yang, L.; Liu, R.H.; Huang, S.B.; Zhou, M.; Li, L.M.; He, J.W. Progress of research on pathogenesis and therapeutic drugs fc rheumatoid arthritis. *China. Pharm.* **2021**, *32*, 2154–2159.
52. Haleagrahara, N.; Miranda-Hernandez, S.; Alim, M.A.; Hayes, L.; Bird, G.; Ketheesan, N. Therapeutic effect of quercetin i collagen-induced arthritis. *Biom. Pharm.* **2017**, *90*, 38–46. [CrossRef]
53. Xiao, P. Study on the Effect of Quercetin on Apoptosis of Human Rheumatoid Arthritis Fibroblast-Like Synoviocytes and It Mechanism. Ph.D. Thesis, Zhengzhou University, Zhengzhou, China, 2013.
54. Yuan, K.; Zhu, Q.Q.; Lu, Q.Y.; Jiang, H.X.; Zhu, M.M.; Li, X.H.; Huang, G.R.; Xu, A.L. Quercetin alleviates rheumatoid arthritis b inhibiting neutrophil inflammatory activities. *J. Nutr. Biochem.* **2020**, *84*, 108454. [CrossRef]
55. Yamaguchi, M.; Weitzmann, M.N. Quercetin, a potent suppressor of NF-κB and Smad activation in osteoblasts. *Int. J. Mol. Me* **2011**, *28*, 521–525.
56. Kim, H.R.; Kim, B.M.; Won, J.Y.; Lee, K.A.; Ko, H.M.; Kang, Y.S.; Lee, S.H.; Kim, K.W. Quercetin, a plant polyphenol, has potentia for the prevention of bone destruction in rheumatoid arthritis. *J. Med. Food* **2019**, *22*, 152–161. [CrossRef]
57. Ji, J.J.; Lin, Y.; Huang, S.S.; Zhang, H.L.; Diao, Y.P.; Li, K. Quercetin: A potential natural drug for adjuvant treatment of rheumatoi arthritis. *Afr. J. Tradit. Complement. Altern. Med.* **2013**, *10*, 418–421. [PubMed]
58. Lin, X.X.; Peng, Q.Z.; Zhang, J.J.; Li, X.; Huang, J.R.; Duan, S.; Zhang, W.S. Quercetin prevents lipopolysaccharide-induce experimental preterm labor in mice and increases offspring survival rate. *Reprod. Sci.* **2020**, *27*, 1047–1057. [CrossRef] [PubMed]
59. Lee, C.J.; Moon, S.J.; Jeong, J.H.; Lee, S.; Lee, M.H.; Yoo, S.M.; Lee, H.S.; Kang, H.C.; Lee, J.Y.; Lee, W.S.; et al. Kaempferol targetin on the fibroblast growth factor receptor 3-ribosomal S6 kinase 2 signaling axis prevents the development of rheumatoid arthriti *Cell Death Dis.* **2018**, *9*, 401. [CrossRef]
60. Tang, M.F. Isolation of Chemical Constituents from *Pantheria camphorata* and Screening of Its In Vitro Anti-Inflammator Immunological Activity. Master's Thesis, Anhui Medical University, Hefei, China, 2010.
61. Tang, X.L.; Liu, J.X.; Dong, W.; Li, P.; Li, L.; Hou, J.C.; Zheng, Y.Q.; Lin, C.R.; Ren, J.G. Protective effect of kaempferol on LPS plu ATP-induced inflammatory response in cardiac fibroblasts. *Inflammation* **2015**, *38*, 94–101. [CrossRef]
62. Zou, Z.J.; Hu, J.X.; Wang, X.M.; Gao, Z.G. Effect of Euphorbiae humifusae herb on expression of VEGF, MMP-9 on H22 cell bearing Mice. *Chin. J. Mod. Med.* **2013**, *23*, 11–14.
63. Zou, Z.J.; Liu, H.Y.; Gao, Z.G.; Wang, X.M. Antitumor effect of Euphorbiae Humifusae Herb on H22 tumor-bearing mice and it mechanism. *Chin. J. Cancer Prev. Treat.* **2014**, *21*, 903–908.

. Zou, Z.J.; Liu, H.Y.; Wang, X.M. Influence of herba Euphorbiae humifusae on anti-tumor effect and apoptosis protein expression in transplanted liver cancer mice. *Chin. J. Exp. Tradit. Med. Formulae* **2013**, *19*, 241–245.

. Hu, J.X.; Xi, X.T.; Wang, X.M.; Zou, Z.J. Mechanism of Humifuse euphorbia in regulating tumor angiogenesis by NF-κB/VEGF signal pathway. *Chin. J. Exp. Tradit. Med. Formulae* **2018**, *24*, 165–170.

. Geng, G.X.; Wei, H.; Chen, H.L.; Shi, M.H.; Li, Q.W. Study on the inhibitory effect of water extract of Dijincao on cervical cancer in mice. *Heilongjiang Anim. Sci. Veter. Med.* **2013**, *12*, 152–154.

. Jiang, T.; Zhou, B.; Fu, D.; Li, M. Effects of *Euphorbia humifusa* on proliferation, apoptosis of Hela cervical cancer cells. *Mod. Chin. Med.* **2017**, *19*, 1713–1716.

. Chu, X.L.; Fan, W.Y.; Lou, L.Y.; Liu, J.H. Pharmacological studies of Dijin. *Lishizhen Med. Mater. Medica Res.* **2001**, *12*, 193–194.

. Dong, P.; Zhou, P.; Guo, L.F. Effect of herba Euphorbiae humifusae on platelet count, aggregation. *Chin. J. Integr. Med.* **1998**, *18*, 330–332.

. Ma, T.J.; Sang, Y.Z.; Si, Z.Z. Alleviation of histopathologic toxicity of hexachlorobenzene by Dijincao in mice. *Chin. J. Mod. Appl. Pharm.* **1987**, *4*, 6–9.

. Ma, B.H.; Gao, Y.S.; Xu, J.H.; Lv, J.; Qi, R.F.; Wang, L.J. Effect of *Euphorbia humifusa* herb on serum myocardial enzyme activity in rats with myocardial ischemia-reperfusion injury. *Chin. J. Ethnomed. Ethnophar.* **2016**, *25*, 44–46.

. Chang, S.J. *Extraction, Isolation, In Vitro Fermentation of Polysaccharides from Euphorbia humifusa and Its Protective Effect on Ulcerative Colitis*; Southern Medical University: Guangzhou, China, 2021.

. Huang, H. Study on the Extraction Technology and Anti-Fatigue of Polysaccharide from Herba Euphorbiae humifusae. Master's Thesis, Agricultural University of Hunan, Changsha, China, 2009.

. Liang, S.L.; Wen, D.W.; Xiao, H.L.; Meng, F.L.; Li, Z.H.; Yu, H.H.; Liu, Y.J. Experimental study on the analgesic effect of *Euphorbia humifusa* decoction in mice. *Chin. Hosp. Pharm. J.* **2009**, *29*, 193–1995.

. Sun, W.J. The Preparation of Polysaccharides from *Euphorbia humifusa* and Its Immune-Enhancing Activity in Mice. Ph.D. Thesis, College of Veterinary Medicine Northwest A & F University, Xianyang, China, 2021.

. Zhang, C.J.; Yang, X.Z.; Gao, Y.C. Acute and subacute toxicity tests of *Euphorbia humifusa* granule. *Shandong Agric. Sci.* **2014**, *46*, 128–130.

MDPI AG
Grosspeteranlage 5
4052 Basel
Switzerland
Tel.: +41 61 683 77 34

Molecules Editorial Office
E-mail: molecules@mdpi.com
www.mdpi.com/journal/molecules

www.ingramcontent.com/pod-product-compliance
Lightning Source LLC
LaVergne TN
LVHW070454061126
851858LV00011B/236